Nečas Center Series

Birkhäuser

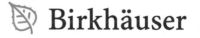

The Nečas Center Series aims to publish high-quality monographs, textbooks, lecture notes, habilitation and Ph.D. theses in the field of mathematics and related areas in the natural and social sciences and engineering. There is no restriction regarding the topic, although we expect that the main fields will include continuum thermodynamics, solid and fluid mechanics, mixture theory, partial differential equations, numerical mathematics, matrix computations, scientific computing and applications. Emphasis will be placed on viewpoints that bridge disciplines and on connections between apparently different fields. Potential contributors to the series are encouraged to contact the editor-in-chief and the manager of the series.

More information about this series at http://www.springer.com/series/16005

Zdeněk Martinec

Principles of Continuum Mechanics

A Basic Course for Physicists

Birkhäuser

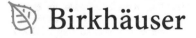

Zdeněk Martinec
Faculty of Mathematics and Physics
Charles University in Prague
Prague, Czech Republic

Dublin Institute for Advanced Studies
Geophysics Section
Dublin, Ireland

ISSN 2523-3343 ISSN 2523-3351 (electronic)
Nečas Center Series
ISBN 978-3-030-05389-5 ISBN 978-3-030-05390-1 (eBook)
https://doi.org/10.1007/978-3-030-05390-1

Mathematics Subject Classification (2010): 74-01, 76-01, 80-01, 86-01

This book is published under the imprint Birkhäuser, www.birkhauser-science.com by the registered company Springer Nature Switzerland AG.
The registered company address is: Gewerbestrasse 11, 6330 Cham, Switzerland

I dedicate this book to my family for their love and support that I will be always thankful for.

Preface

The material in this textbook is suitable for a two-semester course on *Continuum Mechanics*. It is based on lecture notes from an undergraduate course that I have taught over the last two decades. The material is intended for use by undergraduate students of physics with one or more years of university-level calculus behind them.

The literature on *Continuum Mechanics* is very extensive, ranging from books oriented towards the more practical aspects of this discipline to those providing an exact mathematical treatment. The literature listed in *References* was used during the preparation of this book and the preceding lecture series. Moreover, the table given in *Selected References for General Reading* should serve to help the reader to find supplementary literature related to a particular section of the book more quickly.

Like most authors, I am indebted to many people who have assisted in the preparation of this book. In particular, I would like to thank Erik Grafarend, Ctirad Matyska, Jiří Zahradník and Ondřej Čadek, whose interest encouraged me to write this book. It is a pleasure to express my gratitude to those who have made so many helpful comments, among them Ondřej Souček and Vojtěch Patočka. I would also like to thank my oldest son, Zdeněk, who plotted most of figures embedded in the text. In addition, my sincere thanks go to the students who have given feedback from the classroom notes. I would like to acknowledge my indebtedness to Kevin Fleming and Grace Cox, whose thorough proofreading of the entire text is very much appreciated.

I will be grateful to hear from readers regarding errors, omissions and suggestions for improvement.

Prague, Czech Republic | Zdeněk Martinec

Contents

Notation

Scalars

da	Area element in configuration κ_t
dA	Area element in configuration κ_0
dv	Volume element in configuration κ_t
dV	Volume element in configuration κ_0
$g_R(\mathcal{F})$	Symmetry group
h	Heat supply per unit mass
I_A, II_A, III_A	Principal invariants of a tensor $\boldsymbol{A}$
J	Jacobian (determinant of the deformation gradient)
k	Elastic bulk modulus
k_v	Bulk viscosity
K	Total kinetic energy
m	Mass of a body
p	Thermodynamic pressure
P	Total mechanical power
Q_h	Total heat energy
s	Exterior surface of a body in configuration κ_t
S	Exterior surface of a body in configuration κ_0
t	Time
$T(\vec{x}, t)$	Eulerian representation of the absolute temperature
$T(\vec{X}, t)$	Lagrangian representation of the absolute temperature
v	Volume of a body in configuration κ_t
V	Volume of a body in configuration κ_0
W	Strain energy density per unit undeformed volume
z	Entropy supply per unit mass
Z	Total entropy supply
γ	Entropy production density per unit mass
Γ	Total entropy production
$\delta_{kK}, \delta^{-1}_{Kk}$	Shifter symbols

δ_{KL}, δ_{kl}	Kronecker delta symbols
ϵ_{KLM}, ϵ_{klm}	Levi–Cività alternating symbols
ε	Internal energy density per unit mass
E	Total internal energy
η	Entropy density per unit mass
H	Total entropy
λ, μ	Lamé elastic parameters
λ_v, μ_v	Dilatational and shear viscosities
$\theta(\vec{x}, t)$	Eulerian representation of the empirical temperature
$\theta(\vec{X}, t)$	Lagrangian representation of the empirical temperature
κ	Thermal conductivity
κ_t	Present configuration
κ_0	Reference configuration
κ_τ	Relative reference configuration
π	Constraint pressure
ϱ	Mass density in configuration κ_t
$\mathcal{Q}$	Lagrangian representation of ϱ
ϱ^E	Eulerian increment in mass density
$\mathcal{Q}^L$	Lagrangian increment in mass density
ϱ_0	Mass density in configuration κ_0
σ	Mechanical pressure
σ	Eulerian representation of a discontinuity surface
Σ	Lagrangian representation of a discontinuity surface
τ	Normal stress
ϕ	Eulerian representation of gravitational potential
Φ	Lagrangian representation of gravitational potential
ϕ^E	Eulerian increment in gravitational potential
Φ^L	Lagrangian increment in gravitational potential
ψ	Helmholtz free energy density per unit mass

Vectors

$\vec{a}$, a_k	Acceleration in configuration κ_t
$\vec{A}$, A_K	Acceleration in configuration κ_0
$d\vec{a}$, da_k	Oriented area element in configuration κ_t
$d\vec{A}$, dA_K	Oriented area element in configuration κ_0
$\vec{c}^{\,*}$, c_k^*	Translation of Euclidean transformation
$\vec{f}$, f_k	Body force per unit mass
$\vec{F}$, F_K	Lagrangian representation of body force
$\vec{f}^E$	Eulerian increment in body force
$\vec{F}^L$	Lagrangian increment in body force
$\vec{g}$, g_k	Eulerian representation of gravitational attraction

$\vec{G}$, G_K	Lagrangian representation of gravitational attraction
$\vec{g}^E$	Eulerian increment in gravitation
$\vec{G}^L$	Lagrangian increment in gravitation
$\vec{g}_\theta$, $(\vec{g}_\theta)_k$	Eulerian representation of temperature gradient
$\vec{G}_\theta$, $(\vec{G}_\theta)_K$	Lagrangian representation of temperature gradient
$\vec{i}_k$	Cartesian unit base vectors in configuration κ_t
$\vec{I}_K$	Cartesian unit base vectors in configuration κ_0
$\vec{j}$, j_k	Eulerian representation of energy flux
$\vec{n}$, n_k	Outward unit normal vector to a surface in configuration κ_t
$\vec{N}$, N_K	Outward unit normal vector to a surface in configuration κ_0
$\vec{q}$, q_k	Heat flux in configuration κ_t
$\vec{Q}$, Q_K	Heat flux in configuration κ_0
$\vec{s}$, s_k	Entropy flux in configuration κ_t
$\vec{S}$, S_K	Entropy flux in configuration κ_0
$\vec{t}_{(\vec{n})}$	Stress vector on a surface with the unit normal vector $\vec{n}$
$\vec{u}$, u_k	Eulerian representation of displacement
$\vec{U}$, U_K	Lagrangian representation of displacement
$\vec{v}$, v_k	Eulerian representation of velocity
$\vec{V}$, V_K	Lagrangian representation of velocity
$\vec{x}$, x_k	Position of a particle $\mathcal{X}$ in configuration κ_t
$\vec{X}$, X_K	Position of a particle $\mathcal{X}$ in configuration κ_0
$\vec{w}$, w_k	Velocity of a discontinuity surface in configuration κ_t
$\vec{W}$, W_K	Velocity of a discontinuity surface in configuration κ_0
$\vec{\xi}$, ξ_k	Vorticity
$\vec{\chi}$, χ_k	Motion function
$\vec{\chi}_\tau$, $(\vec{\chi}_\tau)_k$	Relative motion function

Tensors

$\boldsymbol{a}_n$, $(\boldsymbol{a}_n)_{kl}$	Rivlin–Ericksen tensor of order n
$\boldsymbol{b}$, b_{kl}	Finger deformation tensor
$\boldsymbol{B}$, B_{KL}	Piola deformation tensor
$\boldsymbol{c}$, c_{kl}	Cauchy deformation tensor
$\boldsymbol{C}$, C_{KL}	Green deformation tensor
$\boldsymbol{c}_t$, $(\boldsymbol{c}_t)_{kl}$	Relative Green deformation tensor
$\boldsymbol{d}$, d_{kl}	Strain-rate tensor
$\boldsymbol{e}$, e_{kl}	Eulerian strain tensor
$\tilde{\boldsymbol{e}}$, $\tilde{e}_{kl}$	Linearised Eulerian strain tensor
$\boldsymbol{E}$, E_{KL}	Lagrangian strain tensor
$\widetilde{\boldsymbol{E}}$, $\widetilde{E}_{KL}$	Linearised Lagrangian strain tensor
$\boldsymbol{F}$, F_{kK}	Deformation gradient tensor
$\boldsymbol{F}_t$, $(\boldsymbol{F}_t)_{kl}$	Relative deformation gradient tensor

$\boldsymbol{H}$, H_{KL}	Displacement gradient tensor
$\boldsymbol{i}$, i_{kl}	Identity tensor in the spatial coordinates
$\boldsymbol{I}$, I_{KL} or I_{kl}	Identity tensor in the referential or spatial coordinates
$\boldsymbol{l}$, l_{kl}	Transposed spatial gradient of the velocity vector
$\boldsymbol{O}$, O_{kl}	Orthogonal transformation of Euclidean transformation
$\boldsymbol{P}$, $P_{\hat{K}L}$	Gradient of transformation of the reference configuration
$\boldsymbol{Q}$, $Q_{\hat{K}L}$	Orthogonal transformation of the reference configuration
$\boldsymbol{R}$, R_{kK}	Rotation tensor
$\widetilde{\boldsymbol{R}}$, $\widetilde{R}_{KL}$	Linearised Lagrangian rotation tensor
$\boldsymbol{t}(\vec{x}, t)$, t_{kl}	Eulerian representation of the Cauchy stress tensor
$\boldsymbol{t}(\vec{X}, t)$, t_{kl}	Lagrangian representation of the Cauchy stress tensor
$\boldsymbol{t}^E$	Eulerian increment in the Cauchy stress tensor
$\boldsymbol{t}^L$	Lagrangian increment in the Cauchy stress tensor
$\boldsymbol{T}^{(1)}$, $T^{(1)}_{Kl}$	First Piola–Kirchhoff stress tensor
$\boldsymbol{T}^{(1),L}$	Lagrangian increment in the first Piola–Kirchhoff stress tensor
$\boldsymbol{T}^{(2)}$, $T^{(2)}_{KL}$	Second Piola–Kirchhoff stress tensor
$\boldsymbol{U}$, U_{KL}	Right-stretch tensor
$\boldsymbol{V}$, V_{kl}	Left-stretch tensor
$\boldsymbol{w}$, w_{kl}	Spin tensor
$\boldsymbol{\kappa}$	Thermal conductivity tensor
$\boldsymbol{\varepsilon}$	Infinitesimal strain tensor
$\boldsymbol{\tau}$	Infinitesimal stress tensor
$\boldsymbol{\Omega}$, Ω_{kl}	Angular velocity tensor of Euclidean transformation

Calligraphics

$\mathcal{B}$	Body
$\mathcal{C}$, $\mathcal{C}_{KLMN}$	Elastic tensor
$\mathcal{E}$	Constitutive functional for the internal energy density
$\mathcal{F}$	Constitutive functional for the Cauchy stress tensor
$\mathcal{Q}$	Constitutive functional for the heat flux vector
$\mathcal{X}$	Material particle

Operators

det	Determinant of a tensor
Div	Divergence in referential coordinates
div	Divergence in spatial coordinates
div_Σ	Surface divergence
Grad	Gradient in referential coordinates

grad	Gradient in spatial coordinates
grad_Σ	Surface gradient
Rot	Rotation in referential coordinates
rot	Rotation in spatial coordinates
sym	Symmetric part of a tensor
tr	Trace of a tensor
$(\cdot)^D$	Deviator of a tensor
$(\cdot)^T$	Transpose of a tensor
$(\cdot)^{-1}$	Inverse of a tensor
$\cdot$	Scalar or dot product
$:$	Double-dot product
$\times$	Cross product
$\dot{\times}$	Dot-cross product
$\otimes$	Dyadic product
$\vec{\nabla}$	Nabla operator
∇^2	Laplace operator
$\frac{D}{Dt}$, $\dot{(\;)}$	Material time derivative
$\partial/\partial t$	Local time derivative

Other Symbols

$:=$	'is defined as'
$\stackrel{!}{=}$	'has to be equal to'
$\approx$	'approximately equal to'

Chapter 1
Geometry of Deformation

A deformation, in general, is a change in the shape or size of a body caused by the application of a force. Within the context of continuum mechanics, deformation is understood as the transformation of a body from one configuration to another. In the first two chapters, we will study deformation in terms of a pure geometrical description of changes in the shape of a body, regardless of the forces causing it. In this chapter, we confine our attention to two configurations without any regard for the sequence by which the second configuration is reached from the first. Both configurations are kept fixed in time and their geometrical relationship is described as a three-dimensional spatial mapping between them.

1.1 Bodies, Configurations and Motion

Materials, such as solids and fluids, are composed of atoms and molecules separated by 'empty' space. On a microscopic scale, materials have a discrete structure. However, experiments have shown that on a macroscopic scale, i.e., a length scale much greater than the size of atoms and inter-atomic distances, certain materials behave as though with a continuous structure, meaning that matter is continuously distributed in the occupied region. This can be accounted for by introducing the notion of a continuum. A *continuum* is an abstract body $\mathcal{B}$ that can be continually sub-divided into infinitesimal elements $\mathcal{X}$, called *particles* or *points*, whose properties are those of the bulk material. Whereas a 'particle' in classical mechanics has an assigned mass, a 'continuum particle' is a point for which a mass density is assigned (see Sect. 3.1). A mathematical study of the mechanics of continua is called *continuum mechanics*.

A *rigid body* is a body in which the distance between any two particles remains constant through time regardless of external forces exerted on it. A *deformable body*

Z. Martinec, *Principles of Continuum Mechanics*, Nečas Center Series,
https://doi.org/10.1007/978-3-030-05390-1_1

is a body in which particles can move relative to each other under the action of external forces. This book deals with deformable bodies.

A body $\mathcal{B}$ is available to us only by its configuration. The *configuration* κ of $\mathcal{B}$ is the specification of the position of all particles of $\mathcal{B}$ in the physical space E^3 (usually the Euclidean space). Different configurations of the body $\mathcal{B}$ correspond to different regions in Euclidean space. Often it is convenient to select one particular configuration, the *reference configuration* κ_0, and refer everything concerning the body to it. Mathematically, the definition of the reference configuration κ_0 is expressed by mapping

$$
\begin{aligned}
\vec{\gamma}_0 :\ & \mathcal{B} \to \mathrm{E}^3 \\
& \mathcal{X} \to \vec{X} = \vec{\gamma}_0(\mathcal{X}),
\end{aligned}
\tag{1.1}
$$

where $\vec{X}$ is the position occupied by the particle $\mathcal{X}$ in the reference configuration κ_0, as shown in Fig. 1.1.

The choice of reference configuration is arbitrary. It may be any simplified representation of the body $\mathcal{B}$, and need not even be a configuration ever occupied by the body. For some choice of κ_0, we may obtain a relatively simple description, just as in geometry one choice of coordinates may lead to a simple equation for a particular figure. However, the reference configuration itself has nothing to do with the motion it is used to describe, just as the coordinate system has nothing to do with the geometric figures themselves. A reference configuration is introduced so as to allow us to employ the mathematical apparatus of Euclidean geometry.

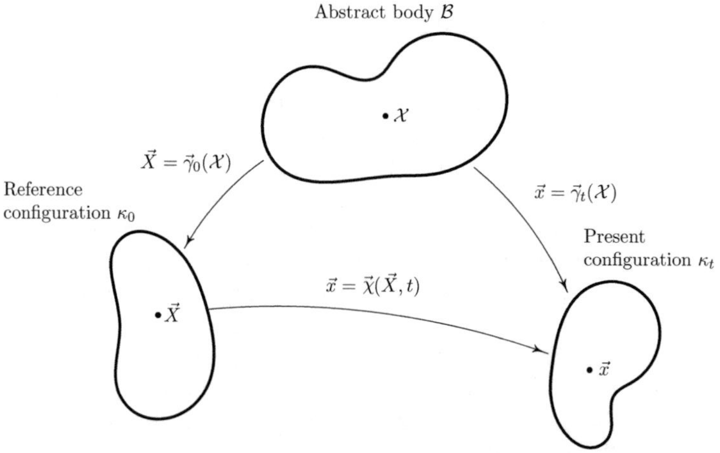

Fig. 1.1 An abstract body, its reference and present configurations

Under the action of external forcing, the body $\mathcal{B}$ deforms, moves and changes its configuration.[1] The configuration of body $\mathcal{B}$ at the present time t, called the *present configuration* κ_t, is defined by mapping

$$\begin{aligned} \vec{\gamma}_t : \ & \mathcal{B} \to \mathrm{E}^3 \\ & \mathcal{X} \to \vec{x} = \vec{\gamma}_t(\mathcal{X}), \end{aligned} \tag{1.2}$$

where $\vec{x}$ is the position occupied by the particle $\mathcal{X}$ in the present configuration κ_t.

The *deformation* of body $\mathcal{B}$ from the reference configuration κ_0 to the present configuration κ_t is described by mapping

$$\begin{aligned} \vec{\chi}(\cdot, t) : \ & \mathrm{E}^3 \to \mathrm{E}^3 \\ & \vec{X} \to \vec{x} = \vec{\chi}(\vec{X}, t), \end{aligned} \tag{1.3}$$

where the time t is held fixed. This equation states that the deformation takes a particle $\mathcal{X}$ from its position $\vec{X}$ in κ_0 to a position $\vec{x}$ in κ_t at a particular time t.

The *motion* of body $\mathcal{B}$ is then a continuous sequence of deformations in time between the reference configuration κ_0 and the present configuration κ_t,

$$\begin{aligned} \vec{\chi} : \ & \mathrm{E}^3 \times \mathbb{R} \to \mathrm{E}^3 \\ & (\vec{X}, t) \to \vec{x} = \vec{\chi}(\vec{X}, t). \end{aligned} \tag{1.4}$$

We assume that the *motion function* $\vec{\chi}$ is continuously differentiable and the Jacobian of the function $\vec{\chi}$ is continuous and non-singular in finite regions of the body or in the entire body. According to the inverse function theorem, the mapping (1.4) is invertible such that

$$\vec{X} = \vec{\chi}^{-1}(\vec{x}, t) \tag{1.5}$$

holds. The functional form (1.4) for a given motion depends on the choice of the reference configuration. If more than one reference configuration is used, it is necessary to label the function accordingly. For example, relative to two reference configurations κ_1 and κ_2, the same motion could be represented symbolically by the two equations $\vec{x} = \vec{\chi}_{\kappa_1}(\vec{X}, t) = \vec{\chi}_{\kappa_2}(\vec{X}, t)$.

It is often convenient to change the reference configuration in the description of motion. To see how the motion is described in a new reference configuration, consider two different configurations κ_τ and κ_t of body $\mathcal{B}$ at two different times,

[1] As introduced, a rigid body cannot deform. However, it can translate and rotate. By Chasles' theorem, the general motion of a rigid body is a combination of a pure rotation about a point of the body, and the translation motion of that point. This point may be arbitrary, but is usually taken to be the centre of mass of the body.

τ and t,

$$\vec{\xi} = \vec{\chi}(\vec{X}, \tau), \qquad \vec{x} = \vec{\chi}(\vec{X}, t), \tag{1.6}$$

that is, $\vec{\xi}$ is the position occupied at time τ by the particle that occupies $\vec{x}$ at time t. Since the function $\vec{\chi}$ is invertible, that is,

$$\vec{X} = \vec{\chi}^{\,-1}(\vec{\xi}, \tau) = \vec{\chi}^{\,-1}(\vec{x}, t), \tag{1.7}$$

we have either

$$\vec{\xi} = \vec{\chi}(\vec{\chi}^{\,-1}(\vec{x}, t), \tau) =: \vec{\chi}_t(\vec{x}, \tau) \tag{1.8}$$

or[2]

$$\vec{x} = \vec{\chi}(\vec{\chi}^{\,-1}(\vec{\xi}, \tau), t) =: \vec{\chi}_\tau(\vec{\xi}, t). \tag{1.9}$$

The map $\vec{\chi}_t(\vec{x}, \tau)$ defines the deformation of the new configuration κ_τ of the body $\mathcal{B}$ relative to the present configuration κ_t, which is considered to be the reference. On the other hand, the map $\vec{\chi}_\tau(\vec{\xi}, t)$ defines the deformation of the new reference configuration κ_τ of the body $\mathcal{B}$ onto the configuration κ_t. Evidently,

$$\vec{\chi}_t(\vec{x}, \tau) = \vec{\chi}_\tau^{\,-1}(\vec{x}, t) \tag{1.10}$$

holds. The functions $\vec{\chi}_t(\vec{x}, \tau)$ and $\vec{\chi}_\tau(\vec{\xi}, t)$ are called the *relative motion functions*. The subscripts t and τ are used to recall which configuration is taken as the reference.

We assume that the functions $\vec{\gamma}_0$, $\vec{\gamma}_t$, $\vec{\chi}$, $\vec{\chi}_t$ and $\vec{\chi}_\tau$ are single-valued and continuous, with continuous derivatives with respect to space and time to whatever order is desired, usually to the second or third, except possibly at some singular points, curves and surfaces. Moreover, each of these functions can be uniquely inverted. This assumption is known as the *axiom of continuity*, which refers to the fact that the matter is *indestructible*. This means that a finite volume of matter cannot be deformed into a zero or infinite volume. Another implication of this axiom is that the matter is *impenetrable*, that is, one portion of matter never penetrates into another. Moreover, this axiom prevents the formation of cavities. In other words, a motion carries every volume into a volume, every surface onto a surface, and every curve onto a curve. There are, however, cases where the axiom of continuity is violated. The formulation presented in this book cannot describe such processes as the creation of new material surfaces, or the cutting, tearing or propagation of cracks. Continuum theories dealing with such processes must weaken the continuity assumption in the neighbourhood of those parts of the body where the map (1.4)

[2]The symbol ':=' means 'is defined as'.

becomes discontinuous. The axiom of continuity is mathematically ensured by the well-known implicit function theorem.

1.2 Descriptions of Motion

Motion can be described in four ways. Under the assumption that the functions $\vec{\kappa}_0$, $\vec{\kappa}_t$, $\vec{\chi}$, $\vec{\chi}_t$ and $\vec{\chi}_\tau$ are differentiable and invertible, all descriptions are equivalent. We refer to them by

- *Material description*, given by the mapping (1.2), whose independent variables are the abstract particle $\mathcal{X}$ and the time t.
- *Referential description*, given by the mapping (1.4), whose independent variables are the position $\vec{X}$ of the particle $\mathcal{X}$ in an arbitrarily chosen reference configuration and the time t. When the reference configuration is chosen to be the actual initial configuration at $t = 0$, the referential description is often called the *Lagrangian description*, although many authors call it the material description, using the particle position $\vec{X}$ in the reference configuration as a label for the particle $\mathcal{X}$ in the material description.
- *Spatial description*, given by the mapping (1.5), whose independent variables are the present position $\vec{x}$ occupied by the particle at the present time t. It is the description most used in fluid mechanics, often called the *Eulerian description*.
- *Relative description*, given by the mapping (1.8), whose independent variables are the present position $\vec{x}$ of the particle and a variable time τ, being the time when the particle has occupied another position $\vec{\xi}$. The motion is then described with $\vec{\xi}$ as a dependent variable. Alternatively, the motion can be described by the mapping (1.9), whose independent variables are the position $\vec{\xi}$ at time τ and the present time t. These two relative descriptions are actually special cases of the referential description, differing from the Lagrangian description in that the reference positions are now denoted by $\vec{x}$ at time t and $\vec{\xi}$ at time τ, respectively, instead of $\vec{X}$ at time $t = 0$.

1.3 Referential and Spatial Coordinates

We consider the body $\mathcal{B}$ that occupies the reference configuration κ_0 at time t_0 and the present configuration κ_t at time t. Let V and v denote the respective volumes and S and s the respective surfaces of the reference and present configurations of the body. The location of particle $\mathcal{X}$ in the reference configuration κ_0 may be described by three Cartesian coordinates X_K, $K = 1, 2, 3$, or by the position vector $\vec{X}$ that extends from the origin O of the coordinate system to the point P, the location of particle $\mathcal{X}$ in the reference configuration, as illustrated in Fig. 1.2. In the present configuration κ_t, the particle $\mathcal{X}$ occupies the point p, which may be located by the

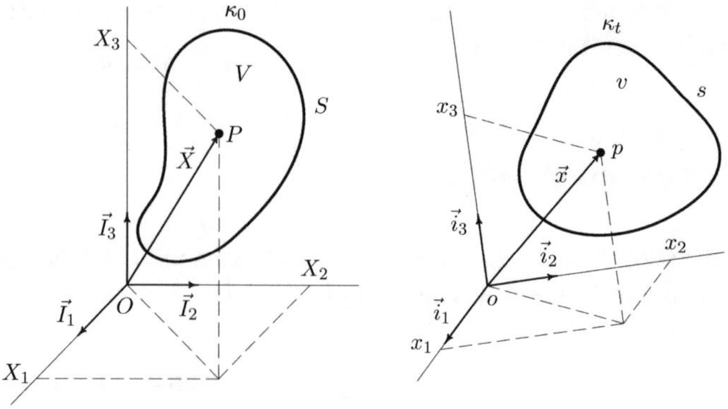

Fig. 1.2 Referential and spatial coordinates for the reference configuration κ_0 and the present configuration κ_t

position vector $\vec{x}$ extending from the origin o of a new set of Cartesian coordinates x_k, $k = 1, 2, 3$. Following the current terminology, we will call X_K the *referential* or *Lagrangian coordinates* and x_k the *spatial* or *Eulerian coordinates.* In following considerations, we assume that these two coordinate systems, one for the reference configuration κ_0 and one for the present configuration κ_t, are non-identical.

The reference position $\vec{X}$ of point P in κ_0 and the present position $\vec{x}$ of p in κ_t, respectively, when referred to the Cartesian coordinates X_K and x_k, are given by[3]

$$\vec{X} = X_K \vec{I}_K, \qquad \vec{x} = x_k \vec{i}_k, \tag{1.11}$$

where $\vec{I}_K$ and $\vec{i}_k$ are the respective *unit base vectors* in Fig. 1.2 (the usual summation convention over repeated indices is employed). Since Cartesian coordinates are employed, the base vectors are mutually orthogonal,

$$\vec{I}_K \cdot \vec{I}_L = \delta_{KL}, \qquad \vec{i}_k \cdot \vec{i}_l = \delta_{kl}, \tag{1.12}$$

where δ_{KL} and δ_{kl} are the *Kronecker delta symbols*, which are equal to 1 when the two indices are equal and zero otherwise; the dot '$\cdot$' denotes the *scalar product of vectors.*[4]

[3]The notions of a position vector and position are interchangeable if we use only one reference origin for the position vector, but 'position' has meaning independent of the choice of origin. The same position may have many different position vectors corresponding to different choices of origin, but the relative position vectors $d\vec{X}$ of neighbouring positions will be the same for all origins.

[4]If a curvilinear coordinate system is employed, the appropriate form of these equations can be obtained by the standard transformation rules. For example, the partial derivatives in a Cartesian coordinate system must be replaced by the partial covariant derivatives. However, for general considerations, we will rely on the already introduced Cartesian coordinate systems.

Let us consider two Cartesian coordinate systems that are related by a rigid (i.e., independent of location) translation and rotation. A unit base vector in one coordinate system can be expressed in terms of its projection onto the other coordinate system as

$$\vec{i}_k = \delta_{kK}\vec{I}_K, \qquad \vec{I}_K = \delta^{-1}_{Kk}\vec{i}_k, \tag{1.13}$$

where the expansion coefficients δ_{kK} and δ^{-1}_{Kk} are called the *shifter symbols*. They are the components of *two-point tensors* (see Sect. 1.5) because they relate the unit base vectors $\vec{i}_k$ and $\vec{I}_K$ in the two Cartesian coordinate systems x_k and X_K. Note that they are *not* Kronecker delta symbols, except when the two coordinate systems coincide.

From the identity

$$\delta_{kl}\,\vec{i}_l = \vec{i}_k = \delta_{kK}\vec{I}_K = \delta_{kK}\delta^{-1}_{Kl}\,\vec{i}_l,$$

we deduce that

$$\delta_{kK}\delta^{-1}_{Kl} = \delta_{kl}, \qquad \delta^{-1}_{Kk}\delta_{kL} = \delta_{KL}. \tag{1.14}$$

To derive a relationship between the two sets of shifter symbols, we express the scalar product of the unit base vectors $\vec{i}_k$ and $\vec{I}_K$ in two different ways by $(1.13)_1$ and $(1.13)_2$, respectively,

$$\vec{i}_k \cdot \vec{I}_K = \delta_{kL}\vec{I}_L \cdot \vec{I}_K = \delta_{kK} \qquad \text{and} \qquad \vec{i}_k \cdot \vec{I}_K = \delta^{-1}_{Kl}\vec{i}_k \cdot \vec{i}_l = \delta^{-1}_{Kk}.$$

Hence,

$$\delta^{-1}_{Kk} = \delta_{kK}, \tag{1.15}$$

while (1.14) takes the form

$$\delta_{kK}\delta_{lK} = \delta_{kl}, \qquad \delta_{kK}\delta_{kL} = \delta_{KL}, \tag{1.16}$$

which shows that δ_{kK} (or δ^{-1}_{Kk}) describe a rigid rotation of the Cartesian coordinate system X_K with respect to the Cartesian coordinate system x_k (see Sect. 5.1). Note that the commutative property of the scalar product does *not* imply the symmetry of the shifter symbols as shown by

$$\delta_{kK} = \vec{i}_k \cdot \vec{I}_K = \vec{I}_K \cdot \vec{i}_k = \vec{I}_K \cdot \delta_{kL}\vec{I}_L = \delta_{kK}.$$

The following notation convention will be such that the quantities associated with the reference configuration κ_0 will be denoted by capital letters, and those associated with the present configuration κ_t by small letters. When these quantities are referred to coordinates X_K, their indices will be upper-case letters, and when they are referred to x_k, their indices will be lower-case letters. For example, a vector $\vec{V}$ in κ_0 referred to X_K will have the components V_K, while when it is referred to x_k, it will have the components V_k, such that

$$V_K = \vec{V} \cdot \vec{I}_K, \qquad V_k = \vec{V} \cdot \vec{i}_k. \tag{1.17}$$

The components V_K can be expressed in terms of the components V_k. Using (1.13) and (1.15), the components V_K and V_k can be related by

$$V_K = V_k \delta_{kK}, \qquad V_k = V_K \delta_{kK}. \tag{1.18}$$

Conversely, consider vector $\vec{v}$ in κ_t that, in general, differs from $\vec{V}$. Its components v_K and v_k when referred to X_K and x_k, respectively, are

$$v_K = \vec{v} \cdot \vec{I}_K, \qquad v_k = \vec{v} \cdot \vec{i}_k. \tag{1.19}$$

Again, using (1.13) and (1.15), the components v_K and v_k can be related by

$$v_K = v_k \delta_{kK}, \qquad v_k = v_K \delta_{kK}. \tag{1.20}$$

1.4 Lagrangian and Eulerian Variables

Every scalar, vector or tensor[5] physical quantity $\mathcal{Q}$ defined for the body $\mathcal{B}$, such as mass density, temperature or velocity, is defined with respect to a particle $\mathcal{X}$ at a certain time t as

$$\mathcal{Q} = \hat{\mathcal{Q}}(\mathcal{X}, t). \tag{1.21}$$

Since the particle $\mathcal{X}$ is available to us in the reference or present configurations, the physical quantity $\mathcal{Q}$ is always considered a function of the position of the particle $\mathcal{X}$ in the reference or present configurations. Assuming that the function (1.1) is invertible, i.e., $\mathcal{X} = \vec{\gamma}_0^{-1}(\vec{X})$, the Lagrangian representation of the quantity $\mathcal{Q}$ is

$$\mathcal{Q} = \hat{\mathcal{Q}}(\mathcal{X}, t) = \hat{\mathcal{Q}}(\vec{\gamma}_0^{-1}(\vec{X}), t) =: Q(\vec{X}, t). \tag{1.22}$$

Alternatively, inverting (1.2), i.e., $\mathcal{X} = \vec{\gamma}_t^{-1}(\vec{x})$, the Eulerian representation of the quantity $\mathcal{Q}$ is

$$\mathcal{Q} = \hat{\mathcal{Q}}(\mathcal{X}, t) = \hat{\mathcal{Q}}(\vec{\gamma}_t^{-1}(\vec{x}), t) =: q(\vec{x}, t). \tag{1.23}$$

[5] Unless specifically stated, we implicitly presume that a tensor denotes a second-order tensor.

Hence, the *Lagrangian* and *Eulerian variables* $Q(\vec{X}, t)$ and $q(\vec{x}, t)$ are referred to the reference configuration κ_0 and the present configuration κ_t of the body $\mathcal{B}$, respectively. In the Lagrangian description, attention is focused on what is happening to the individual particles during the motion, whereas in the Eulerian description, the attention is directed to the events taking place at specific points in space. For example, if $\mathcal{Q}$ is temperature, then $Q(\vec{X}, t)$ gives the temperature recorded by a thermometer attached to a moving particle $\vec{X}$, whereas $q(\vec{x}, t)$ gives the temperature recorded at a fixed point $\vec{x}$ in space, which may be occupied by different particles during time. The relationship between these two descriptions is

$$Q(\vec{X}, t) = q(\vec{\chi}(\vec{X}, t), t), \qquad q(\vec{x}, t) = Q(\vec{\chi}^{-1}(\vec{x}, t), t), \tag{1.24}$$

where the small and capital letters emphasise different functional forms resulting from the change in variables.

As an example, we define the Lagrangian and Eulerian variables for a vector quantity $\vec{\mathcal{V}}$. Let us assume that $\vec{\mathcal{V}}$ in the Eulerian description is

$$\vec{\mathcal{V}} \equiv \vec{v}(\vec{x}, t). \tag{1.25}$$

Vector $\vec{v}$ may be expressed in terms of the referential or spatial components $v_K(\vec{x}, t)$ and $v_k(\vec{x}, t)$, respectively, as

$$\vec{v}(\vec{x}, t) = v_K(\vec{x}, t)\vec{I}_K = v_k(\vec{x}, t)\vec{i}_k. \tag{1.26}$$

The Lagrangian representation of $\vec{\mathcal{V}}$, that is, the vector $\vec{V}(\vec{X}, t)$, is defined by $(1.24)_1$ as

$$\vec{V}(\vec{X}, t) := \vec{v}(\vec{\chi}(\vec{X}, t), t). \tag{1.27}$$

Representing $\vec{V}(\vec{X}, t)$ in terms of the referential or spatial components $V_K(\vec{X}, t)$ and $V_k(\vec{X}, t)$, respectively, that is,

$$\vec{V}(\vec{X}, t) = V_K(\vec{X}, t)\vec{I}_K = V_k(\vec{X}, t)\vec{i}_k, \tag{1.28}$$

the definition (1.27) can be interpreted in two possible component forms:

$$V_K(\vec{X}, t) := v_K(\vec{\chi}(\vec{X}, t), t), \qquad V_k(\vec{X}, t) := v_k(\vec{\chi}(\vec{X}, t), t). \tag{1.29}$$

Expressing the referential components v_K in terms of the spatial components v_k according to (1.20) gives

$$V_K(\vec{X}, t) = v_k(\vec{\chi}(\vec{X}, t), t)\delta_{kK}, \qquad V_k(\vec{X}, t) = v_K(\vec{\chi}(\vec{X}, t), t)\delta_{kK}. \tag{1.30}$$

An analogous consideration may be carried out for the Eulerian variables $v_K(\vec{x}, t)$ and $v_k(\vec{x}, t)$ in the case where $\vec{\mathcal{V}}$ is given in the Lagrangian representation $\vec{\mathcal{V}} \equiv \vec{V}(\vec{X}, t)$.

1.5 Deformation Gradients

The coordinate form of the motion (1.4) is

$$x_k = \chi_k(X_1, X_2, X_3, t), \qquad k = 1, 2, 3, \tag{1.31}$$

or, conversely,

$$X_K = \chi_K^{-1}(x_1, x_2, x_3, t), \qquad K = 1, 2, 3. \tag{1.32}$$

According to the implicit function theorem, the mathematical condition that guarantees the existence of such a unique inversion is that the Jacobian determinant of the matrix of partial derivatives $\partial\chi_k/\partial X_K$ (or, simply, *Jacobian*) does not vanish,

$$J(\vec{X}, t) := \det\left(\frac{\partial\chi_k}{\partial X_K}\right) \neq 0. \tag{1.33}$$

The differentials of (1.31) and (1.32), at a fixed time, are

$$dx_k = \chi_{k,K}\, dX_K, \qquad dX_K = \chi_{K,k}^{-1}\, dx_k, \tag{1.34}$$

where the index after a comma denotes a partial derivative with respect to X_K when the index is a upper-case letter, and with respect to x_k when the index is a lower-case letter,

$$\chi_{k,K} := \frac{\partial\chi_k}{\partial X_K}, \qquad \chi_{K,k}^{-1} := \frac{\partial\chi_K^{-1}}{\partial x_k}. \tag{1.35}$$

The two sets of quantities defined by (1.35) are components of the *deformation gradient* $\boldsymbol{F}$ of the mapping (1.31) and the *deformation gradient* $\boldsymbol{F}^{-1}$ of the inverse mapping (1.32), respectively,

$$\boldsymbol{F}(\vec{X}, t) := \chi_{k,K}(\vec{X}, t)\big(\vec{i}_k \otimes \vec{I}_K\big), \qquad \boldsymbol{F}^{-1}(\vec{x}, t) := \chi_{K,k}^{-1}(\vec{x}, t)\big(\vec{I}_K \otimes \vec{i}_k\big), \tag{1.36}$$

where the symbol $\otimes$ denotes the *dyadic product of vectors*.[6] Alternatively, (1.36) may be written in tensor notation as[7]

$$\boldsymbol{F}(\vec{X}, t) := (\mathrm{Grad}\, \vec{\chi})^T, \qquad \boldsymbol{F}^{-1}(\vec{x}, t) := (\mathrm{grad}\, \vec{\chi}^{-1})^T. \tag{1.37}$$

The deformation gradients $\boldsymbol{F}$ and $\boldsymbol{F}^{-1}$ relate vectors $d\vec{x}$ and $d\vec{X}$ in the present and reference configurations and are therefore expressed in terms of two different base vectors $\vec{i}_k$ and $\vec{I}_K$. They are known as *two-point tensors*, which refers to the fact that their components transform like those of a vector under rotations of only one of two reference axes and like a two-point tensor when the two sets of axes are rotated independently. In tensor notation, (1.34) appears in the form

$$d\vec{x} = dx_k \vec{i}_k = \boldsymbol{F} \cdot d\vec{X}, \qquad d\vec{X} = dX_K \vec{I}_K = \boldsymbol{F}^{-1} \cdot d\vec{x}. \tag{1.38}$$

---

[6] The dyadic product or *tensor product* of two vectors $\vec{u}$ and $\vec{v}$ is defined through its action on an arbitrary vector $\vec{a}$ by

$$(\vec{u} \otimes \vec{v}) \cdot \vec{a} = (\vec{a} \cdot \vec{v})\vec{u}.$$

The dyadic product is a second-order tensor, and is called *dyad*. A dyad is a quantity that has magnitude and *two* associated directions. For example, product of inertia is a measure of how far mass is distributed in two directions. A *dyadic* is a linear combination of dyads. For example, an inertia dyadic describes the mass distribution of a body and is the sum of various dyads associated with products and moments of inertia. In general, the dyadic $\vec{a} \otimes \vec{b} + \vec{c} \otimes \vec{d}$ is not a dyad because it cannot be written as a vector multiplied by a vector. Dyadics differ from vectors in that the sum of two vectors is a vector, whereas the sum of two dyads is not necessarily a dyad. Every second-order tensor can be represented by a dyadic.

Let $\vec{u}$, $\vec{v}$, $\vec{w}$ and $\vec{x}$ be arbitrary vectors, and $\boldsymbol{A}$ be an arbitrary tensor. It can be shown that

$$\begin{aligned}
(\vec{u} \otimes \vec{v})^T &= \vec{v} \otimes \vec{u}, \\
(\vec{u} \otimes \vec{v}) \cdot (\vec{w} \otimes \vec{x}) &= (\vec{v} \cdot \vec{w})(\vec{u} \otimes \vec{x}), \\
(\vec{u} \otimes \vec{v}) \boldsymbol{:} (\vec{w} \otimes \vec{x}) &= (\vec{v} \cdot \vec{w})(\vec{u} \cdot \vec{x}), \\
\boldsymbol{A} \cdot (\vec{u} \otimes \vec{v}) &= (\boldsymbol{A} \cdot \vec{u}) \otimes \vec{v}, \\
(\vec{u} \otimes \vec{v}) \cdot \boldsymbol{A} &= \vec{u} \otimes (\boldsymbol{A}^T \cdot \vec{v}), \\
\boldsymbol{A} \boldsymbol{:} (\vec{u} \otimes \vec{v}) &= \vec{v} \cdot \boldsymbol{A} \cdot \vec{u},
\end{aligned}$$

where the symbol **:** denotes the *double-dot product of dyads*.

[7] The nabla operator $\vec{\nabla}$ in the Cartesian coordinates x_k is defined as $\vec{\nabla} := \vec{i}_k \dfrac{\partial}{\partial x_k}$. With this operator, the gradient of a vector function $\vec{\phi}$ is defined by the left dyadic product of the nabla operator $\vec{\nabla}$ with $\vec{\phi}$, that is, $\mathrm{grad}\,\vec{\phi} := \vec{\nabla} \otimes \vec{\phi}$. The Cartesian components of $\mathrm{grad}\,\vec{\phi}$ are then given by $(\mathrm{grad}\,\vec{\phi})_{kl} = \phi_{l,k}$. Moreover, the gradients 'Grad' and 'grad' denote the gradient operator with respect to referential and spatial coordinates, respectively, the time t being held constant in each case.

One powerful feature of tensor notation is that it describes physical laws in a manner that is independent of any particular coordinate system (or reference frame) used. Such a requirement is clearly necessary for a mathematical description of a physical law to be valid, since the laws of the universe cannot depend on the reference frame used to describe them. In turn, this requirement defines how the components of a tensor transform under a change of reference frame.

The deformation gradient $\boldsymbol{F}$ can thus be thought of as a mapping of the infinitesimal vector $d\vec{X}$ of the reference configuration onto the infinitesimal vector $d\vec{x}$ of the present configuration; the inverse mapping is performed by $\boldsymbol{F}^{-1}$.

In view of (1.31) and (1.32), the chain rule of differentiation yields

$$\chi_{k,K}\overset{-1}{\chi}_{K,l} = \delta_{kl}, \qquad \overset{-1}{\chi}_{K,k}\chi_{k,L} = \delta_{KL}, \tag{1.39}$$

or, in tensor notation,

$$\boldsymbol{F}\cdot\boldsymbol{F}^{-1} = \boldsymbol{i}, \qquad \boldsymbol{F}^{-1}\cdot\boldsymbol{F} = \boldsymbol{I}, \tag{1.40}$$

where $\boldsymbol{i} = \vec{i}_k \otimes \vec{i}_k$ and $\boldsymbol{I} = \vec{I}_K \otimes \vec{I}_K$ are the identity tensors in the referential and spatial coordinates, respectively. We make, however, an exception in the notation and use $\boldsymbol{I}$ for both the identity tensors. Equation (1.40) shows that $\boldsymbol{F}^{-1}$ is indeed the inverse of $\boldsymbol{F}$.

Each of the two systems of nine linear equations (1.39) for the nine unknown $\chi_{k,K}$ or $\overset{-1}{\chi}_{K,k}$ has a unique solution because the Jacobian J is assumed not to equal zero. According to Cramer's rule, the solution of $(1.39)_1$, say, for $\overset{-1}{\chi}_{K,k}$ is

$$\overset{-1}{\chi}_{K,k} = \frac{\text{cofactor}\,(\chi_{k,K})}{J} = \frac{1}{2J}\epsilon_{KLM}\epsilon_{klm}\chi_{l,L}\chi_{m,M}, \tag{1.41}$$

where ϵ_{KLM} and ϵ_{klm} are the Levi–Cività alternating symbols. The Jacobian J, defined by (1.33), can also be expressed with the help of the Levi–Cività symbols as

$$J = \frac{1}{3!}\epsilon_{KLM}\epsilon_{klm}\chi_{k,K}\chi_{l,L}\chi_{m,M}. \tag{1.42}$$

With the help of the basic properties of the Levi–Cività alternating symbols,[8] (1.41) and (1.42) may be written as

$$\begin{aligned} J\epsilon_{KLM}\overset{-1}{\chi}_{K,k} &= \epsilon_{klm}\chi_{l,L}\chi_{m,M}, \\ J\epsilon_{KLM} &= \epsilon_{klm}\chi_{k,K}\chi_{l,L}\chi_{m,M}, \end{aligned} \tag{1.43}$$

[8] The product of two alternating symbols defines a sixth-order tensor with the components

$$\epsilon_{KLM}\epsilon_{RST} = \det\begin{pmatrix} \delta_{KR} & \delta_{LR} & \delta_{MR} \\ \delta_{KS} & \delta_{LS} & \delta_{MS} \\ \delta_{KT} & \delta_{LT} & \delta_{MT} \end{pmatrix}.$$

The successive contraction of indices yields

$$\begin{aligned} \epsilon_{KLM}\epsilon_{KST} &= \delta_{LS}\delta_{MT} - \delta_{LT}\delta_{MS}, \\ \epsilon_{KLM}\epsilon_{KLT} &= 2\delta_{MT}, \\ \epsilon_{KLM}\epsilon_{KLM} &= 6. \end{aligned}$$

A tensor which has the same components in all rotated Cartesian coordinate systems is called an *isotropic tensor*. The Kronecker delta symbols and the Levi–Cività alternating symbols are components of a second-order and third-order isotropic tensors, respectively.

or, in tensor notation,

$$
\begin{aligned}
J\boldsymbol{F}^{-T}\cdot(\vec{A}\times\vec{B}) &= (\boldsymbol{F}\cdot\vec{A})\times(\boldsymbol{F}\cdot\vec{B}),\\
J(\vec{A}\times\vec{B})\cdot\vec{C} &= \big[(\boldsymbol{F}\cdot\vec{A})\times(\boldsymbol{F}\cdot\vec{B})\big]\cdot(\boldsymbol{F}\cdot\vec{C}),
\end{aligned}
\tag{1.44}
$$

which is valid for all vectors $\vec{A}$, $\vec{B}$ and $\vec{C}$.

The determinant of $\boldsymbol{F}$ is[9]

$$
\det\boldsymbol{F} = \frac{[(\boldsymbol{F}\cdot\vec{I}_1)\times(\boldsymbol{F}\cdot\vec{I}_2)]\cdot(\boldsymbol{F}\cdot\vec{I}_3)}{(\vec{I}_1\times\vec{I}_2)\cdot\vec{I}_3},
$$

which, in view of $(1.44)_2$, gives

$$
\det\boldsymbol{F} = J. \tag{1.45}
$$

Note that this relation is valid only in the case where the referential and spatial coordinates are of the same type, as in the case here when both are Cartesian coordinates. If the referential coordinates are of a different type to the spatial coordinates, the volume element in these coordinates is not the same and, consequently, the Jacobian J will differ from the determinant of $\boldsymbol{F}$.

Upon differentiating (1.42) and (1.41), we obtain the following *Jacobi's identities*:

$$
\begin{aligned}
&\frac{dJ}{d\chi_{k,K}} = J\chi^{-1}_{K,k},\\
&\left(J\chi^{-1}_{K,k}\right)_{,K} = 0, \qquad \text{or} \qquad \left(J^{-1}\chi_{k,K}\right)_{,k} = 0.
\end{aligned}
\tag{1.46}
$$

The first identity is proved as follows:

$$
\begin{aligned}
\frac{dJ}{d\chi_{r,R}} &= \frac{1}{3!}\epsilon_{KLM}\epsilon_{klm}\left[\frac{\partial\chi_{k,K}}{\partial\chi_{r,R}}\chi_{l,L}\chi_{m,M} + \chi_{k,K}\frac{\partial\chi_{l,L}}{\partial\chi_{r,R}}\chi_{m,M} + \chi_{k,K}\chi_{l,L}\frac{\partial\chi_{m,M}}{\partial\chi_{r,R}}\right]\\
&= \frac{1}{3!}\left[\epsilon_{RLM}\epsilon_{rlm}\chi_{l,L}\chi_{m,M} + \epsilon_{KRM}\epsilon_{krm}\chi_{k,K}\chi_{m,M} + \epsilon_{KLR}\epsilon_{klr}\chi_{k,K}\chi_{l,L}\right]\\
&= \frac{1}{2}\epsilon_{RLM}\epsilon_{rlm}\chi_{l,L}\chi_{m,M} = \text{cofactor}\,(\chi_{r,R}) = J\chi^{-1}_{R,r}.
\end{aligned}
$$

[9] The determinant of a tensor $\boldsymbol{A}$ is defined by

$$
\det\boldsymbol{A} := \frac{[(\boldsymbol{A}\cdot\vec{a})\times(\boldsymbol{A}\cdot\vec{b})]\cdot(\boldsymbol{A}\cdot\vec{c})}{(\vec{a}\times\vec{b})\cdot\vec{c}}
$$

for any basis $\{\vec{a},\vec{b},\vec{c}\}$. Thus, $|\det\boldsymbol{A}|$ is the ratio of the volume of the parallelepiped defined by vectors $\boldsymbol{A}\cdot\vec{a}$, $\boldsymbol{A}\cdot\vec{b}$ and $\boldsymbol{A}\cdot\vec{c}$ to the volume of the parallelepiped defined by vectors $\vec{a}$, $\vec{b}$ and $\vec{c}$.

Furthermore, differentiating this result with respect to X_K yields

$$\begin{aligned}
\left(J\chi^{-1}_{K,k}\right)_{,K} &= \frac{1}{2}\epsilon_{KLM}\epsilon_{klm}(\chi_{l,LK}\chi_{m,M} + \chi_{l,L}\chi_{m,MK}) \\
&= \epsilon_{KLM}\epsilon_{klm}\chi_{l,LK}\chi_{m,M} \\
&= \frac{1}{2}\epsilon_{klm}\chi_{m,M}\left(\epsilon_{KLM}\chi_{l,LK} + \epsilon_{KLM}\chi_{l,LK}\right) \\
&= \frac{1}{2}\epsilon_{klm}\chi_{m,M}\left(\epsilon_{KLM}\chi_{l,LK} - \epsilon_{LKM}\chi_{l,LK}\right) \\
&= \frac{1}{2}\epsilon_{klm}\chi_{m,M}\left(\epsilon_{KLM}\chi_{l,LK} - \epsilon_{KLM}\chi_{l,KL}\right) \\
&= 0,
\end{aligned}$$

hence proving the second identity in (1.46). In the last step, we have assumed that the order of differentiation with respect to X_K and X_L can be interchanged.

Jacobi's identities can be expressed in tensor notation as[10]

$$\begin{aligned}
&\frac{dJ}{d\boldsymbol{F}} = J\boldsymbol{F}^{-T}, \\
&\text{Div}\left(J\boldsymbol{F}^{-1}\right) = \vec{0}, \qquad \text{or} \qquad \text{div}\left(J^{-1}\boldsymbol{F}\right) = \vec{0}.
\end{aligned} \tag{1.47}$$

The differential operators in the spatial coordinates applied to the Eulerian representation of a physical quantity $\mathcal{Q}$ can be converted to the differential operators in the referential coordinates applied to the Lagrangian representation of the quantity $\mathcal{Q}$ by means of the chain rule of differentiation. For example, the following identities can be verified:

$$\begin{aligned}
\text{grad}\,\bullet &= \boldsymbol{F}^{-T}\cdot\text{Grad}\,\bullet, \\
\text{div}\,\bullet &= J^{-1}\text{Div}\,(J\boldsymbol{F}^{-1}\cdot\bullet), \\
\text{div}\,\text{grad}\,\bullet &= J^{-1}\text{Div}\,(J\boldsymbol{F}^{-1}\cdot\boldsymbol{F}^{-T}\cdot\text{Grad}\,\bullet).
\end{aligned} \tag{1.48}$$

To prove this, let the Euler representation of a tensor $\mathcal{T}$ be denoted by $\boldsymbol{t}(\vec{x},t)$. The corresponding Lagrangian representation of $\mathcal{T}$ is $\boldsymbol{T}(\vec{X},t) = \boldsymbol{t}(\vec{\chi}(\vec{X},t),t)$. Then,

$$\begin{aligned}
\text{grad}\,\boldsymbol{t} &= \vec{i}_k \otimes \frac{\partial \boldsymbol{t}}{\partial x_k} = \vec{i}_k \otimes \frac{\partial \boldsymbol{T}}{\partial X_K}\frac{\partial X_K}{\partial x_k} = \chi^{-1}_{K,k}\vec{i}_k \otimes \frac{\partial \boldsymbol{T}}{\partial X_K} = \left(\vec{I}_K\cdot\boldsymbol{F}^{-1}\right)\otimes\frac{\partial \boldsymbol{T}}{\partial X_K} \\
&= \left(\boldsymbol{F}^{-T}\cdot\vec{I}_K\right)\otimes\frac{\partial \boldsymbol{T}}{\partial X_K} = \boldsymbol{F}^{-T}\cdot\left(\vec{I}_K\otimes\frac{\partial \boldsymbol{T}}{\partial X_K}\right) = \boldsymbol{F}^{-T}\cdot\text{Grad}\,\boldsymbol{T},
\end{aligned}$$

[10] Note that $\boldsymbol{F}^{-T} \equiv \left(\boldsymbol{F}^T\right)^{-1} \equiv \left(\boldsymbol{F}^{-1}\right)^T$.

hence proving the first identity in (1.48). To prove the second identity, we apply the differential identity (A.25) to

$$\mathrm{Div}\,(J\boldsymbol{F}^{-1}\cdot\boldsymbol{T}) = \mathrm{Div}\,(J\boldsymbol{F}^{-1})\cdot\boldsymbol{T} + J\boldsymbol{F}^{-T} : \mathrm{Grad}\,\boldsymbol{T}.$$

The first term is equal to zero due to Jacobi's identity $(1.47)_2$. The double-dot product in the second term can be successively written such that

$$\begin{aligned}\boldsymbol{F}^{-T} : \mathrm{Grad}\,\boldsymbol{T} &= \boldsymbol{F}^{-T} : \vec{I}_K \otimes \frac{\partial \boldsymbol{T}}{\partial X_K} = \chi^{-1}_{K,k}\vec{i}_k \cdot \frac{\partial \boldsymbol{T}}{\partial X_K} = \chi^{-1}_{K,k}\vec{i}_k \cdot \frac{\partial \boldsymbol{t}}{\partial x_m}\frac{\partial x_m}{\partial X_K} \\ &= \chi^{-1}_{K,k}\chi_{m,K}\,\vec{i}_k \cdot \frac{\partial \boldsymbol{t}}{\partial x_m} = \delta_{km}\,\vec{i}_k \cdot \frac{\partial \boldsymbol{t}}{\partial x_m} = \vec{i}_k \cdot \frac{\partial \boldsymbol{t}}{\partial x_k} = \mathrm{div}\,\boldsymbol{t},\end{aligned}$$

which proves the second identity. The last identity in (1.48) derives from the combination of the previous two relationships. Furthermore, the inverse relations to (1.48) are given by

$$\begin{aligned}\mathrm{Grad}\,\bullet &= \boldsymbol{F}^T\cdot\mathrm{grad}\,\bullet, \\ \mathrm{Div}\,\bullet &= J\mathrm{div}\,(J^{-1}\boldsymbol{F}\cdot\bullet), \\ \mathrm{Div}\,\mathrm{Grad}\,\bullet &= J\mathrm{div}\,(J^{-1}\boldsymbol{F}\cdot\boldsymbol{F}^T\cdot\mathrm{grad}\,\bullet).\end{aligned} \tag{1.49}$$

1.6 Polar Decomposition of the Deformation Gradient

Decomposing a deformation into rotation and stretch enables us to assess the local behaviour of deformation. This decomposition, known as the *polar decomposition of the deformation gradient*,[11] states that a non-singular tensor $\boldsymbol{F}$ ($\det\boldsymbol{F} \neq 0$) can be decomposed into two ways:[12]

$$\boldsymbol{F} = \boldsymbol{R}\cdot\boldsymbol{U} = \boldsymbol{V}\cdot\boldsymbol{R}, \tag{1.50}$$

where the tensors $\boldsymbol{R}$, $\boldsymbol{U}$ and $\boldsymbol{V}$ have the following properties:

- $\boldsymbol{U}$ and $\boldsymbol{V}$ are symmetric and positive definite.
- $\boldsymbol{R}$ is *orthogonal*, $\boldsymbol{R}\cdot\boldsymbol{R}^T = \boldsymbol{R}^T\cdot\boldsymbol{R} = \boldsymbol{I}$, i.e., $\boldsymbol{R}$ is either a *rotation tensor* or a *reflection tensor*.[13]

[11] Equation (1.50) is analogous to the polar decomposition of a complex number: $z = re^{i\varphi}$, where $r = (x^2+y^2)^{1/2}$ and $\varphi = \arctan(y/x)$. For this reason, it is referred to as the *polar decomposition*.

[12] The polar decomposition may be applied to every second-order, non-singular tensor as the product of a positive-definite symmetric tensor and an orthogonal tensor.

[13] If for an orthogonal tensor, $\det\boldsymbol{Q} = +1$, $\boldsymbol{Q}$ is said to be a *proper orthogonal tensor*, corresponding to a *rotation*. If $\det\boldsymbol{Q} = -1$, $\boldsymbol{Q}$ is said to be an *improper orthogonal tensor*, corresponding to a *reflection*.

- $\boldsymbol{U}, \boldsymbol{V}$ and $\boldsymbol{R}$ are unique.
- $\boldsymbol{U}$ and $\boldsymbol{V}$ have the same eigenvalues (termed *principal stretches*), but different eigenvectors.

In preparation for proving these statements, we note that an arbitrary non-singular tensor $\boldsymbol{T}$ is positive definite if $\vec{v} \cdot \boldsymbol{T} \cdot \vec{v} > 0$ for all vectors $\vec{v} \neq \vec{0}$. A necessary and sufficient condition for $\boldsymbol{T}$ to be positive definite is that all of its eigenvalues are positive. In this regard, consider the tensor $\boldsymbol{C}$, $\boldsymbol{C} := \boldsymbol{F}^T \cdot \boldsymbol{F}$. Since $\boldsymbol{F}$ is assumed to be non-singular and $\boldsymbol{F} \cdot \vec{v} \neq \vec{0}$ if $\vec{v} \neq \vec{0}$, it follows that $(\boldsymbol{F} \cdot \vec{v}) \cdot (\boldsymbol{F} \cdot \vec{v})$ is a sum of squares and hence greater than zero. Thus,

$$0 < (\boldsymbol{F} \cdot \vec{v}) \cdot (\boldsymbol{F} \cdot \vec{v}) = \vec{v} \cdot \boldsymbol{F}^T \cdot \boldsymbol{F} \cdot \vec{v} = \vec{v} \cdot \boldsymbol{C} \cdot \vec{v},$$

and $\boldsymbol{C}$ is positive definite. In view of the same arguments, the tensor $\boldsymbol{b}$, $\boldsymbol{b} := \boldsymbol{F} \cdot \boldsymbol{F}^T$, is also positive definite. The positive roots of tensors $\boldsymbol{C}$ and $\boldsymbol{b}$ define uniquely two tensors $\boldsymbol{U}$ and $\boldsymbol{V}$ such that[14]

$$\boldsymbol{U}(\vec{X}, t) := \sqrt{\boldsymbol{C}} = \sqrt{\boldsymbol{F}^T \cdot \boldsymbol{F}}, \qquad \boldsymbol{V}(\vec{x}, t) := \sqrt{\boldsymbol{b}} = \sqrt{\boldsymbol{F} \cdot \boldsymbol{F}^T}. \tag{1.51}$$

The tensors $\boldsymbol{U}$ and $\boldsymbol{V}$, called the *right-* and *left-stretch tensors*, are symmetric, positive definite and unique.

Furthermore, two tensors $\boldsymbol{R}$ and $\tilde{\boldsymbol{R}}$,

$$\boldsymbol{R} := \boldsymbol{F} \cdot \boldsymbol{U}^{-1}, \qquad \tilde{\boldsymbol{R}} := \boldsymbol{V}^{-1} \cdot \boldsymbol{F}, \tag{1.52}$$

are orthogonal because

$$\begin{aligned} \boldsymbol{R} \cdot \boldsymbol{R}^T &= \left(\boldsymbol{F} \cdot \boldsymbol{U}^{-1}\right) \cdot \left(\boldsymbol{F} \cdot \boldsymbol{U}^{-1}\right)^T = \boldsymbol{F} \cdot \boldsymbol{U}^{-1} \cdot \boldsymbol{U}^{-1} \cdot \boldsymbol{F}^T = \boldsymbol{F} \cdot \boldsymbol{U}^{-2} \cdot \boldsymbol{F}^T \\ &= \boldsymbol{F} \cdot \left(\boldsymbol{F}^T \cdot \boldsymbol{F}\right)^{-1} \cdot \boldsymbol{F}^T = \boldsymbol{F} \cdot \boldsymbol{F}^{-1} \cdot \boldsymbol{F}^{-T} \cdot \boldsymbol{F}^T = \boldsymbol{I} \end{aligned}$$

and

$$\boldsymbol{R}^T \cdot \boldsymbol{R} = \left(\boldsymbol{F} \cdot \boldsymbol{U}^{-1}\right)^T \cdot \left(\boldsymbol{F} \cdot \boldsymbol{U}^{-1}\right) = \boldsymbol{U}^{-1} \cdot \boldsymbol{F}^T \cdot \boldsymbol{F} \cdot \boldsymbol{U}^{-1} = \boldsymbol{U}^{-1} \cdot \boldsymbol{U}^2 \cdot \boldsymbol{U}^{-1} = \boldsymbol{I}.$$

The orthogonality of $\tilde{\boldsymbol{R}}$ is proved in a similar manner.

Using the orthogonality of $\boldsymbol{R}$ gives

$$\boldsymbol{F} = \boldsymbol{R} \cdot \boldsymbol{U} = \boldsymbol{R} \cdot \boldsymbol{U} \cdot (\boldsymbol{R}^T \cdot \boldsymbol{R}) = (\boldsymbol{R} \cdot \boldsymbol{U} \cdot \boldsymbol{R}^T) \cdot \boldsymbol{R} =: \tilde{\boldsymbol{V}} \cdot \boldsymbol{R}.$$

[14]To find the square roots of a symmetric tensor, three steps need to be made: diagonalise the symmetric tensor by rotation to its principal directions (see Sect. 1.10), take the square roots of the diagonal elements and rotate the result back to the original orientation.

On the other hand, $\boldsymbol{F} = \boldsymbol{V}\cdot\tilde{\boldsymbol{R}}$ in view of $(1.52)_2$. Therefore, it may be concluded that there may be two decompositions of $\boldsymbol{F}$, for which $\sqrt{\boldsymbol{b}} = \boldsymbol{V} = \tilde{\boldsymbol{V}}$ holds. However, since $\boldsymbol{V}$ is unique, $\tilde{\boldsymbol{V}}$ must be equal to $\boldsymbol{V}$, which implies that $\tilde{\boldsymbol{R}} = \boldsymbol{R}$. Thus, $\boldsymbol{R}$ is unique.

To complete the proofs of the statements, let λ and $\vec{v}$ be an eigenvalue and eigenvector of $\boldsymbol{U}$, i.e., $\lambda\vec{v} = \boldsymbol{U}\cdot\vec{v}$. Then, $\lambda\boldsymbol{R}\cdot\vec{v} = (\boldsymbol{R}\cdot\boldsymbol{U})\cdot\vec{v} = (\boldsymbol{V}\cdot\boldsymbol{R})\cdot\vec{v} = \boldsymbol{V}\cdot(\boldsymbol{R}\cdot\vec{v})$, which implies that λ is also an eigenvalue of $\boldsymbol{V}$ and $\boldsymbol{R}\cdot\vec{v}$ is its eigenvector.

It is instructive to write the relation (1.50) in its component form

$$F_{kK} = R_{kL}U_{LK} = V_{kl}R_{lK}, \tag{1.53}$$

which means that $\boldsymbol{R}$ is a two-point tensor, while $\boldsymbol{U}$ and $\boldsymbol{V}$ are ordinary (one-point) tensors. Note that $\boldsymbol{U}$, which is related to the reference configuration, is multiplied by $\boldsymbol{R}$ and $\boldsymbol{R}^T$ to rotate it to the present configuration, to obtain $\boldsymbol{V}$,

$$\boldsymbol{V} = \boldsymbol{R}\cdot\boldsymbol{U}\cdot\boldsymbol{R}^T. \tag{1.54}$$

The equation $d\vec{x} = \boldsymbol{F}\cdot d\vec{X}$ shows that the deformation gradient $\boldsymbol{F}$ can be thought of as the mapping of the infinitesimal vector $d\vec{X}$ of the reference configuration onto the infinitesimal vector $d\vec{x}$ of the present configuration. The polar decomposition replaces the linear transformation $d\vec{x} = \boldsymbol{F}\cdot d\vec{X}$ by rotation and stretch, whose sequence is given by the multiplication of either $\boldsymbol{R}$ and $\boldsymbol{U}$, or $\boldsymbol{V}$ and $\boldsymbol{R}$, as illustrated in Fig. 1.3:

$$d\vec{x} = (\boldsymbol{R}\cdot\boldsymbol{U})\cdot d\vec{X} = (\boldsymbol{V}\cdot\boldsymbol{R})\cdot d\vec{X}. \tag{1.55}$$

It is worth noting that $\boldsymbol{R}$ should not be confused with a rigid-body rotation because, in general, $\boldsymbol{R}$ varies from point to point, $\boldsymbol{R} = \boldsymbol{R}(\vec{X}, t)$, and describes local properties of deformation only.

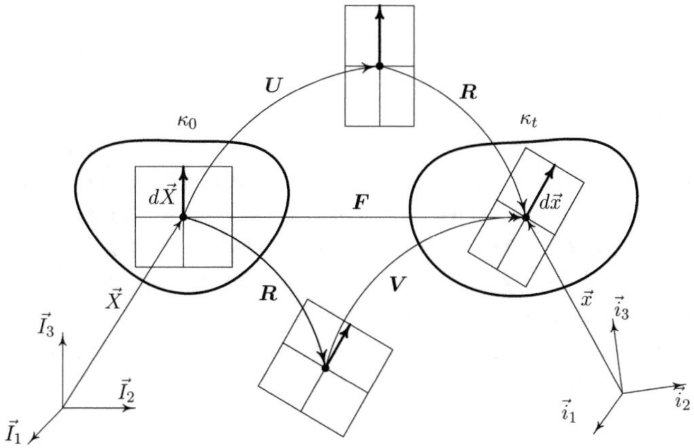

Fig. 1.3 Polar decomposition of the deformation gradient

1.7 Measures of Deformation

Local changes in the geometry of continuous bodies can be described, as usual in differential geometry, by changes in the metric tensor. In Euclidean space, this is particularly simple to accomplish. Consider a point $\vec{X}$ and an infinitesimal vector $d\vec{X}$. Changes in the length of three such linearly independent vectors describe local changes in the geometry.

The infinitesimal vector $d\vec{X}$ in κ_0 is mapped onto the infinitesimal vector $d\vec{x}$ in κ_t. The metric properties of the present configuration κ_t can be described by the square of the length of $d\vec{x}$,

$$ds^2 = d\vec{x} \cdot d\vec{x} = (\boldsymbol{F} \cdot d\vec{X}) \cdot (\boldsymbol{F} \cdot d\vec{X}) = d\vec{X} \cdot \boldsymbol{C} \cdot d\vec{X} = \boldsymbol{C} : (d\vec{X} \otimes d\vec{X}), \tag{1.56}$$

where the *Green deformation tensor* $\boldsymbol{C}$,

$$\boldsymbol{C}(\vec{X}, t) := \boldsymbol{F}^T \cdot \boldsymbol{F}, \tag{1.57}$$

has already been used in the proof of the polar decomposition theorem. Equation (1.56) describes the local geometric property of the present configuration κ_t with respect to the reference configuration κ_0. Alternatively, we can use the inverse tensor expressing the geometry of the reference configuration κ_0 relative to the present configuration κ_t such that

$$dS^2 = d\vec{X} \cdot d\vec{X} = (\boldsymbol{F}^{-1} \cdot d\vec{x}) \cdot (\boldsymbol{F}^{-1} \cdot d\vec{x}) = d\vec{x} \cdot \boldsymbol{c} \cdot d\vec{x} = \boldsymbol{c} : (d\vec{x} \otimes d\vec{x}), \tag{1.58}$$

where $\boldsymbol{c}$ is the *Cauchy deformation tensor*,

$$\boldsymbol{c}(\vec{x}, t) := \boldsymbol{F}^{-T} \cdot \boldsymbol{F}^{-1}. \tag{1.59}$$

Both tensors $\boldsymbol{c}$ and $\boldsymbol{C}$ are symmetric, that is, $\boldsymbol{c} = \boldsymbol{c}^T$ and $\boldsymbol{C} = \boldsymbol{C}^T$. We see that, in contrast to the deformation gradient tensor $\boldsymbol{F}$ with generally nine independent components, the changes in metric properties, following from the changes of configuration, are described by six independent components of the deformation tensors $\boldsymbol{c}$ or $\boldsymbol{C}$.

Apart from the Cauchy deformation tensor $\boldsymbol{c}$ and the Green deformation tensor $\boldsymbol{C}$, we can introduce other equivalent geometrical measures of deformation. Equations (1.56) and (1.58) yield two different expressions for the squares of element of length, ds^2 and dS^2. The difference $ds^2 - dS^2$ is a relative measure of the change of length. When this difference vanishes for any two neighbouring points, the deformation has not changed the distance between the pair. When it is zero for all points in the body, the body has undergone only a *rigid-body motion*, that is, the rigid-body translation and rigid-body rotation. From (1.56) and (1.58)

$$ds^2 - dS^2 = d\boldsymbol{X} \cdot 2\boldsymbol{E} \cdot d\boldsymbol{X} = d\boldsymbol{x} \cdot 2\boldsymbol{e} \cdot d\boldsymbol{x}, \tag{1.60}$$

where we have introduced the *Lagrangian* and *Eulerian strain tensors*, respectively,

$$\boldsymbol{E}(\vec{X}, t) := \frac{1}{2}(\boldsymbol{C} - \boldsymbol{I}), \qquad \boldsymbol{e}(\vec{x}, t) := \frac{1}{2}(\boldsymbol{I} - \boldsymbol{c}). \tag{1.61}$$

Clearly, $ds^2 = dS^2$ when either vanishes. The strain tensors are non-zero only when a strain or stretch arises. Both tensors $\boldsymbol{e}$ and $\boldsymbol{E}$ are symmetric, that is, $\boldsymbol{e} = \boldsymbol{e}^T$ and $\boldsymbol{E} = \boldsymbol{E}^T$. Therefore, in three dimensions there are only six independent components for each of these tensors, for example, E_{11}, E_{22}, E_{33}, $E_{12} = E_{21}$, $E_{13} = E_{31}$ and $E_{23} = E_{32}$. The first three components, E_{11}, E_{22} and E_{33}, are called *normal strains* and the last three, E_{12}, E_{13} and E_{23}, are called *shear strains*. The reason for this will be discussed later in this chapter. Note that, for rigid-body motion, both strain tensors vanish, $\boldsymbol{E} = \boldsymbol{0}$ and $\boldsymbol{e} = \boldsymbol{0}$, but $\boldsymbol{C} = \boldsymbol{U} = \boldsymbol{I}$ and $\boldsymbol{F} = \boldsymbol{R}$.

Two other equivalent measures of deformation are the *reciprocal tensors* $\boldsymbol{b}$ and $\boldsymbol{B}$ (known as the *Finger* and *Piola deformation tensors*, respectively),

$$\boldsymbol{b}(\vec{x}, t) := \boldsymbol{F} \cdot \boldsymbol{F}^T, \qquad \boldsymbol{B}(\vec{X}, t) := \boldsymbol{F}^{-1} \cdot \boldsymbol{F}^{-T}. \tag{1.62}$$

They satisfy the conditions

$$\boldsymbol{b} \cdot \boldsymbol{c} = \boldsymbol{c} \cdot \boldsymbol{b} = \boldsymbol{I}, \qquad \boldsymbol{B} \cdot \boldsymbol{C} = \boldsymbol{C} \cdot \boldsymbol{B} = \boldsymbol{I}, \tag{1.63}$$

which follow from the definitions (1.57), (1.59) and (1.62). Moreover, it holds

$$\boldsymbol{b} = \boldsymbol{R} \cdot \boldsymbol{C} \cdot \boldsymbol{R}^T, \tag{1.64}$$

meaning that the Green deformation tensor $\boldsymbol{C}$, which is related to the reference configuration, is multiplied by $\boldsymbol{R}$ and $\boldsymbol{R}^T$ to rotate it to the present configuration to obtain the Finger deformation tensor b.

We have been using the notion of *tensor* for quantities such as $\boldsymbol{C}$. A tensor refers to a set of quantities that transform according to a certain rule upon *coordinate transformation*. Suppose that the Cartesian coordinates X_K are transformed onto $X'_{K'}$ by the one-to-one mapping

$$X_K = \hat{X}_K(X'_1, X'_2, X'_3). \tag{1.65}$$

Substituting

$$dX_K = \frac{\partial \hat{X}_K}{\partial X'_{K'}} dX'_{K'}$$

into the right-hand side of (1.56) and using the fact that the left-hand side of (1.56) is independent of coordinate transformations gives[15]

$$ds^2 = C_{KL} dX_K dX_L = C_{KL} \frac{\partial \hat{X}_K}{\partial X'_{K'}} \frac{\partial \hat{X}_L}{\partial X'_{L'}} dX'_{K'} dX'_{L'} \stackrel{!}{=} (ds')^2 = C'_{K'L'} dX'_{K'} dX'_{L'}.$$

Since $dX'_{K'}$ is arbitrary, we have

$$C'_{K'L'}(\vec{X}', t) = \frac{\partial \hat{X}_K}{\partial X'_{K'}} \frac{\partial \hat{X}_L}{\partial X'_{L'}} C_{KL}(\vec{X}, t), \tag{1.66}$$

which is called the *transformation rule* between the primed and unprimed components in the two coordinate systems. Thus, knowing C_{KL} in one set of coordinates X_K, we can find the corresponding quantities in another set $X'_{K'}$ once the transformation (1.65) between X_K and X'_K is given. Quantities that transform according to (1.66) are known as *tensors*.

Similarly, the components of the Cauchy deformation tensor $\boldsymbol{c}$ transform under a change of the Cartesian coordinates according to

$$c'_{k'l'}(\vec{x}', t) = \frac{\partial \hat{x}_k}{\partial x'_{k'}} \frac{\partial \hat{x}_l}{\partial x'_{l'}} c_{kl}(\vec{x}, t). \tag{1.67}$$

1.8 Length and Angle Changes

A geometrical meaning of the normal strains E_{11}, E_{22} and E_{33} is provided by considering the length and angle changes that result from deformation. We consider a line element $d\vec{X}$ of length dS that deforms to the element $d\vec{x}$ of length ds. Let $\vec{K}$ be the unit vector along $d\vec{X}$,

$$\vec{K} := \frac{d\vec{X}}{dS}. \tag{1.68}$$

The relative change of length,

$$E_{(\vec{K})} := \frac{ds - dS}{dS}, \tag{1.69}$$

[15] The symbol '$\stackrel{!}{=}$' means 'has to be equal to'.

is called the *extension* or *elongation*.[16] Dividing $(1.60)_1$ by dS^2 gives

$$\frac{ds^2 - dS^2}{dS^2} = \vec{K} \cdot 2\boldsymbol{E} \cdot \vec{K}. \tag{1.70}$$

Expressing the left-hand side in terms of the extension $E_{(\vec{K})}$ leads to the quadratic equation for $E_{(\vec{K})}$,

$$E_{(\vec{K})}\big(E_{(\vec{K})} + 2\big) - \vec{K} \cdot 2\boldsymbol{E} \cdot \vec{K} = 0. \tag{1.71}$$

From the two possible solutions of this equation, we choose the physically admissible option

$$E_{(\vec{K})} = -1 + \sqrt{1 + \vec{K} \cdot 2\boldsymbol{E} \cdot \vec{K}}\,. \tag{1.72}$$

Since $\vec{K}$ is a unit vector, $\vec{K} \cdot 2\boldsymbol{E} \cdot \vec{K} + 1 = \vec{K} \cdot \boldsymbol{C} \cdot \vec{K}$ by $(1.61)_1$. As proved in Sect. 1.6, the Green deformation tensor $\boldsymbol{C}$ is positive definite, that is, $\vec{K} \cdot \boldsymbol{C} \cdot \vec{K} > 0$. Hence, the argument of the square root in (1.72) is positive. Specifically, taking $\vec{K}$ along the X_1 axis, (1.72) simplifies to

$$E_{(1)} = -1 + \sqrt{1 + 2E_{11}}\,. \tag{1.73}$$

Similar expressions hold for the extensions along the X_2 and X_3 axes.

A geometrical meaning of the shear strains E_{12}, E_{13} and E_{23} is provided by considering the angles between two directions $\vec{K}^{(1)}$ and $\vec{K}^{(2)}$,

$$\vec{K}^{(1)} := \frac{d\vec{X}^{(1)}}{dS^{(1)}}, \qquad \vec{K}^{(2)} := \frac{d\vec{X}^{(2)}}{dS^{(2)}}. \tag{1.74}$$

As illustrated in Fig. 1.4, the angle Θ between these vectors in the reference configuration,

$$\cos\Theta = \frac{d\vec{X}^{(1)}}{dS^{(1)}} \cdot \frac{d\vec{X}^{(2)}}{dS^{(2)}}, \tag{1.75}$$

[16] Complementary to the Lagrangian extension $E_{(\vec{K})}$, the Eulerian extension $e_{(\vec{K})}$ can be introduced by

$$e_{(\vec{K})} := \frac{ds - dS}{ds}.$$

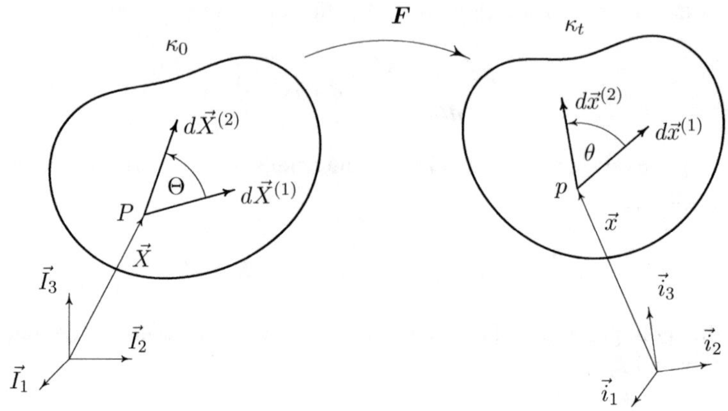

Fig. 1.4 Angle change

is changed by deformation to

$$\cos\theta = \frac{d\vec{x}^{(1)}}{ds^{(1)}} \cdot \frac{d\vec{x}^{(2)}}{ds^{(2)}} = \frac{\boldsymbol{F}\cdot d\vec{X}^{(1)}}{ds^{(1)}} \cdot \frac{\boldsymbol{F}\cdot d\vec{X}^{(2)}}{ds^{(2)}} = \frac{d\vec{X}^{(1)}\cdot\boldsymbol{C}\cdot d\vec{X}^{(2)}}{ds^{(1)}\,ds^{(2)}}. \tag{1.76}$$

From (1.68) and (1.69)

$$\cos\theta = \big(\vec{K}^{(1)}\cdot\boldsymbol{C}\cdot\vec{K}^{(2)}\big)\frac{dS^{(1)}}{ds^{(1)}}\frac{dS^{(2)}}{ds^{(2)}} = \frac{\vec{K}^{(1)}\cdot\boldsymbol{C}\cdot\vec{K}^{(2)}}{\big(E_{(\vec{K}_1)}+1\big)\big(E_{(\vec{K}_2)}+1\big)}, \tag{1.77}$$

which can be rewritten in terms of the Lagrangian strain tensor as

$$\cos\theta = \frac{\vec{K}^{(1)}\cdot(\boldsymbol{I}+2\boldsymbol{E})\cdot\vec{K}^{(2)}}{\sqrt{1+\vec{K}^{(1)}\cdot2\boldsymbol{E}\cdot\vec{K}^{(1)}}\sqrt{1+\vec{K}^{(2)}\cdot2\boldsymbol{E}\cdot\vec{K}^{(2)}}}. \tag{1.78}$$

Specifically, taking $\vec{K}^{(1)}$ along the X_1 axis and $\vec{K}^{(2)}$ along the X_2 axis, (1.78) simplifies to

$$\cos\theta_{(12)} = \frac{2E_{12}}{\sqrt{1+2E_{11}}\sqrt{1+2E_{22}}}. \tag{1.79}$$

1.9 Area and Volume Changes

We now determine a change in area and volume due to deformation. The oriented area element in the reference configuration built on the edge vectors $d\vec{X}^{(1)}$ and $d\vec{X}^{(2)}$,

$$d\vec{A} := d\vec{X}^{(1)} \times d\vec{X}^{(2)}, \tag{1.80}$$

where the cross '$\times$' denotes the *cross product of vectors*, is changed by deformation to the oriented area element with edge vectors $d\vec{x}^{(1)}$ and $d\vec{x}^{(2)}$,

$$d\vec{a} := d\vec{x}^{(1)} \times d\vec{x}^{(2)} = (\boldsymbol{F} \cdot d\vec{X}^{(1)}) \times (\boldsymbol{F} \cdot d\vec{X}^{(2)}) = J\boldsymbol{F}^{-T} \cdot (d\vec{X}^{(1)} \times d\vec{X}^{(2)}),$$

where we have used the identity $(1.44)_1$. Substituting for $d\vec{A}$ gives the relationship between the area elements in the reference and present configurations,

$$d\vec{a} = J\boldsymbol{F}^{-T} \cdot d\vec{A}, \qquad \text{or} \qquad da_k = JX_{K,k}dA_K, \tag{1.81}$$

which is known as *Nanson's formula.*

The oriented area element $d\vec{a}$ can be expressed as the product of the magnitude da of $d\vec{a}$, $da = (d\vec{a} \cdot d\vec{a})^{1/2}$, and the unit vector $\vec{n}$ normal to the plane area $d\vec{x}^{(1)} \times d\vec{x}^{(2)}$, that is, $d\vec{a} = \vec{n}da$. Similarly, $d\vec{A} = \vec{N}dA$. Using Nanson's formula (1.81) then gives

$$da = J\sqrt{\vec{N} \cdot \boldsymbol{B} \cdot \vec{N}}\, dA, \tag{1.82}$$

where $\boldsymbol{B}$ is the Piola deformation tensor defined by $(1.62)_2$. Dividing $(1.81)_1$ by da and using (1.82) yields

$$\vec{n} = \frac{\vec{N} \cdot \boldsymbol{F}^{-1}}{\sqrt{\vec{N} \cdot \boldsymbol{B} \cdot \vec{N}}}. \tag{1.83}$$

To find the change of the volume element by deformation, an infinitesimal rectilinear parallelepiped in the reference configuration defined by the vectors $d\vec{X}^{(1)}, d\vec{X}^{(2)}$ and $d\vec{X}^{(3)}$ is considered (see Fig. 1.5). Its volume is given by the scalar product of $d\vec{X}^{(3)}$ with $d\vec{X}^{(1)} \times d\vec{X}^{(2)}$,

$$dV := \big(d\vec{X}^{(1)} \times d\vec{X}^{(2)}\big) \cdot d\vec{X}^{(3)}. \tag{1.84}$$

In the present configuration, the volume of the parallelepiped is

$$\begin{aligned} dv &:= \big(d\vec{x}^{(1)} \times d\vec{x}^{(2)}\big) \cdot d\vec{x}^{(3)} = \big[(\boldsymbol{F} \cdot d\vec{X}^{(1)}) \times (\boldsymbol{F} \cdot d\vec{X}^{(2)})\big] \cdot (\boldsymbol{F} \cdot d\vec{X}^{(3)}) \\ &= J\big(d\vec{X}^{(1)} \times d\vec{X}^{(2)}\big) \cdot d\vec{X}^{(3)}, \end{aligned}$$

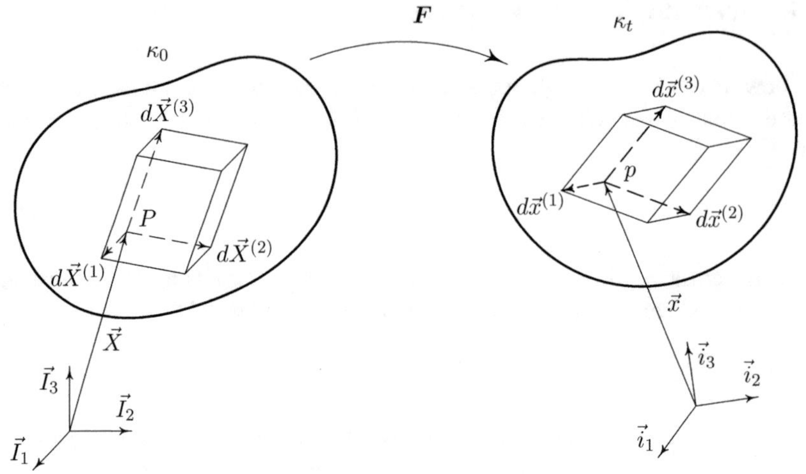

Fig. 1.5 Deformation of an infinitesimal rectilinear parallelepiped

where we have used the identity $(1.44)_2$. Substituting for dV gives the relationship between the volume elements in the reference and present configurations,

$$dv = J dV. \tag{1.85}$$

Consequently, the Jacobian J measures the volume changes of infinitesimal elements. Since volumes cannot be negative and $J \neq 0$, J must be positive at all points of the body at all times,

$$J > 0. \tag{1.86}$$

1.10 Strain Invariants and Principal Strains

In this section, a brief review of the invariants of a second-order symmetric tensor is given. The Lagrangian strain tensor $\boldsymbol{E}$ is considered as a typical example of this group. It is of interest to determine, at a given point $\vec{X}$, the directions $\vec{V}$ for which the expression $\vec{V} \cdot \boldsymbol{E} \cdot \vec{V}$ takes extremal values. For this we must differentiate $\vec{V} \cdot \boldsymbol{E} \cdot \vec{V}$ with respect to $\vec{V}$ subject to the condition $\vec{V} \cdot \vec{V} = 1$. Using Lagrange's method of multipliers, we set

$$\frac{\partial}{\partial V_K}\left[\vec{V} \cdot \boldsymbol{E} \cdot \vec{V} - \lambda(\vec{V} \cdot \vec{V} - 1)\right] = 0, \tag{1.87}$$

where λ is the unknown Lagrange multiplier, which gives

$$(E_{KL} - \lambda\delta_{KL})\, V_L = 0. \tag{1.88}$$

A non-trivial solution of the homogeneous equations (1.88) exists only if the *characteristic determinant* vanishes,

$$\det(\boldsymbol{E} - \lambda\boldsymbol{I}) = 0. \tag{1.89}$$

Upon expanding this determinant, we obtain a cubic algebraic equation in λ, known as the *characteristic equation* of the tensor $\boldsymbol{E}$:

$$-\lambda^3 + I_E\,\lambda^2 - II_E\,\lambda + III_E = 0, \tag{1.90}$$

where

$$\begin{aligned} I_E &:= E_{11} + E_{22} + E_{33} \equiv \operatorname{tr}\boldsymbol{E}, \\ II_E &:= E_{11}E_{22} + E_{11}E_{33} + E_{22}E_{33} - E_{12}^2 - E_{13}^2 - E_{23}^2 \\ &= \frac{1}{2}\left[(\operatorname{tr}\boldsymbol{E})^2 - \operatorname{tr}(\boldsymbol{E}^2)\right], \\ III_E &:= \det\boldsymbol{E}. \end{aligned} \tag{1.91}$$

The quantities I_E, II_E and III_E are known as the *principal invariants* of the tensor $\boldsymbol{E}$. These quantities remain invariant under orthogonal transformations of $\boldsymbol{E}$, $\boldsymbol{E}^* = \boldsymbol{Q}\cdot\boldsymbol{E}\cdot\boldsymbol{Q}^T$, where $\boldsymbol{Q}$ is an orthogonal tensor. This is deduced from the equivalence of the characteristic equations for $\boldsymbol{E}^*$ and $\boldsymbol{E}$:

$$\begin{aligned} 0 = \det(\boldsymbol{E}^* - \lambda\boldsymbol{I}) &= \det(\boldsymbol{Q}\cdot\boldsymbol{E}\cdot\boldsymbol{Q}^T - \lambda\boldsymbol{I}) = \det\left[\boldsymbol{Q}\cdot(\boldsymbol{E} - \lambda\boldsymbol{I})\cdot\boldsymbol{Q}^T\right] \\ &= \det\boldsymbol{Q}\det(\boldsymbol{E} - \lambda\boldsymbol{I})\det\boldsymbol{Q}^T = (\det\boldsymbol{Q})^2\det(\boldsymbol{E} - \lambda\boldsymbol{I}) = \det(\boldsymbol{E} - \lambda\boldsymbol{I}). \end{aligned}$$

A tensor $\boldsymbol{E}$ possesses only three independent invariants. That is, all other invariants of $\boldsymbol{E}$ can be shown to be functions of the three principal invariants. For instance, three other invariants are

$$\tilde{I}_E := \operatorname{tr}\boldsymbol{E}, \qquad \tilde{I}\tilde{I}_E := \operatorname{tr}\boldsymbol{E}^2, \qquad I\tilde{I}I_E := \operatorname{tr}\boldsymbol{E}^3. \tag{1.92}$$

The relationships between these and I_E, II_E and III_E are

$$\begin{aligned} I_E &= \tilde{I}_E, \\ II_E &= \frac{1}{2}\left(\tilde{I}_E^2 - \tilde{I}\tilde{I}_E\right), \\ III_E &= \frac{1}{3}\left(I\tilde{I}I_E - \frac{3}{2}\tilde{I}_E\,\tilde{I}\tilde{I}_E + \frac{1}{2}\tilde{I}_E^3\right), \end{aligned} \tag{1.93}$$

or, conversely,

$$
\begin{aligned}
\tilde{I}_E &= I_E, \\
\tilde{II}_E &= I_E^2 - 2II_E, \\
\tilde{III}_E &= I_E^3 - 3I_E\, II_E + 3III_E.
\end{aligned}
\tag{1.94}
$$

The roots λ_α, $\alpha = 1, 2, 3$, of the characteristic equation (1.90) are called the *characteristic roots*. If $\boldsymbol{E}$ is the Lagrangian or Eulerian strain tensor, λ_α are called the *principal strains*. For each of the characteristic roots, we determine a *principal direction* $\vec{V}_\alpha$, $\alpha = 1, 2, 3$, by solving the equation

$$
\boldsymbol{E} \cdot \vec{V}_\alpha = \lambda_\alpha \vec{V}_\alpha \tag{1.95}
$$

together with the normalising condition $\vec{V}_\alpha \cdot \vec{V}_\alpha = 1$. For a symmetric tensor $\boldsymbol{E}$, it is not difficult to show that (i) all characteristic roots are real, and (ii) the principal directions corresponding to two distinct characteristic roots are unique and mutually orthogonal. If, however, there is a pair of equal roots, say $\lambda_1 = \lambda_2$, then only the direction associated with λ_3 will be unique. In this case, any other two directions which are orthogonal to $\vec{V}_3$, and to one another so as to form a right-handed system, may be taken as principal directions. If $\lambda_1 = \lambda_2 = \lambda_3$, every set of right-handed orthogonal axes qualifies as principal axes, and every direction is said to be a principal direction. Thus, at each point $\vec{X}$ it is always possible to find at least three mutually orthogonal directions for which the expression $\vec{V} \cdot \boldsymbol{E} \cdot \vec{V}$ takes the stationary values.

A tensor $\boldsymbol{E}$ takes a particularly simple form when the reference coordinate system is selected to coincide with the principal directions. Let the components of tensor $\boldsymbol{E}$ be given initially with respect to arbitrary Cartesian axes X_K with the base vectors $\vec{I}_K$, and let the principal axes of $\boldsymbol{E}$ be designated by X_α with the base vectors $\vec{I}_\alpha \equiv \vec{V}_\alpha$. The tensor $\boldsymbol{E}$ can then be represented in terms of base vectors $\vec{I}_K$ and $\vec{V}_\alpha$, respectively,

$$
\boldsymbol{E} = E_{KL}(\vec{I}_K \otimes \vec{I}_L) = E_{\alpha\beta}(\vec{V}_\alpha \otimes \vec{V}_\beta), \tag{1.96}
$$

where the diagonal elements $E_{\alpha\alpha}$ are equal to the principal values λ_α, while the off-diagonal elements $E_{\alpha\beta}$, $\alpha \neq \beta$, are zero. The projection of the vector $\vec{I}_K$ onto the base of the vectors $\vec{V}_\alpha$ is

$$
\vec{I}_K = (\vec{I}_K \cdot \vec{V}_\alpha)\vec{V}_\alpha, \tag{1.97}
$$

where $\vec{I}_K \cdot \vec{V}_\alpha$ are the direction cosines between the two established sets of axes X_K and X_α. Substituting (1.97) into (1.96) gives

$$
E_{\alpha\beta} = (\vec{I}_M \cdot \vec{V}_\alpha)(\vec{I}_N \cdot \vec{V}_\beta)E_{MN}. \tag{1.98}
$$

Hence, the determination of the principal directions $\vec{V}_\alpha$ and the characteristic roots λ_α of a tensor $\boldsymbol{E}$ is equivalent to finding the Cartesian coordinate system in which the matrix of elements E_{KL} takes the diagonal form.

1.11 The Displacement Vector

The geometrical measures of deformation can also be expressed in terms of the *displacement vector* $\vec{u}$ that extends from a point $\vec{X}$ in the reference configuration κ_0 to its present position $\vec{x}$ in the present configuration κ_t, as illustrated in Fig. 1.6:

$$\vec{u} := \vec{x} - \vec{X} + \vec{b}, \tag{1.99}$$

where $\vec{b}$ is the position vector of the origin o of the Cartesian coordinate system x_k relative to the origin O of the Cartesian coordinate system X_K. The definition (1.99) may be interpreted by either the Lagrangian or Eulerian representations of $\vec{u}$,

$$\vec{U}(\vec{X}, t) = \vec{\chi}(\vec{X}, t) - \vec{X} + \vec{b}, \qquad \vec{u}(\vec{x}, t) = \vec{x} - \vec{\chi}^{-1}(\vec{x}, t) + \vec{b}. \tag{1.100}$$

Taking the scalar product of both sides of $(1.100)_1$ and $(1.100)_2$ by $\vec{i}_k$ and $\vec{I}_K$, and using $(1.18)_1$ and $(1.20)_1$, respectively, the Lagrangian and Eulerian displacements can be expressed in terms of the spatial and referential components as

$$U_k(\vec{X}, t) = \chi_k - \delta_{kL} X_L + b_k, \qquad u_K(\vec{x}, t) = \delta_{lK} x_l - \chi_K^{-1} + b_K, \tag{1.101}$$

where $b_k = \vec{b} \cdot \vec{i}_k$ and $b_K = \vec{b} \cdot \vec{I}_K$. We again see here the appearance of shifter symbols. Differentiating $(1.101)_1$ and $(1.101)_2$ with respect to X_K and x_k,

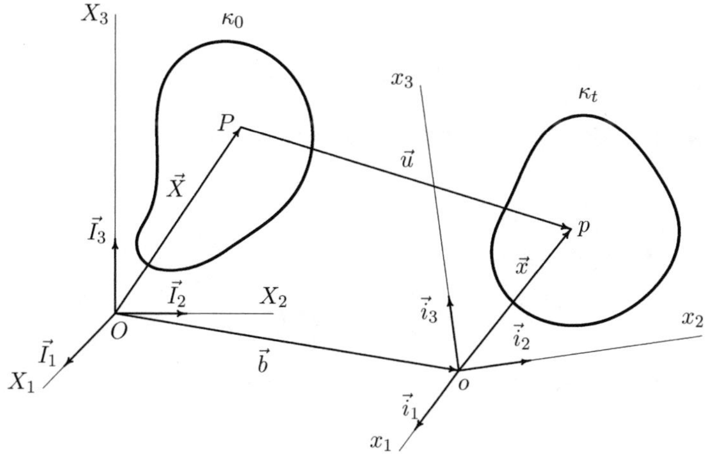

Fig. 1.6 Displacement vector

respectively, results in

$$U_{k,K}(\vec{X},t) = \chi_{k,K} - \delta_{kL}\delta_{LK}, \qquad u_{K,k}(\vec{x},t) = \delta_{lK}\delta_{lk} - \overset{-1}{\chi}_{K,k},$$

which simplifies to

$$U_{k,K}(\vec{X},t) = \chi_{k,K} - \delta_{kK}, \qquad u_{K,k}(\vec{x},t) = \delta_{kK} - \overset{-1}{\chi}_{K,k}, \tag{1.102}$$

due to the sifting property of the Kronecker delta symbols δ_{KL} and δ_{kl}.

In principle, all physical quantities representing the measures of deformation can be expressed in terms of the displacement vector and its gradient. For instance, the Lagrangian strain tensor $(1.61)_1$ is

$$2E_{KL} = C_{KL} - \delta_{KL} = \chi_{k,K}\chi_{k,L} - \delta_{KL} = (\delta_{kK} + U_{k,K})(\delta_{kL} + U_{k,L}) - \delta_{KL},$$

which simplifies to

$$2E_{KL} = U_{K,L} + U_{L,K} + U_{k,K}U_{k,L}, \tag{1.103}$$

in view of (1.16) and $(1.18)_1$. Similarly, the Eulerian strain tensor $(1.61)_2$ is

$$2e_{kl} = \delta_{kl} - c_{kl} = \delta_{kl} - \overset{-1}{\chi}_{K,k}\overset{-1}{\chi}_{K,l} = \delta_{kl} - (\delta_{kK} - u_{K,k})(\delta_{lK} - u_{K,l}),$$

which simplifies to

$$2e_{kl} = u_{k,l} + u_{l,k} - u_{K,k}u_{K,l}. \tag{1.104}$$

It is often convenient not to distinguish between the referential coordinates X_K and the spatial coordinates x_k. In such a case, $\vec{b} = \vec{0}$, the shifter symbols δ_{kL} reduce to the Kronecker delta symbols δ_{KL} and the displacement vector $(1.100)_1$ simplifies to

$$\vec{U}(\vec{X},t) = \vec{\chi}(\vec{X},t) - \vec{X}. \tag{1.105}$$

Consequently, $(1.102)_1$ written in tensor notation is

$$\boldsymbol{F} = \boldsymbol{I} + \boldsymbol{H}^T, \tag{1.106}$$

where $\boldsymbol{H}$ is the *displacement gradient tensor*,

$$\boldsymbol{H}(\vec{X},t) := \operatorname{Grad}\vec{U}(\vec{X},t). \tag{1.107}$$

Furthermore, the tensor notation of the Lagrangian strain (1.103) is

$$\boldsymbol{E} = \frac{1}{2}(\boldsymbol{H} + \boldsymbol{H}^T + \boldsymbol{H}\cdot\boldsymbol{H}^T). \tag{1.108}$$

1.12 Geometrical Linearisation

Provided that the numerical values of all components of the displacement gradient tensor are very small compared to one, we may neglect the squares and products of these quantities in comparison to the gradients themselves. A convenient measure of smallness of deformation is the magnitude of the displacement gradient,

$$|\boldsymbol{H}| \ll 1, \tag{1.109}$$

where

$$|\boldsymbol{H}|^2 = H_{kl}H_{kl} = \boldsymbol{H} : \boldsymbol{H}^T. \tag{1.110}$$

In the following, the term *small deformation* will correspond to the case of small displacement gradients. *Geometrical linearisation* or *kinematical linearisation* is the process of developing all kinematic variables correct to first order in $|\boldsymbol{H}|$ and neglecting all terms of orders higher than $O(|\boldsymbol{H}|)$. In its geometrical interpretation, a small value of $|\boldsymbol{H}|$ implies small strains as well as small rotations.

1.12.1 Linearised Analysis of Deformation

Let us decompose the transposed displacement gradient into the symmetric and skew-symmetric parts,[17]

$$\boldsymbol{H}^T = \widetilde{\boldsymbol{E}} + \widetilde{\boldsymbol{R}}, \tag{1.111}$$

where

$$\widetilde{\boldsymbol{E}} := \frac{1}{2}(\boldsymbol{H} + \boldsymbol{H}^T), \qquad \widetilde{\boldsymbol{R}} := -\frac{1}{2}(\boldsymbol{H} - \boldsymbol{H}^T). \tag{1.112}$$

The symmetric tensor $\widetilde{\boldsymbol{E}}$ and the skew-symmetric tensor $\widetilde{\boldsymbol{R}}$ are called the *linearised Lagrangian strain tensor* and *linearised Lagrangian rotation tensor*, respectively.

[17] Every second-order tensor $\boldsymbol{A}$ can be written as the sum of a symmetric tensor $\boldsymbol{S}$, $\boldsymbol{S}^T = \boldsymbol{S}$, and a skew-symmetric tensor $\boldsymbol{R}$, $\boldsymbol{R}^T = -\boldsymbol{R}$, in the form

$$\boldsymbol{A} = \boldsymbol{S} + \boldsymbol{R},$$

where

$$\boldsymbol{S} = \frac{1}{2}(\boldsymbol{A} + \boldsymbol{A}^T), \qquad \boldsymbol{R} = \frac{1}{2}(\boldsymbol{A} - \boldsymbol{A}^T).$$

Carrying this decomposition into (1.108) gives

$$\boldsymbol{E} = \widetilde{\boldsymbol{E}} + \frac{1}{2}(\widetilde{\boldsymbol{E}} - \widetilde{\boldsymbol{R}}) \cdot (\widetilde{\boldsymbol{E}} + \widetilde{\boldsymbol{R}}). \tag{1.113}$$

It is now clear that for $\boldsymbol{E} \approx \widetilde{\boldsymbol{E}}$, not only the strains $\widetilde{\boldsymbol{E}}$ must be small, but also the rotations $\widetilde{\boldsymbol{R}}$, so that products such as $\widetilde{\boldsymbol{E}}^T \cdot \widetilde{\boldsymbol{E}}$, $\widetilde{\boldsymbol{E}}^T \cdot \widetilde{\boldsymbol{R}}$ and $\widetilde{\boldsymbol{R}}^T \cdot \widetilde{\boldsymbol{R}}$ will be negligible compared to $\widetilde{\boldsymbol{E}}$.

In view of the decomposition (1.111), the deformation gradient is

$$\boldsymbol{F} = \boldsymbol{I} + \widetilde{\boldsymbol{E}} + \widetilde{\boldsymbol{R}}. \tag{1.114}$$

Thus, the multiplicative decomposition of the deformation gradient into orthogonal and positive-definite factors is approximated for small deformations by the additive decomposition into symmetric and skew-symmetric parts.

In the geometrical linearisation, the other deformation and rotation tensors take the form

$$\begin{aligned}
\boldsymbol{F}^{-1} &= \boldsymbol{I} - \boldsymbol{H}^T + O(|\boldsymbol{H}|^2), \\
\det \boldsymbol{F} &= 1 + \operatorname{tr} \boldsymbol{H} + O(|\boldsymbol{H}|^2), \\
\boldsymbol{C} &= \boldsymbol{I} + \boldsymbol{H} + \boldsymbol{H}^T + O(|\boldsymbol{H}|^2), \\
\boldsymbol{B} &= \boldsymbol{I} - \boldsymbol{H} - \boldsymbol{H}^T + O(|\boldsymbol{H}|^2), \\
\boldsymbol{U} &= \boldsymbol{I} + \frac{1}{2}(\boldsymbol{H} + \boldsymbol{H}^T) + O(|\boldsymbol{H}|^2), \\
\boldsymbol{V} &= \boldsymbol{I} + \frac{1}{2}(\boldsymbol{H} + \boldsymbol{H}^T) + O(|\boldsymbol{H}|^2), \\
\boldsymbol{R} &= \boldsymbol{I} - \frac{1}{2}(\boldsymbol{H} - \boldsymbol{H}^T) + O(|\boldsymbol{H}|^2), \\
\boldsymbol{b} &= \boldsymbol{I} + \boldsymbol{H} + \boldsymbol{H}^T + O(|\boldsymbol{H}|^2).
\end{aligned} \tag{1.115}$$

The transformation (1.48) between the differential operators applied to the Eulerian variables and those applied to the Lagrangian variables can be linearised as

$$\begin{aligned}
\operatorname{grad} \bullet &= \operatorname{Grad} \bullet - \boldsymbol{H} \cdot \operatorname{Grad} \bullet + O(|\boldsymbol{H}|^2), \\
\operatorname{div} \bullet &= \operatorname{Div} \bullet - \boldsymbol{H} : \operatorname{Grad} \bullet + O(|\boldsymbol{H}|^2), \\
\operatorname{div} \operatorname{grad} \bullet &= \operatorname{Div} \operatorname{Grad} \bullet - 2\boldsymbol{H} : \operatorname{Grad} \operatorname{Grad} \bullet - \operatorname{Div} \boldsymbol{H} \cdot \operatorname{Grad} \bullet + O(|\boldsymbol{H}|^2),
\end{aligned} \tag{1.116}$$

or, conversely,

$$\begin{aligned} \operatorname{Grad} \bullet &= \operatorname{grad} \bullet + \boldsymbol{H} \cdot \operatorname{grad} \bullet + O(|\boldsymbol{H}|^2), \\ \operatorname{Div} \bullet &= \operatorname{div} \bullet + \boldsymbol{H} : \operatorname{grad} \bullet + O(|\boldsymbol{H}|^2), \\ \operatorname{Div} \operatorname{Grad} \bullet &= \operatorname{div} \operatorname{grad} \bullet + 2\boldsymbol{H} : \operatorname{grad} \operatorname{grad} \bullet + \operatorname{Div} \boldsymbol{H} \cdot \operatorname{grad} \bullet + O(|\boldsymbol{H}|^2). \end{aligned} \tag{1.117}$$

Equation (1.104) shows that the geometrical linearisation of the Eulerian strain tensor $\boldsymbol{e}$ results in the *linearised Eulerian strain tensor* $\tilde{\boldsymbol{e}}$,

$$\boldsymbol{e} \approx \tilde{\boldsymbol{e}} = \frac{1}{2}\Big[\operatorname{grad} \vec{u} + (\operatorname{grad} \vec{u})^T\Big]. \tag{1.118}$$

In view of $(1.116)_1$, the tensor $\tilde{\boldsymbol{e}}$ is, correct to first order in $|\boldsymbol{H}|$, equal to the tensor $\widetilde{\boldsymbol{E}}$,

$$\tilde{\boldsymbol{e}} = \widetilde{\boldsymbol{E}} + O(|\boldsymbol{H}|^2). \tag{1.119}$$

Thus, the distinction between the Lagrangian and Eulerian strain tensors disappears in the linearised theory.

1.12.2 Length and Angle Changes

Expanding the square root in (1.73) by the binomial theorem and neglecting the square and higher powers of E_{11} results in

$$E_{(1)} \approx E_{11} \approx \widetilde{E}_{11}. \tag{1.120}$$

Similar results are of course valid for E_{22} and E_{33}, which indicates that the infinitesimal normal strains are approximately the extensions along the coordinate axes in the reference configuration. Likewise, the geometrical linearisation of (1.79) yields

$$\cos \theta_{(12)} \approx 2E_{12} \approx 2\widetilde{E}_{12}. \tag{1.121}$$

Hence, writing $\cos \theta_{(12)} = \sin(\pi/2 - \theta_{(12)}) = \sin \Gamma_{(12)} \approx \Gamma_{(12)}$, where $\Gamma_{(12)} = \pi/2 - \theta_{(12)}$, yields

$$\Gamma_{(12)} \approx 2\widetilde{E}_{12}. \tag{1.122}$$

Similar results are valid for E_{13} and E_{23}. This provides geometrical meaning for shear strains; the infinitesimal shear strains are approximately one half of the angle change between the coordinate axes in the reference configuration.

1.12.3 Area and Volume Changes

Substituting the linearised forms $(1.115)_{1,2}$ for the inverse deformation gradient $\boldsymbol{F}^{-1}$ and the Jacobian J into Nanson's formula (1.81) gives the linearised relationship between $d\vec{a}$ and $d\vec{A}$,

$$d\vec{a} = \big[\boldsymbol{I} + (\operatorname{tr}\boldsymbol{H})\boldsymbol{I} - \boldsymbol{H}\big] \cdot d\vec{A} + O(|\boldsymbol{H}|^2). \tag{1.123}$$

We may also linearise the separate contributions to $d\vec{a}$. Using the linearised form $(1.115)_4$ for the Piola deformation tensor $\boldsymbol{B}$ gives

$$\frac{1}{\sqrt{\vec{N} \cdot \boldsymbol{B} \cdot \vec{N}}} = \frac{1}{\sqrt{1 - 2\vec{N} \cdot \boldsymbol{H} \cdot \vec{N}}} = 1 + \vec{N} \cdot \boldsymbol{H} \cdot \vec{N} + O(|\boldsymbol{H}|^2), \tag{1.124}$$

where $\vec{N}$ is the unit vector normal to the area element dA in the reference configuration. Employing this and again the linearised forms $(1.115)_{1,2}$ in (1.82) and (1.83), the magnitude da of the area element $d\vec{a}$ and the unit vector $\vec{n}$ normal to da may, within the framework of linear approximation, be written as

$$da = \big(1 + \operatorname{tr}\boldsymbol{H} - \vec{N} \cdot \boldsymbol{H} \cdot \vec{N}\big) dA + O(|\boldsymbol{H}|^2), \tag{1.125}$$

$$\vec{n} = (1 + \vec{N} \cdot \boldsymbol{H} \cdot \vec{N})\vec{N} - \boldsymbol{H} \cdot \vec{N} + O(|\boldsymbol{H}|^2). \tag{1.126}$$

The first equation accounts for the change in the magnitude of area element, while the second gives the deflection of the unit vector normal to da.

The linearised relationship between the area elements can also be expressed in terms of the surface gradient operator $\operatorname{Grad}_\Sigma$ and the surface divergence operator $\operatorname{Div}_\Sigma$ defined in Appendix B. For example, the surface gradient and the surface divergence of the displacement vector $\vec{U}$ are given by (B.27) and (B.28),

$$\operatorname{Grad}_\Sigma \vec{U} = \boldsymbol{H} - \vec{N} \otimes (\vec{N} \cdot \boldsymbol{H}), \tag{1.127}$$

$$\operatorname{Div}_\Sigma \vec{U} = \operatorname{tr}\boldsymbol{H} - \vec{N} \cdot \boldsymbol{H} \cdot \vec{N}, \tag{1.128}$$

where $\boldsymbol{H} = \operatorname{Grad}\vec{U}$ and $\operatorname{tr}\boldsymbol{H} = \operatorname{Div}\vec{U}$. Equations (1.123), (1.125) and (1.126), expressed in terms of the surface displacement gradient and divergence, become

$$d\vec{a} = \big[\boldsymbol{I} + (\operatorname{Div}_\Sigma \vec{U})\boldsymbol{I} - \operatorname{Grad}_\Sigma \vec{U}\big] \cdot d\vec{A} + O(|\boldsymbol{H}|^2), \tag{1.129}$$

$$da = \big(1 + \operatorname{Div}_\Sigma \vec{U}\big) dA + O(|\boldsymbol{H}|^2), \tag{1.130}$$

$$\vec{n} = (\boldsymbol{I} - \operatorname{Grad}_\Sigma \vec{U}) \cdot \vec{N} + O(|\boldsymbol{H}|^2). \tag{1.131}$$

Equation (1.130) is the surface analogue of the volumetric relation (1.85); the volume elements dv and dV in the present and reference configurations are related by

$$dv = (1 + \operatorname{Div} \vec{U})dV + O(|\boldsymbol{H}|^2). \tag{1.132}$$

Chapter 2
Basic Kinematics

In the preceding chapter, we discussed the geometric properties of the present configuration κ_t under the assumption that the parameter t describing temporal changes of the body is kept fixed. That is, the motion function $\vec{\chi}$ is considered to be a deformation map $\vec{\chi}(\cdot, t)$. Now, we turn our attention to the problem in the case of motion, that is, where the function $\vec{\chi}$ is treated as the map $\vec{\chi}(\vec{X}, \cdot)$ for a fixed point $\vec{X}$. We assume that the mapping $\vec{\chi}(\vec{X}, \cdot)$ is twice differentiable.

2.1 Material and Spatial Time Derivatives

If we focus our attention on a specific particle $\mathcal{X}^P$ having the position $\vec{X}^P$, (1.4) takes the form

$$\vec{x}^P = \vec{\chi}(\vec{X}^P, t), \tag{2.1}$$

which describes the *path* or *trajectory* of that particle as a function of time. The *velocity* $\vec{V}^P$ of the particle along this path is defined as the time rate of change of position, or

$$\vec{V}^P := \frac{d\vec{x}^P}{dt} = \left.\frac{\partial \vec{\chi}}{\partial t}\right|_{\vec{X}=\vec{X}^P}, \tag{2.2}$$

where the subscript $\vec{X}$ accompanying a vertical bar indicates that $\vec{X}$ is held fixed (equal to $\vec{X}^P$) in the differentiation of $\vec{\chi}$. In an obvious generalisation, we define the *velocity* of an arbitrary particle $\mathcal{X}$ with the position vector $\vec{X}$ as

$$\vec{V}(\vec{X}, t) := \left.\frac{d\vec{x}}{dt}\right|_{\vec{X}} = \left.\frac{\partial \vec{\chi}}{\partial t}\right|_{\vec{X}}. \tag{2.3}$$

Z. Martinec, *Principles of Continuum Mechanics*, Nečas Center Series,
https://doi.org/10.1007/978-3-030-05390-1_2

This is the Lagrangian representation of velocity (or, simply, the Lagrangian velocity). The time rate of change of a physical quantity with respect to a fixed, but moving particle is called the *material time derivative*. Hence, the Lagrangian velocity of the particle is defined as the material time derivative of its position at time t. Similarly, the material time derivative of the Lagrangian velocity defines the Lagrangian representation of acceleration,

$$\vec{A}(\vec{X},t) := \left.\frac{d\vec{V}}{dt}\right|_{\vec{X}} = \left.\frac{\partial\vec{V}}{\partial t}\right|_{\vec{X}} = \left.\frac{\partial^2\vec{\chi}(\vec{X},t)}{\partial^2 t}\right|_{\vec{X}} . \tag{2.4}$$

Substituting $\vec{\chi}(\vec{X},t)$ from $(1.100)_1$ gives

$$\vec{V}(\vec{X},t) = \left.\frac{\partial\vec{U}}{\partial t}\right|_{\vec{X}}, \qquad \vec{A}(\vec{X},t) = \left.\frac{\partial^2\vec{U}}{\partial^2 t}\right|_{\vec{X}} . \tag{2.5}$$

A particle with a given velocity or acceleration is, in the Lagrangian representation, identifiable, such that both the velocity and acceleration fields are defined with respect to the reference configuration κ_0. This is frequently not convenient, for instance, in classical fluid mechanics. When the present configuration κ_t is chosen as the reference configuration, the motion function $\vec{\chi}(\vec{X},t)$ cannot be specified. Thus, in the Eulerian description, the velocity and acceleration at time t and spatial point $\vec{x}$ are known, but the particle occupying this point is not known.

Since the fundamental laws of continuum dynamics involve the acceleration of particles, and a Lagrangian description of velocity may not be available, acceleration must be calculated from an Eulerian description of velocity. To that end, let us again consider a particle $\mathcal{X}^P$ moving along the trajectory $\vec{x}^P = \vec{\chi}(\vec{X}^P,t)$. The Eulerian representation of velocity (or, simply, the Eulerian velocity) of this particle is defined by

$$\vec{v}(\vec{x}^P,t)\Big|_{\vec{x}^P=\vec{\chi}(\vec{X}^P,t)} := \vec{V}(\vec{X}^P,t), \tag{2.6}$$

which may be termed the *fundamental principle of kinematics*. In other words, the Eulerian velocity $\vec{v}$ at a fixed Eulerian position $\vec{x}^P$ is the velocity of the particle that happens to be at that position at that instant in time. To find the Eulerian representation of acceleration, we differentiate the left-hand side of (2.6) with respect to time and use the chain rule:

$$\vec{a}(\vec{x}^P,t) = \left.\frac{\partial\vec{v}}{\partial t}\right|_{\vec{x}^P} + \frac{\partial\vec{v}}{\partial x_k^P}\frac{\partial\chi_k(\vec{X}^P,t)}{\partial t} . \tag{2.7}$$

In view of (2.2) and (2.6), the last term on the right is equal to $v_k(\vec{x}^P,t)$. Dropping the particle label P from the notation, (2.7) simplifies to the desired equation for the Eulerian representation of acceleration, expressed in terms of the Eulerian representation of velocity,

$$\vec{a}(\vec{x}, t) = \left.\frac{\partial \vec{v}}{\partial t}\right|_{\vec{x}} + \vec{v} \cdot \operatorname{grad} \vec{v}, \tag{2.8}$$

where differentiation in the gradient operator is taken with respect to the spatial coordinates. In this equation, the first term on the right gives the time rate of change of velocity at a fixed position $\vec{x}$, known as the *local rate of change* or *spatial time derivative*. The second term results from the particles changing position in space and is referred to as the *convective term*. Note that the convective term can equivalently be expressed as

$$\vec{v} \cdot \operatorname{grad} \vec{v} = \operatorname{grad} \frac{v^2}{2} - \vec{v} \times \operatorname{rot} \vec{v}. \tag{2.9}$$

The material time derivative of any other field quantity can be calculated in the same way if its Lagrangian or Eulerian representation is known. This allows the introduction of the *material time derivative operator*

$$\frac{D}{Dt} \equiv (\dot{\ }) := \left.\frac{d}{dt}\right|_{\vec{X}} = \begin{cases} \left.\dfrac{\partial}{\partial t}\right|_{\vec{X}} & \text{for a field in the Lagrangian representation,} \\ \left.\dfrac{\partial}{\partial t}\right|_{\vec{x}} + \vec{v} \cdot \operatorname{grad} & \text{for a field in the Eulerian representation,} \end{cases} \tag{2.10}$$

which can be applied to any field quantity given in the Lagrangian or Eulerian representation.

2.2 Time Derivatives of Some Geometric Quantities

All of the geometric quantities that we discussed in Chap. 1 can be calculated from the deformation gradient $\boldsymbol{F}$. For this reason, we begin with an investigation of the material time derivative of $\boldsymbol{F}$. In index notation, we can write

$$\frac{D\chi_{k,K}}{Dt} = \frac{D}{Dt}\left(\frac{\partial \chi_k}{\partial X_K}\right) = \left.\frac{\partial}{\partial t}\left(\frac{\partial \chi_k}{\partial X_K}\right)\right|_{\vec{X}} = \left.\frac{\partial}{\partial X_K}\left(\frac{\partial \chi_k}{\partial t}\right)\right|_{\vec{X}} = \frac{\partial V_k}{\partial X_K} = V_{k,K} = v_{k,l}\chi_{l,K}, \tag{2.11}$$

where we have used the fact that X_K are kept fixed in the material time derivative, so that the material time derivative D/Dt can be interchanged with the material gradient $\partial/\partial X_K$. In addition, we have employed the Lagrangian and Eulerian representations of velocity (2.3) and (2.6), respectively. In tensor notation, we have

$$\dot{\boldsymbol{F}} = \boldsymbol{l} \cdot \boldsymbol{F}, \qquad \text{or} \qquad \boldsymbol{l} = \dot{\boldsymbol{F}} \cdot \boldsymbol{F}^{-1}, \tag{2.12}$$

where $\boldsymbol{l}$ is the transposed *velocity gradient*,

$$\boldsymbol{l}(\vec{x}, t) := \operatorname{grad}^T \vec{v}(\vec{x}, t), \qquad \text{or} \qquad l_{kl}(\vec{x}, t) := \frac{\partial v_k(\vec{x}, t)}{\partial x_l}. \tag{2.13}$$

A consequence of (2.12) is

$$(\boldsymbol{F}^{-1})^{\cdot} = -\boldsymbol{F}^{-1} \cdot \boldsymbol{l}. \tag{2.14}$$

To show this, we take the material time derivative of $\boldsymbol{F} \cdot \boldsymbol{F}^{-1} = \boldsymbol{I}$,

$$\dot{\boldsymbol{F}} \cdot \boldsymbol{F}^{-1} + \boldsymbol{F} \cdot (\boldsymbol{F}^{-1})^{\cdot} = \boldsymbol{0},$$

and substitute $\dot{\boldsymbol{F}}$ from (2.12) to obtain (2.14).

The material time derivative of the Jacobian is

$$\dot{J} = J \operatorname{div} \vec{v} = J \operatorname{tr} \boldsymbol{l}, \tag{2.15}$$

which is shown using Jacobi's identity $(1.46)_1$ and

$$\dot{J} = \left(\det \chi_{k,K}\right)^{\cdot} = \frac{\partial J}{\partial \chi_{k,K}} (\chi_{k,K})^{\cdot} = \frac{\partial J}{\partial \chi_{k,K}} v_{k,l} x_{l,K}.$$

The velocity gradient $\boldsymbol{l}$ determines the material time derivatives of the line element $d\vec{x}$, the oriented area element $d\vec{a}$ and the volume element dv according to the formulae

$$(d\vec{x})^{\cdot} = \boldsymbol{l} \cdot d\vec{x}, \tag{2.16}$$

$$(d\vec{a})^{\cdot} = \left[(\operatorname{div} \vec{v})\boldsymbol{I} - \boldsymbol{l}^T\right] \cdot d\vec{a}, \tag{2.17}$$

$$(dv)^{\cdot} = \operatorname{div} \vec{v}\, dv. \tag{2.18}$$

The proofs are straightforward. We take the material time derivative of $(1.38)_1$ and substitute $\dot{\boldsymbol{F}}$ from (2.12) to obtain

$$(d\vec{x})^{\cdot} = \dot{\boldsymbol{F}} \cdot d\vec{X} = \boldsymbol{l} \cdot \boldsymbol{F} \cdot d\vec{X}.$$

Replacing $d\vec{X}$ by $d\vec{x}$ proves (2.16). Taking the material time derivative of Nanson's formula (1.81) and using (2.14) and (2.15) gives

$$(d\vec{a})^{\cdot} = \left[\dot{J}\, \boldsymbol{F}^{-T} + J(\boldsymbol{F}^{-T})^{\cdot}\right] \cdot d\vec{A} = \left[J \operatorname{div} \vec{v}\, \boldsymbol{F}^{-T} - J\, \boldsymbol{l}^T \cdot \boldsymbol{F}^{-T}\right] \cdot d\vec{A}.$$

Replacing $d\vec{A}$ by $d\vec{a}$ results in (2.17). The third statement can be verified by differentiating (1.85) with respect to time and using (2.15),

$$(dv)^{\cdot} = \dot{J}\,dV = J\,\mathrm{div}\vec{v}\,dV = \mathrm{div}\vec{v}\,dv,$$

which completes the proofs of these statements.

The velocity gradient $\boldsymbol{l}$, as any other tensor, can be uniquely decomposed into symmetric and skew-symmetric parts,

$$\boldsymbol{l} = \boldsymbol{d} + \boldsymbol{w}, \tag{2.19}$$

where

$$\boldsymbol{d} := \frac{1}{2}(\boldsymbol{l} + \boldsymbol{l}^T), \qquad \boldsymbol{w} := \frac{1}{2}(\boldsymbol{l} - \boldsymbol{l}^T). \tag{2.20}$$

The symmetric tensor $\boldsymbol{d}$ is called the *strain-rate tensor* and the skew-symmetric tensor $\boldsymbol{w}$ is called the *spin tensor*.

To highlight the meaning of the spin tensor $\boldsymbol{w}$, we readily see that every skew-symmetric tensor $\boldsymbol{w}$ has only three independent scalar components. Within a sign convention, $\boldsymbol{w}$ may be represented as

$$\boldsymbol{w} = \begin{pmatrix} 0 & -\xi_3 & \xi_2 \\ \xi_3 & 0 & -\xi_1 \\ -\xi_2 & \xi_1 & 0 \end{pmatrix} (\vec{i}_k \otimes \vec{i}_l),$$

where ξ_i are components of the so-called *vorticity vector* $\vec{\xi}$. The components of $\boldsymbol{w}$ and $\vec{\xi}$ are related by

$$w_{kl} = \varepsilon_{lkm}\xi_m, \qquad \text{or} \qquad \xi_k = \frac{1}{2}\varepsilon_{klm}w_{ml}.$$

Since $w_{kl} = \frac{1}{2}(v_{k,l} - v_{l,k})$, we find that the vorticity vector $\vec{\xi}$ is equal to one-half of the curl of the velocity vector $\vec{v}$,

$$\vec{\xi} = \frac{1}{2}\mathrm{rot}\,\vec{v}, \tag{2.21}$$

hence the name *spin* or *vorticity* is given to the tensor $\boldsymbol{w}$. Furthermore, the scalar product of $\boldsymbol{w}$ with vector $\vec{a}$ can be expressed as

$$\boldsymbol{w} \cdot \vec{a} = \vec{\xi} \times \vec{a}, \tag{2.22}$$

giving (2.16) in the form

$$(d\vec{x})^{\cdot} = \boldsymbol{d} \cdot d\vec{x} + \vec{\xi} \times d\vec{x}. \tag{2.23}$$

The physical significance of the spin tensor may be seen by considering the case where the strain-rate tensor is zero, i.e., $\boldsymbol{d} = \boldsymbol{0}$. Then, $(d\vec{x})^{\cdot} = \vec{\xi} \times d\vec{x}$, according to which the spin tensor $\boldsymbol{w}$ describes an instantaneous local rigid-body rotation about an axis passing through a point $\vec{x}$. The corresponding angular velocity, the direction and the sense of this rotation is described by the vorticity vector $\vec{\xi}$. If, on the other hand, the vorticity $\vec{\xi}$ (or, equivalently, the spin tensor $\boldsymbol{w}$) is everywhere zero in a region occupied by particles of a moving body, the motion is said to be *irrotational* in that region.

Taking the material time derivative of the polar decomposition (1.50) of the deformation gradient $\boldsymbol{F}$, i.e., $\dot{\boldsymbol{F}} = \dot{\boldsymbol{R}} \cdot \boldsymbol{U} + \boldsymbol{R} \cdot \dot{\boldsymbol{U}}$, and substituting this into $(2.12)_2$ yields

$$\boldsymbol{l} = \dot{\boldsymbol{R}} \cdot \boldsymbol{R}^T + \boldsymbol{R} \cdot (\dot{\boldsymbol{U}} \cdot \boldsymbol{U}^{-1}) \cdot \boldsymbol{R}^T. \tag{2.24}$$

The spin tensor $\boldsymbol{w}$ is then expressed as

$$\boldsymbol{w} = \dot{\boldsymbol{R}} \cdot \boldsymbol{R}^T + \frac{1}{2}\boldsymbol{R} \cdot \left(\dot{\boldsymbol{U}} \cdot \boldsymbol{U}^{-1} - \boldsymbol{U}^{-1} \cdot \dot{\boldsymbol{U}}\right) \cdot \boldsymbol{R}^T, \tag{2.25}$$

where $\boldsymbol{R} \cdot (\boldsymbol{R}^T)^{\cdot} = -\dot{\boldsymbol{R}} \cdot \boldsymbol{R}^T$ has been used, which follows from the time differentiation of $\boldsymbol{R} \cdot \boldsymbol{R}^T = \boldsymbol{I}$. In the special case of a rigid-body motion, for which $\boldsymbol{U} = \boldsymbol{I}$, (2.25) reduces to

$$\boldsymbol{w} = \dot{\boldsymbol{R}} \cdot \boldsymbol{R}^T. \tag{2.26}$$

Hence, the first term in (2.24) represents an instantaneous local rigid-body rotation about an axis passing through a point $\vec{x}$, while the second term in (2.24) represents the contribution due to deformation because $\dot{\boldsymbol{U}} = \boldsymbol{0}$ if there is no deformation.

The material time derivative of the Lagrangian strain tensor is

$$\dot{\boldsymbol{E}} = \frac{1}{2}\,\dot{\boldsymbol{C}} = \boldsymbol{F}^T \cdot \boldsymbol{d} \cdot \boldsymbol{F}, \tag{2.27}$$

which follows from

$$\begin{aligned}\dot{\boldsymbol{C}} &= \left(\boldsymbol{F}^T \cdot \boldsymbol{F}\right)^{\cdot} = (\boldsymbol{F}^T)^{\cdot} \cdot \boldsymbol{F} + \boldsymbol{F}^T \cdot \dot{\boldsymbol{F}} = \boldsymbol{F}^T \cdot \boldsymbol{l}^T \cdot \boldsymbol{F} + \boldsymbol{F}^T \cdot \boldsymbol{l} \cdot \boldsymbol{F} \\ &= \boldsymbol{F}^T \cdot (\boldsymbol{l} + \boldsymbol{l}^T) \cdot \boldsymbol{F} = 2\boldsymbol{F}^T \cdot \boldsymbol{d} \cdot \boldsymbol{F}.\end{aligned}$$

The material time derivative of the strain tensor $\boldsymbol{E}$ is determined by the tensor $\boldsymbol{d}$, but not by the tensor $\boldsymbol{w}$. This is why $\boldsymbol{d}$ is called the *strain-rate* tensor.

The material time derivative of the Eulerian strain tensor is

$$\dot{\boldsymbol{e}} = -\frac{1}{2}\,\dot{\boldsymbol{c}} = \boldsymbol{d} - (\boldsymbol{e}\cdot\boldsymbol{l} + \boldsymbol{l}^T\cdot\boldsymbol{e}), \tag{2.28}$$

which follows from

$$\begin{aligned}\dot{\boldsymbol{c}} &= \left(\boldsymbol{F}^{-T}\cdot\boldsymbol{F}^{-1}\right)^{\cdot} = (\boldsymbol{F}^{-T})^{\cdot}\cdot\boldsymbol{F}^{-1} + \boldsymbol{F}^{-T}\cdot(\boldsymbol{F}^{-1})^{\cdot}\\ &= -\boldsymbol{l}^T\cdot\boldsymbol{F}^{-T}\cdot\boldsymbol{F}^{-1} - \boldsymbol{F}^{-T}\cdot\boldsymbol{F}^{-1}\cdot\boldsymbol{l} = -\boldsymbol{l}^T\cdot\boldsymbol{c} - \boldsymbol{c}\cdot\boldsymbol{l}\end{aligned}$$

and substituting $\boldsymbol{c}$ from $(1.61)_2$. Note that $\dot{\boldsymbol{e}}$ does not equal $\boldsymbol{d}$.

Similarly, it can be readily shown that the material time derivative of the Piola deformation tensor is

$$\dot{\boldsymbol{B}} = -\boldsymbol{F}^{-1}\cdot 2\boldsymbol{d}\cdot\boldsymbol{F}^{-T}. \tag{2.29}$$

Thus, the material time derivative of the unit vector $\vec{n}$ normal to the area element da is

$$(\vec{n})^{\cdot} = -\vec{n}\cdot\boldsymbol{l} + (\vec{n}\cdot\boldsymbol{l}\cdot\vec{n})\vec{n}. \tag{2.30}$$

To show this, we take the material time derivative of (1.83), use (2.14) and (2.29),

$$\begin{aligned}(\vec{n})^{\cdot} &= \frac{\vec{N}\cdot(\boldsymbol{F}^{-1})^{\cdot}}{(\vec{N}\cdot\boldsymbol{B}\cdot\vec{N})^{1/2}} - \frac{\vec{N}\cdot\boldsymbol{F}^{-1}}{2(\vec{N}\cdot\boldsymbol{B}\cdot\vec{N})^{3/2}}(\vec{N}\cdot\dot{\boldsymbol{B}}\cdot\vec{N})\\ &= -\frac{\vec{N}\cdot\boldsymbol{F}^{-1}\cdot\boldsymbol{l}}{(\vec{N}\cdot\boldsymbol{B}\cdot\vec{N})^{1/2}} + \frac{\vec{N}\cdot\boldsymbol{F}^{-1}}{(\vec{N}\cdot\boldsymbol{B}\cdot\vec{N})^{3/2}}(\vec{N}\cdot\boldsymbol{F}^{-1}\cdot\boldsymbol{d}\cdot\boldsymbol{F}^{-T}\cdot\vec{N}),\end{aligned}$$

and substitute for $\vec{n}$ from (1.83).

The material time derivative of the square of the length of $d\vec{x}$ is

$$(ds^2)^{\cdot} = 2\,d\vec{x}\cdot\boldsymbol{d}\cdot d\vec{x}. \tag{2.31}$$

To show this, we take the material time derivative of (1.56),

$$\begin{aligned}(ds^2)^{\cdot} &= (d\vec{x}\cdot d\vec{x})^{\cdot} = (d\vec{X}\cdot\boldsymbol{C}\cdot d\vec{X})^{\cdot} = d\vec{X}\cdot\dot{\boldsymbol{C}}\cdot d\vec{X} = (\boldsymbol{F}^{-1}\cdot d\vec{x})\cdot\dot{\boldsymbol{C}}\cdot(\boldsymbol{F}^{-1}\cdot d\vec{x})\\ &= d\vec{x}\cdot\boldsymbol{F}^{-T}\cdot\dot{\boldsymbol{C}}\cdot\boldsymbol{F}^{-1}\cdot d\vec{x},\end{aligned}$$

and use (2.27).

Higher-order material time derivatives of ds^2 can be carried out by repeated differentiation. Therefore, the nth material time derivative of ds^2 can be expressed as

$$(ds^2)^{(n)} = d\vec{x}\cdot\boldsymbol{a}_n\cdot d\vec{x}, \tag{2.32}$$

where

$$\boldsymbol{a}_n(\vec{x}, t) := \boldsymbol{F}^{-T} \cdot \overset{(n)}{\boldsymbol{C}} \cdot \boldsymbol{F}^{-1} \tag{2.33}$$

are known as the *Rivlin–Ericksen* tensors of order n. The proof of (2.32) is similar to that of (2.31). Note that $\overset{(n)}{\boldsymbol{C}}$ is the nth material time derivative of the Green deformation tensor $\boldsymbol{C}$. The first two Rivlin–Ericksen tensors are

$$\boldsymbol{a}_0 = \boldsymbol{I}, \qquad \boldsymbol{a}_1 = 2\boldsymbol{d}. \tag{2.34}$$

The Rivlin–Ericksen tensors are used in the formulation of a constitutive equation for a fluid, as will be presented in Sects. 6.14 and 6.15.

2.3 Reynolds Transport Theorem

In this section, we will prove that the material time derivative of a volume integral of a scalar or vector field ϕ over volume $v(t)$ of the present configuration is

$$\frac{D}{Dt}\int_{v(t)} \phi\, dv = \int_{v(t)} \left(\frac{D\phi}{Dt} + \phi \operatorname{div} \vec{v} \right) dv, \tag{2.35}$$

which is often referred to as the *Reynolds transport theorem*.

To prove this theorem, we first transform the integral over $v(t)$ to the integral over volume V of the reference configuration. Under the assumption of the existence of the mapping (1.4), and using (1.85) gives

$$\frac{D}{Dt}\int_{v(t)} \phi\, dv = \frac{D}{Dt}\int_{V} \Phi J dV,$$

where $\Phi(\vec{X}, t) = \phi(\vec{x}(\vec{X}, t), t)$. Since the volume V of the reference configuration is time independent, the differentiation D/Dt can be interchanged with the integration over V, and the differentiation D/Dt can be performed inside the integral sign,

$$\begin{aligned} \frac{D}{Dt}\int_{V} \Phi J dV &= \int_{V} \frac{D}{Dt}(\Phi J)\, dV = \int_{V} \left(\frac{D\Phi}{Dt} J + \Phi \frac{DJ}{Dt} \right) dV \\ &= \int_{V} \left(\frac{D\Phi}{Dt} + \Phi \operatorname{div} \vec{v} \right) J dV. \end{aligned}$$

Converting this back to the Eulerian description by (1.85) proves (2.35).

The transport theorem may be expressed in an alternative form. We first determine the material time derivative of a scalar field ϕ,

$$\frac{D\phi}{Dt} = \frac{\partial \phi}{\partial t} + \vec{v} \cdot \operatorname{grad} \phi, \tag{2.36}$$

and substitute it into (2.35). With the product rule of the form

$$\operatorname{div}(\vec{v}\phi) = \phi \operatorname{div}\vec{v} + \vec{v}\cdot\operatorname{grad}\phi, \tag{2.37}$$

which is valid for a scalar field ϕ, we then arrive at

$$\frac{D}{Dt}\int_{v(t)} \phi\, dv = \int_{v(t)} \left(\frac{\partial\phi}{\partial t} + \operatorname{div}(\vec{v}\phi)\right) dv.$$

Arranging the second term on the right according to the Gauss theorem

$$\int_v \operatorname{div}\vec{u}\, dv = \int_s \vec{n}\cdot\vec{u}\, da, \tag{2.38}$$

where $\vec{u}$ is a vector field continuously differentiable in v, s is the surface bounding volume v and $\vec{n}$ is the outward unit normal to s, we obtain the equivalent form of the Reynolds transport theorem

$$\frac{D}{Dt}\int_{v(t)} \phi\, dv = \int_{v(t)} \frac{\partial\phi}{\partial t} dv + \int_{s(t)} (\vec{n}\cdot\vec{v})\phi\, da, \tag{2.39}$$

where both ϕ and $\vec{v}$ are again required to be continuously differentiable in v.

The form (2.39) of the Reynolds theorem holds also for a vector field ϕ. The only difference is that the product rule

$$\operatorname{div}(\vec{v}\otimes\phi) = \phi \operatorname{div}\vec{v} + \vec{v}\cdot\operatorname{grad}\phi \tag{2.40}$$

is used instead of (2.37) and the Gauss theorem for a second-order tensor $\boldsymbol{A}$

$$\int_v \operatorname{div}\boldsymbol{A}\, dv = \int_s \vec{n}\cdot\boldsymbol{A}\, da \tag{2.41}$$

replaces (2.38). To prove (2.41), we employ the Gauss theorem (2.38) for vector $\boldsymbol{A}\cdot\vec{c}$, where $\vec{c}$ is a vector with a constant magnitude and constant, but arbitrary, direction

$$\int_v \operatorname{div}(\boldsymbol{A}\cdot\vec{c})\, dv = \int_s \vec{n}\cdot\boldsymbol{A}\cdot\vec{c}\, da.$$

Using the identity $\operatorname{div}(\boldsymbol{A}\cdot\vec{c}) = \operatorname{div}\boldsymbol{A}\cdot\vec{c}$, valid for a constant vector $\vec{c}$, gives

$$\left[\int_v \operatorname{div}\boldsymbol{A}\, dv - \int_s \vec{n}\cdot\boldsymbol{A}\, da\right]\cdot\vec{c} = 0.$$

Since $|\vec{c}| \neq 0$ and its direction is arbitrary, meaning that the cosine of the included angle cannot **always** vanish, the term in brackets must vanish, which verifies (2.41).

The material time derivative of the flux of vector $\vec{q}$ passing across surface $s(t)$ is

$$\frac{D}{Dt}\int_{s(t)} \vec{q}\cdot d\vec{a} = \int_{s(t)} \left(\frac{D\vec{q}}{Dt} + (\operatorname{div}\vec{v})\vec{q} - \vec{q}\cdot \operatorname{grad}\vec{v}\right)\cdot d\vec{a}. \tag{2.42}$$

To prove this form of the transport theorem, the surface integral over surface $s(t)$ of the present configuration is first transformed into the integral over time-independent surface S of the reference configuration by (1.81), the derivative D/Dt can then be moved inside the integral and the time differentiation of the integrand is done in a similar manner as in the proof of (2.17).

2.4 Modified Reynolds Transport Theorem

In the preceding section, we derived the Reynolds transport theorem under the assumption that the field quantities ϕ and $\vec{v}$ are continuously differentiable within volume $v(t)$. This assumption is also implemented in the Gauss theorems (2.38) and (2.41) for field variables $\vec{u}$ and $\boldsymbol{A}$, respectively. When a field variable does not satisfy the continuity conditions at a surface intersecting volume $v(t)$, the two integral theorems must be modified.

We distinguish two particular surfaces: (1) the surface within a body occupied by the same particle at all times is referred to as a *material surface*; (2) the surface within a body across which a physical quantity undergoes a jump is referred to as a *discontinuity surface*.

Let $\sigma(t)$ be a discontinuity surface, not necessarily material, across which a physical variable is discontinuous, while the variable is assumed to be continuously differentiable in the remaining part of a body. Let us assume that $\sigma(t)$ moves with velocity $\vec{w}$, which is not necessarily equal to the velocity $\vec{v}$ of material particles. We will now find a condition for the time evolution of $\sigma(t)$.

The moving discontinuity surface $\sigma(t)$ can be defined implicitly by the scalar equation

$$f_\sigma(\vec{x}, t) = 0 \qquad \vec{x} \in \sigma(t), \tag{2.43}$$

where f_σ is differentiable. Let us assume that there is a surface Σ in the reference configuration such that surfaces $\sigma(t)$ and Σ are related by a one-to-one mapping of the form

$$\vec{x} = \vec{\chi}_\sigma(\vec{X}, t) \qquad \vec{x} \in \sigma(t),\ \vec{X} \in \Sigma. \tag{2.44}$$

Since $\sigma(t)$ moves with velocity $\vec{w}$, which generally differs from the material velocity $\vec{v}$, the mapping $\vec{x} = \vec{\chi}_\sigma(\vec{X}, t)$ may differ from the motion $\vec{x} = \vec{\chi}(\vec{X}, t)$.

The Lagrangian and Eulerian representations of velocity given by (2.3) and (2.6), respectively, are now applied to the mapping (2.44) for the particles located on $\sigma(t)$,

$$\vec{W}(\vec{X},t) = \left.\frac{\partial \vec{\chi}_\sigma}{\partial t}\right|_{\vec{X}}, \qquad \left.\vec{w}(\vec{x},t)\right|_{\vec{x}=\vec{\chi}_\sigma(\vec{X},t)} = \vec{W}(\vec{X},t). \tag{2.45}$$

In view of (2.10), the material time derivative of (2.43) is

$$\frac{\partial f_\sigma}{\partial t} + \vec{w}\cdot \operatorname{grad} f_\sigma = 0, \tag{2.46}$$

where the differentiation in the gradient operator is taken with respect to $\vec{x}$. Introducing the unit normal $\vec{n}$ to $\sigma(t)$ by

$$\vec{n}(\vec{x},t) = \frac{\operatorname{grad} f_\sigma}{|\operatorname{grad} f_\sigma|} \qquad \vec{x}\in\sigma(t), \tag{2.47}$$

we obtain

$$\frac{\partial f_\sigma}{\partial t} + |\operatorname{grad} f_\sigma|(\vec{n}\cdot\vec{w}) = 0. \tag{2.48}$$

The condition (2.46) or (2.48) is called the *kinematic condition* for the time evolution of the moving surface $\sigma(t)$. The respective velocities $\vec{W}$ and $\vec{w}$ of the discontinuity surface $\sigma(t)$ in the reference and present configurations are known as the *displacement velocity* and the *propagation velocity*, respectively.

Consider a closed surface $\sigma(t)$ in volume v across which a tensor field $\boldsymbol{A}$ undergoes a jump. The surface $\sigma(t)$ divides the volume v into two parts, namely, v^+, into which the unit normal $\vec{n}$ is pointing, and v^- on the other side. Figure 2.1

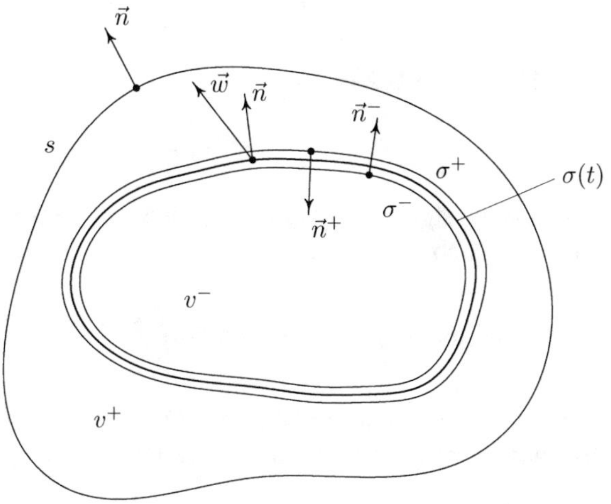

Fig. 2.1 The discontinuity surface $\sigma(t)$ and the sign convention followed

illustrates this concept and the sign convention followed. The Gauss theorem (2.41) is then modified to become

$$\int_{v-\sigma} \operatorname{div} \boldsymbol{A}\, dv = \int_{s} \vec{n} \cdot \boldsymbol{A}\, da - \int_{\sigma} \vec{n} \cdot [\boldsymbol{A}]_{-}^{+}\, da. \tag{2.49}$$

The volume integral over $v - \sigma$ refers to the volume v of the body excluding the points located on the discontinuity surface σ, i.e.,

$$v - \sigma := v^{+} + v^{-}. \tag{2.50}$$

The brackets indicate the *jump* of the enclosed quantity across $\sigma(t)$,

$$[\boldsymbol{A}]_{-}^{+} := \boldsymbol{A}^{+} - \boldsymbol{A}^{-}, \tag{2.51}$$

where the superscripts $\pm$ denote evaluation for the particles lying on the positive and negative sides of the discontinuity $\sigma(t)$. To prove (2.49), we apply the Gauss theorem (2.41) to the two volumes v^{+} and v^{-} bounded by $s + \sigma^{+}$ and σ^{-}, respectively,

$$\int_{v^{+}} \operatorname{div} \boldsymbol{A}\, dv = \int_{s} \vec{n} \cdot \boldsymbol{A}\, da + \int_{\sigma^{+}} \vec{n}^{+} \cdot \boldsymbol{A}^{+}\, da,$$

$$\int_{v^{-}} \operatorname{div} \boldsymbol{A}\, dv = \int_{\sigma^{-}} \vec{n}^{-} \cdot \boldsymbol{A}^{-}\, da,$$

where $\vec{n}^{+}$ and $\vec{n}^{-}$ are the outward unit normal vectors to σ^{+} and σ^{-}, respectively. Adding these two equations gives

$$\int_{v^{+}+v^{-}} \operatorname{div} \boldsymbol{A}\, dv = \int_{s} \vec{n} \cdot \boldsymbol{A}\, da + \int_{\sigma^{+}} \vec{n}^{+} \cdot \boldsymbol{A}^{+}\, da + \int_{\sigma^{-}} \vec{n}^{-} \cdot \boldsymbol{A}^{-}\, da.$$

From Fig. 2.1 we deduce that

$$\vec{n}^{+} = -\vec{n}^{-} = -\vec{n}.$$

The negative sign at $\vec{n}$ appears since the unit normal vector on the discontinuity surface is pointing to its positive side. Finally, letting σ^{+} and σ^{-} approach σ yields

$$\int_{\sigma^{+}} \vec{n}^{+} \cdot \boldsymbol{A}^{+}\, da + \int_{\sigma^{-}} \vec{n}^{-} \cdot \boldsymbol{A}^{-}\, da = \int_{\sigma} \vec{n} \cdot \left(\boldsymbol{A}^{-} - \boldsymbol{A}^{+}\right)\, da = -\int_{\sigma} \vec{n} \cdot [\boldsymbol{A}]_{-}^{+}\, da.$$

The Reynolds transport theorem (2.35) also needs to be modified once the discontinuity surface $\sigma(t)$ moves with velocity $\vec{w}$, which, as stated before, differs from the material velocity $\vec{v}$. The modified (2.35) is

$$\frac{D}{Dt}\int_{v-\sigma} \phi dv = \int_{v-\sigma}\left(\frac{D\phi}{Dt} + \phi \operatorname{div} \vec{v}\right) dv + \int_{\sigma} \vec{n}\cdot\left[(\vec{v}-\vec{w})\otimes\phi\right]_{-}^{+} da. \quad (2.52)$$

Both ϕ and $\vec{v}$ are required to be continuously differentiable in $v-\sigma$. To prove (2.52), we apply (2.39) to the two volumes v^{+} and v^{-} bounded by $s+\sigma^{+}$ and σ^{-}, respectively,

$$\frac{D}{Dt}\int_{v^+} \phi\, dv = \int_{v^+} \frac{\partial\phi}{\partial t} dv + \int_{s} (\vec{n}\cdot\vec{v})\,\phi\, da + \int_{\sigma^+} (\vec{n}^{+}\cdot\vec{w})\,\phi^{+}\, da,$$

$$\frac{D}{Dt}\int_{v^-} \phi\, dv = \int_{v^-} \frac{\partial\phi}{\partial t} dv + \int_{\sigma^-} (\vec{n}^{-}\cdot\vec{w})\,\phi^{-}\, da.$$

Adding these two equations, letting σ^{+} and σ^{-} approach σ and using $\vec{n}^{+} = -\vec{n}^{-} = -\vec{n}$ gives

$$\frac{D}{Dt}\int_{v^+ + v^-} \phi\, dv = \int_{v^+ + v^-} \frac{\partial\phi}{\partial t} dv + \int_{s} (\vec{n}\cdot\vec{v})\,\phi\, da - \int_{\sigma} (\vec{n}\cdot\vec{w})\,[\phi]_{-}^{+}\, da.$$

Replacing the second term on the right from the Gauss theorem (2.49) applied to $\boldsymbol{A} = \vec{v}\otimes\phi$, we obtain

$$\frac{D}{Dt}\int_{v^+ + v^-} \phi\, dv = \int_{v^+ + v^-} \left(\frac{\partial\phi}{\partial t} + \operatorname{div}\,(\vec{v}\otimes\phi)\right) dv + \int_{\sigma} \vec{n}\cdot\left[(\vec{v}-\vec{w})\otimes\phi\right]_{-}^{+} da.$$

To complete the proof of (2.52), the first term on the right is reformulated using (2.36) and (2.40).

Chapter 3
Measures of Stress

In this chapter, we abandon the purely kinematic aspects of the motion of bodies and will focus on the actual forces causing the motion. Two distinct types of forces are recognised in continuum mechanics: body forces are conceived as acting on the particles of a body, and the surface (or contact) forces as arising from the action of one part of a body on an adjacent part across a separating surface. Just as there are many different strain measures, there are several different measures of the surface forces. We will see that the surface forces can be described as a second-order tensor. Thus, the surface forces can always be quantified by a set of nine numbers, and the various different definitions are all equivalent.

3.1 Mass and Density

Mass is an intrinsic property of $\mathcal{B}$. Intuitively, mass is a measure of the amount of material contained in a body. It is a positive real scalar quantity that is additive, i.e., the mass of a body is the sum of the masses of its parts. In continuum mechanics, we assume the existence of a scalar function ϱ, assigned to each particle $\mathcal{X}$, such that the mass of a body $\mathcal{B}$ occupying the finite volume $v(\mathcal{B})$ is determined by

$$m(\mathcal{B}) = \int_{v(\mathcal{B})} \varrho \, dv, \tag{3.1}$$

where ϱ is the *mass density* of the material composing $\mathcal{B}$. Thus, the mass density measures the mass of material per unit volume. We will be dealing with a continuous

Z. Martinec, *Principles of Continuum Mechanics*, Nečas Center Series,
https://doi.org/10.1007/978-3-030-05390-1_3

mass distribution[1] in which (3.1) is valid, which implies that $m(\mathcal{B}) \to 0$ as $v(\mathcal{B}) \to 0$ such that

$$0 < \varrho < \infty. \tag{3.2}$$

A body $\mathcal{B}$ composed of a material with mass density function ϱ is called a *material body*, and a particle $\mathcal{X}$ with an assigned density ϱ is called a *material particle*.

3.2 Body and Surface Forces

The macroscopic forces that act on a continuum, or between portions of it, may be divided into *long-range* and *short-range* forces. This division comes from classifying interactions between continuum particles into two types: (1) those that attenuate 'slowly' with increasing distance such that they are still effective over distances comparable to the macroscopic length scale L of the body; (2) those that attenuate 'rapidly' over distances comparable to a characteristic linear dimension ℓ of a volume element and whose effects do not penetrate the macroscopic volume. Thus, the distinction between long-range and short-range effects is related to the validity of the asymptotic limit $\ell/L \to 0$.

Long-range forces are comprised of *gravitational, electromagnetic* and *inertial* forces. As stated, these forces decrease gradually with increasing distance between interacting particles. As a result, long-range forces act uniformly on all matter contained within a sufficiently small volume such that they are proportional to the volume size involved. In continuum mechanics, long-range forces are referred to as *body* or *volume forces*. A body force acting on the material body $\mathcal{B}$ is specified by the vector field $\vec{f}$, which is a measure of the body force per unit mass of $\mathcal{B}$. The total body force acting on $\mathcal{B}$ currently occupying volume $v(\mathcal{B})$ is

$$\vec{F}(\mathcal{B}) = \int_{v(\mathcal{B})} \varrho \vec{f} \, dv. \tag{3.3}$$

Short-range forces comprise several types of *molecular* forces. Their characteristic feature is that they decrease extremely rapidly with increasing distance between the interacting particles. Hence, they are of consequence only when this distance does not exceed molecular dimensions. As a result, if matter inside a volume is acted upon by short-range forces originating from interactions with matter outside this volume, these forces only act upon a thin layer immediately below its surface. In continuum mechanics, short-range forces, called *surface* or *contact* forces, are specified more closely by constitutive equations in Chap. 6.

[1]Discrete mass distributions, such as concentrated masses, are not considered in this book.

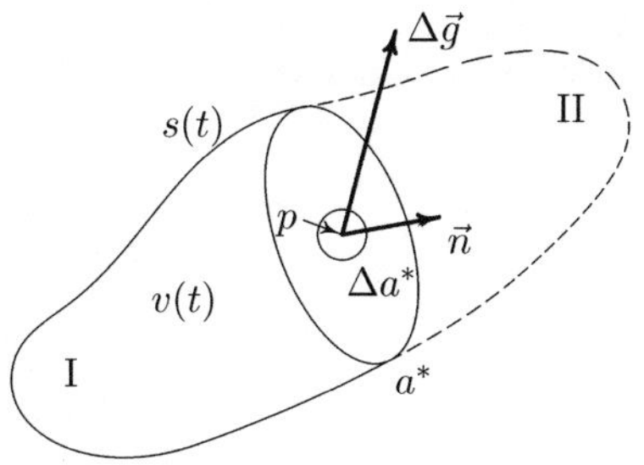

Fig. 3.1 Surface force $\Delta\vec{g}$ on surface area element Δa^*. Body force $\vec{f}$ and surface force $\vec{g}$ acting in $v(t)$ and on $s(t)$, respectively, are not drawn

3.3 Cauchy Traction Principle

A mathematical description of surface forces originates from the Cauchy traction principle. Let a body of a current volume $v(t)$ that is bounded by surface $s(t)$ be subject to body force $\vec{f}$ and surface force $\vec{g}$. Let p be an interior point of $v(t)$ and let a plane surface a^* pass through point p and cut the body into two parts labelled by I and II in Fig. 3.1. Point p is lying in the area element Δa^* of the cutting plane with the unit normal vector $\vec{n}$ pointing in the direction from part I to part II. The action of the material occupying part II on the material occupying part I is represented by a surface force $\Delta\vec{g}$ distributed on Δa^*. Note that the force $\Delta\vec{g}$ is not, in general, in the direction of the unit normal vector $\vec{n}$, as illustrated in Fig. 3.1. The *Cauchy traction principle* postulates that the limit when the area Δa^* shrinks to zero, with p remaining an interior point, exists such that

$$\vec{t}_{(\vec{n})} := \lim_{\Delta a^* \to 0} \frac{\Delta\vec{g}}{\Delta a^*}. \tag{3.4}$$

Obviously, this limit is meaningful only if Δa^* degenerates only into a point p. The vector $\vec{t}_{(\vec{n})}$ is called the *Cauchy stress vector* or the *Cauchy traction vector* (force per unit area). It is important to note that, in general, $\vec{t}_{(\vec{n})}$ depends not only on the position of p on Δa^*, but also on the orientation of surface area Δa^*, that is, on its outward unit normal vector $\vec{n}$. This dependence is therefore indicated by the subscript $\vec{n}$.[2] Thus, for the infinity of cutting planes imaginable through point p, each identified by a specific $\vec{n}$, there is also an infinity of associated stress vectors $\vec{t}_{(\vec{n})}$ for a given loading of the body.

We should here mention that a continuous distribution of surface forces acting across some surface is, in general, equivalent to a resultant force and a resultant torque. In (3.4) we have made the assumption that, in the limit at p, the torque per unit area vanishes and therefore there is no remaining contact torque. A material

[2]The assumption that the stress vector $\vec{t}_{(\vec{n})}$ depends only on the outward unit normal vector $\vec{n}$ and not on the differential geometric property of the surface, such as the curvature, was introduced by Cauchy and is referred to as the Cauchy assumption.

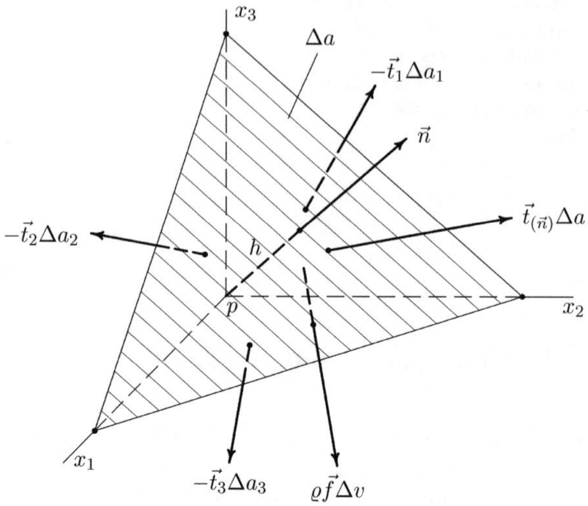

Fig. 3.2 Equilibrium of an infinitesimal tetrahedron

that admits contact forces but not contact torques is called a *non-polar material*. The following text considers non-polar materials only. For a discussion of polar materials, the reader is referred to Eringen (1967).

3.4 Cauchy Stress Formula

To find the dependence of the stress vector on the unit normal vector $\vec{n}$, we set up an infinitesimal tetrahedron of volume Δv, as shown in Fig. 3.2. Its vertex is at p, it has three coordinate surfaces Δa_k and the base Δa with an oriented unit normal vector $\vec{n}$. The stress vector[3] on the coordinate surface $x_k =$ const. is denoted by $-\vec{t}_k$. We now apply Newton's second law of motion to this tetrahedron to yield

$$\int_{\Delta v} \varrho \vec{f}\, dv - \int_{\Delta a_k} \vec{t}_k\, da_k + \int_{\Delta a} \vec{t}_{(\vec{n})}\, da = \frac{D}{Dt}\int_{\Delta v} \varrho \vec{v}\, dv.$$

The surface and volume integrals are evaluated by the mean value theorem such that

$$\varrho^* \vec{f}^* \Delta v - \vec{t}_k^* \Delta a_k + \vec{t}_{(\vec{n})}^* \Delta a = \frac{D}{Dt}\left(\varrho^* \vec{v}^* \Delta v\right), \tag{3.5}$$

[3]Since the outward normal vector to a coordinate surface $x_k =$ const. is in the direction of $-x_k$, without loss of generality, we denote the stress vector acting on this coordinate surface by $-\vec{t}_k$ rather than $\vec{t}_k$.

where ϱ^*, $\vec{f}^*$ and $\vec{v}^*$ are, respectively, the values of ϱ, $\vec{f}$ and $\vec{v}$ at some interior points of the tetrahedron, and $\vec{t}^*_{(\vec{n})}$ and $\vec{t}^*_k$ are the values of $\vec{t}_{(\vec{n})}$ and $\vec{t}_k$ on the surface Δa and on coordinate surfaces Δa_k. The volume of the tetrahedron is

$$\Delta v = \frac{1}{3} h \Delta a, \tag{3.6}$$

where h is the perpendicular distance from point p to the base Δa. Moreover, the area vector $\Delta\vec{a}$ is equal to the sum of three area elements of coordinate surfaces,

$$\Delta\vec{a} = \vec{n}\Delta a = \Delta a_k \vec{i}_k, \tag{3.7}$$

which gives

$$\Delta a_k = n_k \Delta a. \tag{3.8}$$

Substituting (3.6) and (3.8) into (3.5) and canceling the common factor Δa yields

$$\frac{1}{3}\varrho^* \vec{f}^* h - \vec{t}^*_k n_k + \vec{t}^*_{(\vec{n})} = \frac{1}{3}\frac{D}{Dt}\left(\varrho^* \vec{v}^* h\right). \tag{3.9}$$

Now, letting the tetrahedron shrink to point p by taking the limit $h \to 0$ and noting that the starred quantities take the values at point p, we obtain

$$\vec{t}_{(\vec{n})} = \vec{t}_k n_k, \tag{3.10}$$

which is the *Cauchy stress formula*. Equation (3.10) allows the determination of the Cauchy stress vector at a point acting across an arbitrarily inclined plane, if the Cauchy stress vectors acting across the three coordinate surfaces through that point are known.

The stress vectors $\vec{t}_k$ are, by definition, independent of $\vec{n}$. From (3.10) it therefore follows that

$$-\vec{t}_{(-\vec{n})} = \vec{t}_{(\vec{n})}. \tag{3.11}$$

The stress vector acting on a surface with the unit normal vector $\vec{n}$ is equal to the negative stress vector acting on the corresponding surface with the unit normal vector $-\vec{n}$. In Newtonian mechanics this statement is known as Newton's third law of action and reaction.

We now introduce the definition of the Cauchy stress tensor. The t_{kl} component of the *Cauchy stress tensor* $\boldsymbol{t}$ is given by the lth component of the stress vector $\vec{t}_k$ acting on the positive side of the kth coordinate surface:

$$\vec{t}_k = t_{kl}\vec{i}_l, \qquad \text{or} \qquad t_{kl} = \vec{t}_k \cdot \vec{i}_l. \tag{3.12}$$

The first subscript in t_{kl} denotes the coordinate surface on which the stress vector $\vec{t}_k$ acts, while the second subscript denotes the direction in which $\vec{t}_k$ acts. For example, t_{23} is the x_3 component of the stress vector $\vec{t}_2$ acting on the coordinate surface $x_2 =$ const. Now, if the outward normal vector to $x_2 =$ const. points in the positive direction of the x_2 axis, t_{23} points in the positive direction of the x_3 axis. If the outward normal vector to $x_2 =$ const. is in the negative direction of the x_2 axis, t_{23} is directed in the negative direction of the x_3 axis. Figure 3.3 illustrates the positive stress components on the faces of a parallelepiped built on the coordinate surfaces. For clarity, only the stress components on two pairs of parallel faces are shown. The nine components t_{kl} of the Cauchy stress tensor $\boldsymbol{t}$ may be represented in matrix form as

$$\boldsymbol{t} = \begin{pmatrix} t_{11} & t_{12} & t_{13} \\ t_{21} & t_{22} & t_{23} \\ t_{31} & t_{32} & t_{33} \end{pmatrix} (\vec{i}_k \otimes \vec{i}_l). \tag{3.13}$$

In view of (3.12), the Cauchy stress formula (3.10) becomes

$$\vec{t}_{(\vec{n})} = \vec{n} \cdot \boldsymbol{t}, \tag{3.14}$$

which says that the Cauchy stress vector acting on any plane through a point is fully characterised as a linear function of the stress tensor at that point. The normal component of the stress vector,

$$\tau := \vec{t}_{(\vec{n})} \cdot \vec{n} = \vec{n} \cdot \boldsymbol{t} \cdot \vec{n}, \tag{3.15}$$

called the *normal stress*, is said to be *tensile* when positive and *compressive* when negative. The stress vector directed tangentially to surface has the form

$$\vec{t}_t := \vec{t}_{(\vec{n})} - \tau\vec{n} = \vec{n} \cdot \boldsymbol{t} - (\vec{n} \cdot \boldsymbol{t} \cdot \vec{n})\vec{n}. \tag{3.16}$$

The size of $\vec{t}_t$ is known as the *shear stress*. For example, the components t_{11}, t_{22} and t_{33} in Fig. 3.3 are the normal stresses and the mixed components t_{12}, t_{13}, etc., are the shear stresses.

It still remains to verify that the Cauchy stress tensor $\boldsymbol{t}$ is indeed a second-order tensor. In view of the concept of tensors, we should verify that the stress components t_{kl} satisfy the transformation rule (1.67). Suppose that the Cartesian coordinates x_k are transformed onto x'_k by the one-to-one mapping

$$x_k = \hat{x}_k(x'_1, x'_2, x'_3) \iff x'_{k'} = \hat{x}'_{k'}(x_1, x_2, x_3). \tag{3.17}$$

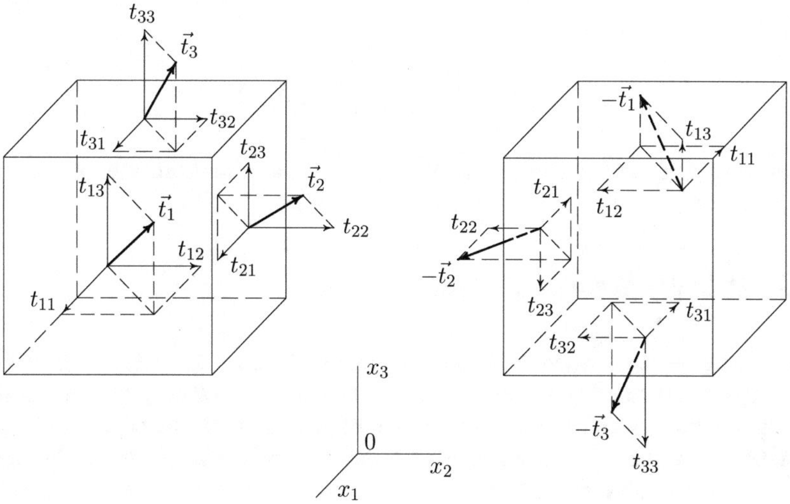

Fig. 3.3 The positive stress components t_{kl} on the front (left) and back (right) faces of an parallelepiped

The unit normal vector $\vec{n}(\vec{x}, t)$ to surface $\sigma(\vec{x}, t)$ that is represented by the implicit equation $f_\sigma(\vec{x}, t) = 0$ is given by (2.47) and its component form is[4]

$$n_k(\vec{x}) = \frac{1}{|\text{grad}\, f_\sigma(\vec{x})|} \frac{\partial f_\sigma(\vec{x})}{\partial x_k} = \frac{1}{|\text{grad}\, f_\sigma(\vec{x})|} \frac{\partial f'_\sigma(\vec{x}')}{\partial x'_{k'}} \frac{\partial \hat{x}'_{k'}}{\partial x_k},$$

where $f'_\sigma(\vec{x}') := f_\sigma(\vec{\hat{x}}(\vec{x}'))$. Using the invariance property of the scalar function $|\text{grad}\, f_\sigma(\vec{x})|$ under the *coordinate transformation* (3.17), i.e., $|\text{grad}\, f_\sigma(\vec{x})| = |\text{grad}'\, f'_\sigma(\vec{x}')|$, yields

$$n_k(\vec{x}) = \frac{\partial \hat{x}'_{k'}}{\partial x_k} n'_{k'}(\vec{x}'). \tag{3.18}$$

Since the normal component of stress t_n is independent of the coordinate transformation, we have

$$t_n = n_k n_l t_{kl} = \frac{\partial \hat{x}'_{k'}}{\partial x_k} \frac{\partial \hat{x}'_{l'}}{\partial x_l} n'_{k'} n'_{l'} t_{kl} \stackrel{!}{=} t'_{n'} = n'_{k'} n'_{l'} t'_{k'l'}.$$

[4]The argument t is omitted to shorten the notation.

Hence,

$$t'_{k'l'}(\vec{x}', t) = \frac{\partial \hat{x}'_{k'}}{\partial x_k} \frac{\partial \hat{x}'_{l'}}{\partial x_l} t_{kl}(\vec{x}', t) \tag{3.19}$$

because $n'_{k'}$ is arbitrary. Thus, t_{kl} transforms such as a second-order tensor.

3.5 Other Measures of Stress

So far, we have represented short-range intermolecular contact forces in terms of the Cauchy stress vector $\vec{t}_k$ or tensor $\boldsymbol{t}$. There are, however, two other ways of measuring or representing these forces, each of which plays a role in the theory of continuum mechanics. The Cauchy stress tensor gives the surface force acting on the deformed elementary area da such that

$$d\vec{g} = \vec{t}_{(\vec{n})} da = (\vec{n} \cdot \boldsymbol{t}) da = d\vec{a} \cdot \boldsymbol{t}. \tag{3.20}$$

The Cauchy stress tensor, as any other variable, has both an Eulerian and a Lagrangian representation. The relationship between the Lagrangian representation of the Cauchy stress tensor (or, simply, the *Lagrangian Cauchy stress tensor*) $\boldsymbol{T}(\vec{X}, t)$ and the Eulerian representation of the Cauchy stress tensor (or, simply, the *Eulerian Cauchy stress tensor*) $\boldsymbol{t}(\vec{x}, t)$ is $\boldsymbol{T}(\vec{X}, t) := \boldsymbol{t}(\vec{x}(\vec{X}, t), t)$. We make, however, an exception in the notation and use $\boldsymbol{t}(\vec{X}, t)$ for the Lagrangian Cauchy stress tensor. The Eulerian Cauchy stress tensor $\boldsymbol{t}(\vec{x}, t)$ arises naturally in the Eulerian form of the principle of conservation of linear momentum. The corresponding Lagrangian form of this principle cannot, however, be readily expressed in terms of the Lagrangian Cauchy stress tensor $\boldsymbol{t}(\vec{X}, t)$.

A simple Lagrangian form of the conservation of linear momentum can be obtained if a stress measure is referred to an area in the reference configuration. This can be achieved by introducing the so-called *first Piola–Kirchhoff stress tensor* $\boldsymbol{T}^{(1)}$ as a stress measure referred to the undeformed area element $d\vec{A}$,

$$d\vec{g} = d\vec{a} \cdot \boldsymbol{t} =: d\vec{A} \cdot \boldsymbol{T}^{(1)}. \tag{3.21}$$

Here, the tensor $\boldsymbol{T}^{(1)}$ gives the surface force acting on the deformed area element $d\vec{a}$ at $\vec{x}$ in terms of the corresponding undeformed area element $d\vec{A}$ at the reference point $\vec{X}$. Thus, $\boldsymbol{T}^{(1)}$ is a measure of the force per unit undeformed area, whereas both the Eulerian and Lagrangian Cauchy stress tensors $\boldsymbol{t}(\vec{x}, t)$ and $\boldsymbol{t}(\vec{X}, t)$, respectively, are measures of the force per unit deformed area. The relationship between $\boldsymbol{T}^{(1)}$ and $\boldsymbol{t}$ is found using Nanson's formula (1.81). The result can be expressed in either of the following two equivalent forms:

$$\boldsymbol{T}^{(1)}(\vec{X}, t) = J\boldsymbol{F}^{-1} \cdot \boldsymbol{t}(\vec{X}, t), \qquad \boldsymbol{t}(\vec{X}, t) = J^{-1}\boldsymbol{F} \cdot \boldsymbol{T}^{(1)}(\vec{X}, t). \tag{3.22}$$

The surface force $d\vec{g}$ in (3.21) acts upon the present point $\vec{x}$, whereas the area element vector $d\vec{A}$ is referred to the point $\vec{X}$. The first Piola–Kirchhoff stress tensor $\boldsymbol{T}^{(1)}$ is therefore a two-point tensor. This can also be deduced from the component form of (3.22),

$$T^{(1)}_{Kl}(\vec{X},t) = J X_{K,k} t_{kl}(\vec{X},t), \qquad t_{kl}(\vec{X},t) = J^{-1} x_{k,K} T^{(1)}_{Kl}(\vec{X},t). \tag{3.23}$$

The constitutive equations for simple materials (see Sect. 6.3) are expressed most conveniently in terms of another measure of stress, known as the *second Piola–Kirchhoff stress tensor*, denoted by $\boldsymbol{T}^{(2)}$. This quantity gives, instead of the actual surface force $d\vec{g}$ acting on the deformed area element $d\vec{a}$, a force $d\vec{G}$ related to $d\vec{g}$ in the same way as the differential $d\vec{X}$ is related to the differential $d\vec{x}$. That is,

$$d\vec{G} = \boldsymbol{F}^{-1} \cdot d\vec{g}, \tag{3.24}$$

in the same manner as $d\vec{X} = \boldsymbol{F}^{-1} \cdot d\vec{x}$. Defining $\boldsymbol{T}^{(2)}$,

$$d\vec{G} =: d\vec{A} \cdot \boldsymbol{T}^{(2)}, \tag{3.25}$$

the first and second Piola–Kirchhoff stresses are related by

$$\boldsymbol{T}^{(2)} = \boldsymbol{T}^{(1)} \cdot \boldsymbol{F}^{-T}, \qquad \boldsymbol{T}^{(1)} = \boldsymbol{T}^{(2)} \cdot \boldsymbol{F}^{T}. \tag{3.26}$$

Comparing this result with (3.22) yields the relationship between the second Piola–Kirchhoff stress tensor and the Lagrangian Cauchy stress tensor,

$$\boldsymbol{T}^{(2)}(\vec{X},t) = J\boldsymbol{F}^{-1} \cdot \boldsymbol{t}(\vec{X},t) \cdot \boldsymbol{F}^{-T}, \qquad \boldsymbol{t}(\vec{X},t) = J^{-1}\boldsymbol{F} \cdot \boldsymbol{T}^{(2)}(\vec{X},t) \cdot \boldsymbol{F}^{T}. \tag{3.27}$$

Since the transformed surface force $d\vec{G}$ may be considered to act at the reference position $\vec{X}$ rather than at the present position $\vec{x}$, the second Piola–Kirchhoff stress tensor is an ordinary (a one-point) rather than a two-point tensor. This can also be seen from the component form of (3.27)

$$T^{(2)}_{KL}(\vec{X},t) = J X_{K,k} X_{L,l} t_{kl}(\vec{X},t), \qquad t_{kl}(\vec{X},t) = J^{-1} x_{k,K} x_{l,L} T^{(2)}_{KL}(\vec{X},t). \tag{3.28}$$

The previous expressions may be linearised provided that the displacement gradient $\boldsymbol{H}$ is much smaller than unity. To this end, we carry the linearised forms $(1.115)_{1,2}$ into $(3.22)_1$ and $(3.27)_1$, and obtain

$$\begin{aligned} \boldsymbol{T}^{(1)} &= (1 + \operatorname{tr}\boldsymbol{H})\,\boldsymbol{t} - \boldsymbol{H}^T \cdot \boldsymbol{t} + O(|\boldsymbol{H}|^2), \\ \boldsymbol{T}^{(2)} &= (1 + \operatorname{tr}\boldsymbol{H})\,\boldsymbol{t} - \boldsymbol{H}^T \cdot \boldsymbol{t} - \boldsymbol{t} \cdot \boldsymbol{H} + O(|\boldsymbol{H}|^2). \end{aligned} \tag{3.29}$$

In Sect. 4.3.3, we will show that the tensors $\boldsymbol{t}$ and $\boldsymbol{T}^{(2)}$ are symmetric. The last equation then demonstrates that the symmetry of tensor $\boldsymbol{T}^{(2)}$ has not been violated by the linearisation process. Conversely,

$$\begin{aligned} \boldsymbol{t} &= (1 - \operatorname{tr} \boldsymbol{H})\, \boldsymbol{T}^{(1)} + \boldsymbol{H}^T \cdot \boldsymbol{T}^{(1)} + O(|\boldsymbol{H}|^2) \\ &= (1 - \operatorname{tr} \boldsymbol{H})\, \boldsymbol{T}^{(2)} + \boldsymbol{H}^T \cdot \boldsymbol{T}^{(2)} + \boldsymbol{T}^{(2)} \cdot \boldsymbol{H} + O(|\boldsymbol{H}|^2). \end{aligned} \tag{3.30}$$

Supposing, in addition, that stresses are small compared to unity (the *infinitesimal deformation and stress theory*), then

$$\boldsymbol{T}^{(1)} \cong \boldsymbol{T}^{(2)} \cong \boldsymbol{t}, \tag{3.31}$$

showing that, when considering infinitesimal deformation and stress, a distinction between the Cauchy and the Piola–Kirchhoff stresses is not necessary.

Chapter 4
Fundamental Conservation Principles

The fundamental principles of continuum mechanics are those that deal with the conservation of some physical quantity. Conservation means that these quantities cannot change over time, unless an external force intervenes. These *conservation principles*, as they are often called, are postulated for all material continua, irrespective of the material's constitution and geometry, and result in equations that must always be satisfied. They deal with mass, linear and angular momenta, energy and entropy. The conservation principles are also valid for all bodies subject to thermomechanical effects.

4.1 Global Conservation Principles

The conservation principles are usually formulated in a global (integral) form derived on the basis of the conservation of some properties of the body as a whole. The global equations may then be used to develop associated *field equations* that are valid at all points within the body and on its surface.

Principle 1 (Conservation of Mass) *The total mass of a body does not change with motion.*[1]

This principle assumes that the mass production and supply is zero, and then postulates that the mass of a material body is independent of motion and remains constant in every configuration,

$$\int_V \varrho_0 \, dV = \int_{v(t)} \varrho \, dv, \tag{4.1}$$

[1] However, the mass density can change as the volume of the body changes while in motion.

Z. Martinec, *Principles of Continuum Mechanics*, Nečas Center Series,
https://doi.org/10.1007/978-3-030-05390-1_4

where V and $v(t)$ are the respective material volumes and $\varrho_0(\vec{X})$ and $\varrho(\vec{x}, t)$ are the respective mass densities of the reference and present configurations of the body. Applying the material time derivative to (4.1) results in the alternative form

$$\frac{D}{Dt} \int_{v(t)} \varrho \, dv = 0. \tag{4.2}$$

Principle 2 (Conservation of Linear Momentum) *The time rate of change of the total linear momentum of a body equals the total force acting on the body.*

Let a body of a current volume $v(t)$ that is bounded by surface $s(t)$ with outward unit normal vector $\vec{n}$ be subject to surface traction $\vec{t}_{(\vec{n})}$ (force per unit area) and body force $\vec{f}$ (force per unit mass). The total force acting on the body is

$$\int_{s(t)} \vec{t}_{(\vec{n})} \, da + \int_{v(t)} \varrho \vec{f} \, dv.$$

In addition, let the particles of the body be in motion under the velocity field $\vec{v}(\vec{x}, t)$. The *linear momentum* of the body is defined by the vector

$$\int_{v(t)} \varrho \vec{v} \, dv,$$

where $\varrho \vec{v}$ is the linear momentum (per unit volume) of a material particle. Thus, the principle of conservation of linear momentum states that

$$\frac{D}{Dt} \int_{v(t)} \varrho \vec{v} \, dv = \int_{s(t)} \vec{t}_{(\vec{n})} \, da + \int_{v(t)} \varrho \vec{f} \, dv. \tag{4.3}$$

Principle 3 (Conservation of Angular Momentum) *The time rate of change of the total angular momentum of a body equals the total moment of all forces acting on the body,*

$$\frac{D}{Dt} \int_{v(t)} \vec{x} \times \varrho \vec{v} \, dv = \int_{s(t)} \vec{x} \times \vec{t}_{(\vec{n})} \, da + \int_{v(t)} \vec{x} \times \varrho \vec{f} \, dv, \tag{4.4}$$

where the left-hand side is the time rate of change of the total *angular momentum* about the origin, which is also frequently called the *moment of momentum*. The surface and volume integrals on the right are the moment of the surface tractions about the origin and the total moment of body forces about the origin, respectively.

Principle 4 (Conservation of Energy) *The time rate of change of the kinetic energy K and the internal energy E equals the mechanical power P done by the* *surface and body forces and all other energies Q entering or leaving the body,*

$$\frac{D}{Dt}(K+E) = P + Q. \tag{4.5}$$

The total *kinetic energy* of the body is

$$K = \frac{1}{2}\int_{v(t)} \varrho\vec{v}\cdot\vec{v}\,dv, \tag{4.6}$$

where $(\vec{v}\cdot\vec{v})/2$ is the kinetic energy (per unit mass) of a material particle of the body. In continuum mechanics, the *internal energy density* ε (energy density per unit mass) is postulated to exist such that

$$E = \int_{v(t)} \varrho\varepsilon\,dv. \tag{4.7}$$

The *mechanical power* of the surface traction $\vec{t}_{(\vec{n})}$ and body forces $\vec{f}$ is

$$P = \int_{s(t)} \vec{t}_{(\vec{n})}\cdot\vec{v}\,da + \int_{v(t)} \varrho\vec{f}\cdot\vec{v}\,dv. \tag{4.8}$$

Other energies Q entering or leaving the body may include heat, electromagnetic, chemical or some other energies. In this book, we consider only mechanical and heat energies, so that the energy transfer is only due to work and heat. The *heat energy* consists of the *heat flux* $\vec{q}$ (per unit area) entering or leaving the body across the surface and the *heat supply* h (per unit mass) generated by internal sources (for example, radioactive decay or chemical reactions),

$$Q \equiv Q_h := -\int_{s(t)} \vec{q}\cdot\vec{n}\,da + \int_{v(t)} \varrho h\,dv. \tag{4.9}$$

The negative sign at the surface integral indicates that the heat flux vector $\vec{q}$ is pointing from higher temperatures towards lower temperatures, so that $\int_{s(t)}\vec{q}\cdot\vec{n}\,da$ is the total outward heat flux. Thus, the principle of conservation of energy becomes

$$\frac{D}{Dt}\int_{v(t)} (\varrho\varepsilon + \frac{1}{2}\varrho\vec{v}\cdot\vec{v})\,dv = \int_{s(t)} (\vec{t}_{(\vec{n})}\cdot\vec{v} - \vec{q}\cdot\vec{n})\,da + \int_{v(t)} (\varrho\vec{f}\cdot\vec{v} + \varrho h)\,dv, \tag{4.10}$$

which is a statement of the *first law of thermodynamics*.

If the body is not subject to body forces or heat supply, the volume integral on the right-hand side of (4.10) vanishes. Moreover, with the aid of (3.14), the integrand of the surface integral can be put into the form

$$\vec{t}_{(\vec{n})}\cdot\vec{v} - \vec{q}\cdot\vec{n} = \vec{n}\cdot(\boldsymbol{t}\cdot\vec{v} - \vec{q}) =: \vec{j}\cdot\vec{n}. \tag{4.11}$$

Equation (4.10) then reduces to

$$\frac{D}{Dt}\int_{v(t)}(\varrho\varepsilon + \frac{1}{2}\varrho\vec{v}\cdot\vec{v})\,dv = -\int_{s(t)}\vec{j}\cdot\vec{n}\,da. \tag{4.12}$$

Hence, the time rate of change of the total energy (i.e., kinetic energy plus internal energy) of the material occupying an arbitrary volume $v(t)$ is equal to the total flux of vector $\vec{j}$ through the surface $s(t)$. In view of this result, $\vec{j}$ is known as the *energy flux vector.*

Principle 5 (Entropy Inequality) *The total entropy production Γ is non-negative for all thermodynamic processes,*

$$\Gamma \geq 0, \tag{4.13}$$

where

$$\Gamma := \frac{DH}{Dt} - Z + \int_{s(t)}\vec{s}\cdot\vec{n}\,da, \tag{4.14}$$

H is the total entropy of the body, Z is the total entropy supply generated by internal sources and $\vec{s}$ is the *entropy flux* across the surface of the body. In continuum mechanics, the existence of the *entropy density* η (entropy per unit mass) and the *entropy supply* z (per unit mass) is postulated such that

$$H = \int_{v(t)}\varrho\eta\,dv, \qquad Z = \int_{v(t)}\varrho z\,dv. \tag{4.15}$$

The entropy inequality (4.13) then becomes

$$\frac{D}{Dt}\int_{v(t)}\varrho\eta\,dv - \int_{v(t)}\varrho z\,dv + \int_{s(t)}\vec{s}\cdot\vec{n}\,da \geq 0, \tag{4.16}$$

which is a statement of the *second law of thermodynamics*. The inequality is valid for an irreversible process; the equality holds for a reversible process.

It is not yet clear why the entropy density η and entropy supply z are introduced as additional variables for the description of thermomechanical phenomena. However, real physical processes are directional. They proceed in certain time sequences, but are not necessarily reversible. This *principle of irreversibility* can be expressed mathematically by introducing the entropy principle, which requires that the entropy production is non-negative for all thermodynamic processes. More precise specifications of this concept will be given in Chap. 7. Hence, we postulate that entropy and temperature exist, an assumption that may be viewed as a part of the entropy principle.

The five global principles postulated above are valid for all bodies, irrespective of their internal constitutions. To find the corresponding local forms, additional principles are necessary to postulate, as will be discussed in the next two sections.

4.2 Local Conservation Principles in the Eulerian Form

4.2.1 Continuity Equation

Applying the Reynolds transport theorem (2.52) for $\phi = \varrho$ to (4.2) gives

$$\int_{v(t)-\sigma(t)} \left(\frac{D\varrho}{Dt} + \varrho \operatorname{div} \vec{v}\right) dv + \int_{\sigma(t)} \vec{n} \cdot \left[\varrho(\vec{v} - \vec{w})\right]_{-}^{+} da = 0. \tag{4.17}$$

We now assume that the integrands of the volume and surface integrals in (4.17) are continuous functions of x_k in the volume $v(t) - \sigma(t)$ and along the discontinuity surface $\sigma(t)$, respectively. Moreover, we postulate that *all global conservation principles hold for an arbitrary part of the volume and of the discontinuity surface* (*the additive principle*).[2] The two assumptions then imply that the integrands of each integral must vanish separately, which is the statement of the *localisation theorem*,

$$\frac{D\varrho}{Dt} + \varrho \operatorname{div} \vec{v} = 0 \qquad \text{in } v(t) - \sigma(t), \tag{4.18}$$

$$\vec{n} \cdot \left[\varrho(\vec{v} - \vec{w})\right]_{-}^{+} = 0 \qquad \text{on } \sigma(t). \tag{4.19}$$

These are the equations of local conservation of mass and the jump condition in the Eulerian form. Equation (4.18) is often called the *continuity equation*. Writing the material time derivative of ϱ as

$$\frac{D\varrho}{Dt} = \frac{\partial \varrho}{\partial t} + \vec{v} \cdot \operatorname{grad} \varrho$$

allows (4.18) to be expressed in another form

$$\frac{\partial \varrho}{\partial t} + \operatorname{div}(\varrho \vec{v}) = 0 \qquad \text{in } v(t) - \sigma(t). \tag{4.20}$$

Equation (4.19) in turn means that the flux of mass (the amount of mass per unit area and unit time) entering the discontinuity surface must leave it on the other side, that is, there is no mass accumulation on the discontinuity surface. In other words,

[2]This principle is not used in non-local continuum theories, and only the global conservation principles, valid for the entire body, are considered to hold.

the quantity in square brackets in (4.19) is the amount of mass passing through a moving discontinuity surface in relation to the motion of the body behind and ahead of the surface. According to (4.19), they must be equal.

4.2.2 Equation of Motion

Substituting the Cauchy stress vector $\vec{t}_{(\vec{n})}$ from (3.14) into the global principle of conservation of linear momentum (4.3) leads to

$$\frac{D}{Dt}\int_{v(t)-\sigma(t)} \varrho\vec{v}\,dv = \int_{s(t)-\sigma(t)} \vec{n}\cdot\boldsymbol{t}\,da + \int_{v(t)-\sigma(t)} \varrho\vec{f}\,dv. \tag{4.21}$$

Applying the modified Gauss theorem (2.49) to the surface integral on the right gives

$$\frac{D}{Dt}\int_{v(t)-\sigma(t)} \varrho\vec{v}\,dv = \int_{v(t)-\sigma(t)} (\operatorname{div}\boldsymbol{t} + \varrho\vec{f})\,dv + \int_{\sigma(t)} \vec{n}\cdot[\boldsymbol{t}]_-^+\,da, \tag{4.22}$$

which upon using the Reynolds transport theorem (2.52) for $\phi = \varrho\vec{v}$ yields

$$\int_{v(t)-\sigma(t)} \left(\frac{D(\varrho\vec{v})}{Dt} + \varrho\vec{v}\operatorname{div}\vec{v} - \operatorname{div}\boldsymbol{t} - \varrho\vec{f}\right) dv$$
$$+ \int_{\sigma(t)} \vec{n}\cdot\left[\varrho(\vec{v}-\vec{w})\otimes\vec{v} - \boldsymbol{t}\right]_-^+ da = \vec{0}. \tag{4.23}$$

Adopting the additive principle, the last equation is postulated to be valid for all parts of the body. Thus, the integrands vanish separately, resulting in

$$\operatorname{div}\boldsymbol{t} + \varrho\vec{f} = \varrho\frac{D\vec{v}}{Dt} \qquad \text{in } v(t)-\sigma(t) \tag{4.24}$$

$$\vec{n}\cdot\left[\varrho(\vec{v}-\vec{w})\otimes\vec{v} - \boldsymbol{t}\right]_-^+ = \vec{0} \qquad \text{on } \sigma(t) \tag{4.25}$$

in view of (4.18). Equation (4.24) is known as *Cauchy's equation of motion*, expressing the local conservation of linear momentum in the Eulerian form, and (4.25) is the associated jump condition on the discontinuity surface σ.

4.2.3 Symmetry of the Cauchy Stress Tensor

The angular momentum of the surface tractions about the origin occurring in the global principle of conservation of angular momentum (4.4) can be rewritten by employing the Cauchy stress formula (3.14), the tensor identity

$$\vec{v} \times (\vec{w} \cdot \boldsymbol{A}) = -\vec{w} \cdot (\boldsymbol{A} \times \vec{v}), \tag{4.26}$$

where $\vec{v}$, $\vec{w}$ are vectors and $\boldsymbol{A}$ is a second-order tensor, and the modified Gauss theorem (2.49) as

$$\int_{s(t)-\sigma(t)} \vec{x} \times \vec{t}_{(\vec{n})}\, da = -\int_{v(t)-\sigma(t)} \operatorname{div}(\boldsymbol{t} \times \vec{x})\, dv - \int_{\sigma(t)} \vec{n} \cdot \left[\boldsymbol{t} \times \vec{x}\right]_-^+ da. \tag{4.27}$$

Using two differential identities,

$$\operatorname{div}(\boldsymbol{t} \times \vec{v}) = \operatorname{div} \boldsymbol{t} \times \vec{v} + \boldsymbol{t}^T \dot{\times} \operatorname{grad} \vec{v}, \tag{4.28}$$

$$\operatorname{grad} \vec{x} = \boldsymbol{I}, \tag{4.29}$$

where $\dot{\times}$ denotes the *dot-cross product of dyads*[3] and $\boldsymbol{I}$ is the second-order identity tensor, we obtain

$$\int_{v(t)-\sigma(t)} \operatorname{div}(\boldsymbol{t} \times \vec{x})\, dv = \int_{v(t)-\sigma(t)} (\operatorname{div} \boldsymbol{t} \times \vec{x} + \boldsymbol{t}^T \dot{\times} \boldsymbol{I})\, dv.$$

Substituting this and (4.27) into the principle of conservation of angular momentum (4.4) and using the Reynolds transport theorem (2.52) for $\phi = \vec{x} \times \varrho\vec{v}$ yields

$$\int_{v(t)-\sigma(t)} \left(\frac{D(\varrho\vec{x} \times \vec{v})}{Dt} + (\varrho\vec{x} \times \vec{v}) \operatorname{div} \vec{v} + \operatorname{div} \boldsymbol{t} \times \vec{x} + \boldsymbol{t}^T \dot{\times} \boldsymbol{I} - \varrho\vec{x} \times \vec{f} \right) dv$$

$$+ \int_{\sigma(t)} \vec{n} \cdot \left[(\vec{v} - \vec{w}) \otimes (\vec{x} \times \varrho\vec{v})\right]_-^+ da + \int_{\sigma(t)} \vec{n} \cdot \left[\boldsymbol{t} \times \vec{x}\right]_-^+ da = \vec{0},$$

which can be rearranged to the form

$$\int_{v(t)-\sigma(t)} \left[(\vec{x} \times \vec{v}) \left(\frac{D\varrho}{Dt} + \varrho \operatorname{div} \vec{v} \right) + \varrho \frac{D\vec{x}}{Dt} \times \vec{v} + \vec{x} \times \left(\varrho \frac{D\vec{v}}{Dt} - \operatorname{div} \boldsymbol{t} - \varrho \vec{f} \right) \right.$$

$$\left. + \boldsymbol{t}^T \dot{\times} \boldsymbol{I} \right] dv - \int_{\sigma(t)} \vec{n} \cdot \left[\varrho(\vec{v} - \vec{w}) \otimes \vec{v} - \boldsymbol{t}\right]_-^+ \times \vec{x}\, da = \vec{0}. \tag{4.30}$$

---

[3] We define

$$(\vec{u} \otimes \vec{v}) \dot{\times} (\vec{w} \otimes \vec{x}) := (\vec{v} \cdot \vec{w})(\vec{u} \times \vec{x}).$$

Considering

$$\frac{D\vec{x}}{Dt} \times \vec{v} = \vec{v} \times \vec{v} = \vec{0}$$

along with the local principles of conservation of mass (4.18), linear momentum (4.24) and the associated jump condition (4.25), (4.30) reduces to

$$\int_{v(t)-\sigma(t)} \boldsymbol{t}^T \dot{\times} \boldsymbol{I} \, dv = \vec{0}. \tag{4.31}$$

Again, postulating that this is valid for all parts of $v(t) - \sigma(t)$, the integrand must vanish such that

$$\boldsymbol{t}^T \dot{\times} \boldsymbol{I} = \vec{0}, \qquad \text{or} \qquad \boldsymbol{t}^T = \boldsymbol{t} \qquad \text{in } v(t) - \sigma(t). \tag{4.32}$$

Thus, the necessary and sufficient condition for satisfying the local conservation of angular momentum is the symmetry of the Cauchy stress tensor $\boldsymbol{t}$. We have seen that the associated jump condition for the angular momentum is satisfied identically.

Note that in formulating the principle of conservation of angular momentum (4.4), we have assumed that no body torques or contact torques act on the body. If any such concentrated moments do act, the material is said to be a *polar* material and the symmetry property of the Cauchy stress tensor $\boldsymbol{t}$ no longer holds. However, this is a rather special situation, which is not considered here.

4.2.4 Energy Equation

The same methodology is now applied to the principle of conservation of energy (4.10). The integrand of the surface integral on the right-hand side of (4.10) is expressed using the Cauchy stress formula (3.14). The surface integral is then converted to the volume integral by the modified Gauss theorem (2.49), resulting in

$$\int_{s(t)-\sigma(t)} (\vec{t}_{(\vec{n})} \cdot \vec{v} - \vec{q} \cdot \vec{n}) \, da = \int_{v(t)-\sigma(t)} (\text{div}\,(\boldsymbol{t} \cdot \vec{v}) - \text{div}\,\vec{q}) \, dv + \int_{\sigma(t)} \vec{n} \cdot \left[\boldsymbol{t} \cdot \vec{v} - \vec{q}\right]_-^+ da. \tag{4.33}$$

In addition, the divergence of vector $\boldsymbol{t} \cdot \vec{v}$ is expressed using the identity as

$$\text{div}\,(\boldsymbol{t} \cdot \vec{v}) = \text{div}\,\boldsymbol{t} \cdot \vec{v} + \boldsymbol{t}^T : \text{grad}\,\vec{v}. \tag{4.34}$$

The Reynolds transport theorem (2.52) for $\phi = \varrho\varepsilon + \frac{1}{2}\varrho\vec{v}\cdot\vec{v}$ is used to arrange the left-hand side of the principle of conservation of energy (4.10) as

$$
\begin{aligned}
&\frac{D}{Dt}\int_{v(t)-\sigma(t)} (\varrho\varepsilon + \frac{1}{2}\varrho\vec{v}\cdot\vec{v})\,dv \\
&= \int_{v(t)-\sigma(t)} \left[\frac{D}{Dt}(\varrho\varepsilon + \frac{1}{2}\varrho\vec{v}\cdot\vec{v}) + (\varrho\varepsilon + \frac{1}{2}\varrho\vec{v}\cdot\vec{v})\operatorname{div}\vec{v}\right] dv \\
&\quad + \int_{\sigma(t)} \vec{n}\cdot\left[(\vec{v}-\vec{w})(\varrho\varepsilon + \frac{1}{2}\varrho\vec{v}\cdot\vec{v})\right]_-^+ da \\
&= \int_{v(t)-\sigma(t)} \left[(\varepsilon + \frac{1}{2}\vec{v}\cdot\vec{v})\left(\frac{D\varrho}{Dt} + \varrho\operatorname{div}\vec{v}\right) + \varrho\frac{D\varepsilon}{Dt} + \varrho\frac{D\vec{v}}{Dt}\cdot\vec{v}\right] dv \\
&\quad + \int_{\sigma(t)} \vec{n}\cdot\left[(\vec{v}-\vec{w})(\varrho\varepsilon + \frac{1}{2}\varrho\vec{v}\cdot\vec{v})\right]_-^+ da,
\end{aligned}
$$

which, in view of the principle of conservation of mass (4.18), reduces to

$$
\begin{aligned}
\frac{D}{Dt}\int_{v(t)-\sigma(t)} (\varrho\varepsilon + \frac{1}{2}\varrho\vec{v}\cdot\vec{v})dv &= \int_{v(t)-\sigma(t)} \varrho\left(\frac{D\varepsilon}{Dt} + \frac{D\vec{v}}{Dt}\cdot\vec{v}\right) dv \\
&\quad + \int_{\sigma(t)} \vec{n}\cdot\left[(\vec{v}-\vec{w})(\varrho\varepsilon + \frac{1}{2}\varrho\vec{v}\cdot\vec{v})\right]_-^+ da. \qquad (4.35)
\end{aligned}
$$

Using (4.33)–(4.35), the equation of motion (4.24), the symmetry of the Cauchy stress tensor (4.32) and upon setting the integrand of the result equal to zero, we obtain

$$\varrho\frac{D\varepsilon}{Dt} = \boldsymbol{t} : \boldsymbol{l} - \operatorname{div}\vec{q} + \varrho h \qquad \text{in } v(t)-\sigma(t), \quad (4.36)$$

$$\vec{n}\cdot\left[(\vec{v}-\vec{w})(\varrho\varepsilon + \frac{1}{2}\varrho\vec{v}\cdot\vec{v}) - \boldsymbol{t}\cdot\vec{v} + \vec{q}\right]_-^+ = 0 \qquad \text{on } \sigma(t), \quad (4.37)$$

where $\boldsymbol{l}$ is the transposed velocity gradient tensor defined by (2.13). In view of the symmetry of the Cauchy stress tensor $\boldsymbol{t}$, (4.36) can be written in the alternative form

$$\varrho\frac{D\varepsilon}{Dt} = \boldsymbol{t} : \boldsymbol{d} - \operatorname{div}\vec{q} + \varrho h \qquad \text{in } v(t)-\sigma(t), \quad (4.38)$$

where $\boldsymbol{d}$ is the strain-rate tensor defined by $(2.20)_1$. The term $\boldsymbol{t} : \boldsymbol{d}$, called the *stress power* (per unit volume), represents the work done by the internal forces. Equation (4.38) is the energy equation for a thermomechanical continuum in the Eulerian form and (4.37) is the associated jump condition on the discontinuity surface σ.

4.2.5 Entropy Inequality

The same methodology can be applied to the global entropy principle (4.16) to derive its local form. Using again the Reynolds transport theorem, the Cauchy stress formula and the Gauss theorem, and assuming that (4.16) is valid for any part of the body, we obtain the local form of the entropy inequality

$$\varrho \frac{D\eta}{Dt} + \operatorname{div}\vec{s} - \varrho z \geq 0 \qquad \text{in } \upsilon(t) - \sigma(t), \tag{4.39}$$

$$\vec{n} \cdot \left[\varrho\eta(\vec{v} - \vec{w}) + \vec{s}\right]_-^+ \geq 0 \qquad \text{on } \sigma(t). \tag{4.40}$$

4.2.6 Overview of Local Conservation Principles in the Eulerian Form

In summary, all local conservation principles expressed in the Eulerian description are as follows:

(i) *Conservation of mass*

$$\frac{\partial \varrho}{\partial t} + \operatorname{div}(\varrho\vec{v}) = 0 \qquad \text{in } \upsilon(t) - \sigma(t), \tag{4.41}$$

$$\vec{n} \cdot \left[\varrho(\vec{v} - \vec{w})\right]_-^+ = 0 \qquad \text{on } \sigma(t). \tag{4.42}$$

(ii) *Conservation of linear momentum*

$$\operatorname{div}\boldsymbol{t} + \varrho\vec{f} = \varrho \frac{D\vec{v}}{Dt} \qquad \text{in } \upsilon(t) - \sigma(t), \tag{4.43}$$

$$\vec{n} \cdot \left[\varrho(\vec{v} - \vec{w}) \otimes \vec{v} - \boldsymbol{t}\right]_-^+ = \vec{0} \qquad \text{on } \sigma(t). \tag{4.44}$$

(iii) *Conservation of angular momentum*

$$\boldsymbol{t} = \boldsymbol{t}^T \qquad \text{in } \upsilon(t) - \sigma(t). \tag{4.45}$$

(iv) *Conservation of energy*

$$\varrho \frac{D\varepsilon}{Dt} = \boldsymbol{t} : \boldsymbol{d} - \operatorname{div}\vec{q} + \varrho h \qquad \text{in } \upsilon(t) - \sigma(t), \tag{4.46}$$

$$\vec{n} \cdot \left[(\vec{v} - \vec{w})(\varrho\varepsilon + \frac{1}{2}\varrho\vec{v} \cdot \vec{v}) - \boldsymbol{t} \cdot \vec{v} + \vec{q}\right]_-^+ = 0 \qquad \text{on } \sigma(t). \tag{4.47}$$

(v) *Entropy inequality*

$$\varrho \frac{D\eta}{Dt} + \operatorname{div} \vec{s} - \varrho z \geq 0 \qquad \text{in } v(t) - \sigma(t), \quad (4.48)$$

$$\vec{n} \cdot \left[\varrho\eta(\vec{v} - \vec{w}) + \vec{s} \right]_-^+ \geq 0 \qquad \text{on } \sigma(t). \quad (4.49)$$

The jump conditions (4.42), (4.44), (4.47) and (4.49) associated with the principles of conservation of mass, momentum and energy, and the entropy inequality, are termed the *dynamic jump conditions*.

4.2.7 Jump Conditions in Special Cases

If there is a moving discontinuity surface $\sigma(t)$ sweeping particles of the body with velocity $\vec{w}$ in the direction of the unit normal vector $\vec{n}$ of $\sigma(t)$, then the dynamic jump conditions must be satisfied along the surface $\sigma(t)$. Some of these jump conditions will now be specified for two special cases.

(i) The discontinuity surface is a material surface, which is a surface occupied by the same material particles at all times. In this case, $\vec{w} = \vec{v}$, (4.42) is satisfied identically, and (4.44) and (4.47) reduce to

$$\vec{n} \cdot [\boldsymbol{t}]_-^+ = \vec{0}, \quad (4.50)$$

$$\vec{n} \cdot \left[\boldsymbol{t} \cdot \vec{v} - \vec{q} \right]_-^+ = 0, \quad (4.51)$$

which can be combined to obtain

$$\vec{n} \cdot \boldsymbol{t} \cdot \left[\vec{v} \right]_-^+ - \vec{n} \cdot \left[\vec{q} \right]_-^+ = 0. \quad (4.52)$$

Hence, along a material discontinuity surface, the surface traction $\vec{n} \cdot \boldsymbol{t}$ is continuous, and the jump of the power of tractions across this discontinuity surface is balanced by the jump of the normal component of the heat flux vector.

The dynamic jump conditions (4.50) and (4.51) must be supplemented by the *kinematic jump conditions*. Along a *welded discontinuity surface*, such as a discontinuity surface between two solids, there is no tangential slip and the Eulerian velocity passes across this discontinuity surface continuously,

$$\left[\vec{v} \right]_-^+ = \vec{0}. \quad (4.53)$$

Consequently, the jump condition (4.52) further reduces to

$$\vec{n} \cdot \left[\vec{q} \right]_-^+ = 0, \quad (4.54)$$

which implies the continuity of the normal component of the heat flux vector across the discontinuity surface $\sigma(t)$.

Tangential slip is allowed along a discontinuity surface between a solid and an inviscid fluid, or it may also occur along a fault surface separating two solids. Along such a *slipping discontinuity surface*, (4.53) is replaced by

$$[\vec{n} \cdot \vec{v}]_{-}^{+} = 0, \tag{4.55}$$

which guarantees that there is no separation or interpenetration of the two materials.

A *frictionless discontinuity surface* between two materials is a material discontinuity surface along which the motion from one of its sides runs without friction. This means that the shear stresses of the Cauchy stress tensor $\boldsymbol{t}$ are equal to zero from one side of the discontinuity surface (e.g. with superscript '−'),

$$\vec{n}^{-} \cdot \boldsymbol{t}^{-} \cdot \left(\boldsymbol{I} - \vec{n}^{-} \otimes \vec{n}^{-}\right) = \vec{0}, \qquad \text{or} \qquad \vec{n}^{-} \cdot \boldsymbol{t}^{-} = \left(\vec{n}^{-} \cdot \boldsymbol{t}^{-} \cdot \vec{n}^{-}\right) \vec{n}^{-}. \tag{4.56}$$

Substituting this into (4.50) and considering $\vec{n} = \vec{n}^{-} = -\vec{n}^{+}$ along $\sigma(t)$, the stress vector $\vec{n} \cdot \boldsymbol{t}$ from both sides of a frictionless discontinuity surface is

$$\vec{n} \cdot \boldsymbol{t} = \tau \vec{n}, \tag{4.57}$$

where τ is the normal stress, $\tau = \vec{n} \cdot \boldsymbol{t} \cdot \vec{n}$. The condition (4.50) is then reduced to the continuity of τ across $\sigma(t)$

$$[\tau]_{-}^{+} = 0. \tag{4.58}$$

(ii) The material discontinuity surface coincides with the surface bounding a material body. The jump conditions (4.50) and (4.51) are again valid, but $\vec{n} \cdot \boldsymbol{t}^{+}$ is now interpreted as the surface traction of external forces and $\boldsymbol{t}^{+} \cdot \vec{v}^{+}$ is its power. In the special case where no external force acts on the bounding surface, that is, when $\vec{n} \cdot \boldsymbol{t}^{+} = \vec{0}$, (4.50) implies that

$$\vec{n} \cdot \boldsymbol{t}^{-} = \vec{0}, \tag{4.59}$$

and the condition (4.51) involves the heat flux alone

$$\vec{n} \cdot [\vec{q}]_{-}^{+} = 0. \tag{4.60}$$

4.3 Local Conservation Principles in the Lagrangian Form

In the preceding section, the local conservation principles were expressed in the Eulerian form. These equations may also be represented in the Lagrangian form, which we now introduce.

4.3.1 Continuity Equation

The Lagrangian form of the continuity equation can be derived from (4.1) by using the transformation rule $dv = J dV$ such that

$$\int_V (\varrho_0 - \mathcal{Q} J)\, dV = 0, \tag{4.61}$$

where V is the volume of the body in the reference configuration and $\mathcal{Q}(\vec{X}, t)$ is the Lagrangian representation of mass density,

$$\mathcal{Q}(\vec{X}, t) := \varrho(\vec{\chi}(\vec{X}, t), t). \tag{4.62}$$

Assuming that the integrand is a continuous function of the referential coordinates X_K in the volume $V - \Sigma$, where Σ is a discontinuity surface, and adopting the additive principle leads to

$$\varrho_0 = \mathcal{Q} J \qquad \text{in } V - \Sigma, \tag{4.63}$$

which is equivalent to

$$\frac{D\varrho_0}{Dt} = 0 \qquad \text{in } V - \Sigma. \tag{4.64}$$

Equations (4.63) and (4.64) express the Lagrangian forms of the continuity equation.

The relation (4.63) can be considered as a general solution to the continuity equation (4.18). To show this, let a body occupy the reference configuration κ_0 at time $t = 0$ and the present configuration κ_t at time t. Using the relationship for the material time derivative of the Jacobian, $\dot{J} = J \operatorname{div} \vec{v}$, (4.18) is written as

$$\frac{\dot{\varrho}}{\varrho} + \frac{\dot{J}}{J} = 0.$$

Integrating this equation with respect to time from $t = 0$ to t and realising that $J = 1$ for $t = 0$, we find that

$$\ln \frac{\varrho}{\varrho_0} + \ln J = 0,$$

which gives (4.63) after some algebraic manipulation. Hence, in the case that the deformation gradient $\boldsymbol{F}$ is given as the solution to the other field equations, the mass density ϱ can be calculated from (4.63) after solving these field equations. It means that (4.63) does not have to be included in the set of the governing equations. Such a situation appears in the Lagrangian description of solids. This is, however, not the case if the Eulerian description is employed. Since the initial configuration is not the reference configuration, neither $\boldsymbol{F}$ nor J follow from the field equations, and the continuity equation (4.18) must be included in the set of the governing equations in order to specify ϱ.

To express the Eulerian jump condition (4.19) in the Lagrangian form, we first modify (4.19) such that

$$\left[\vec{n}da \cdot \varrho(\vec{v} - \vec{w})\right]_-^+ = 0 \qquad \text{on } \sigma(t), \tag{4.65}$$

where the surface area element da is now also involved in (4.19) since it appears in the surface integral (4.17). Moreover, the unit normal vector $\vec{n}$ is, in addition, involved in the square brackets assuming that $\vec{n}$ points towards the $+$ side of the discontinuity surface $\sigma(t)$ on both sides of $\sigma(t)$. Since the area element da of $\sigma(t)$ changes continuously across $\sigma(t)$, that is, $da^+ = da^-$, the jump condition (4.65) is equivalent to the form (4.19). Using Nanson's formula (1.81), (4.65) transforms to

$$\left[\varrho_0(\vec{N}dA \cdot \vec{W})\right]_-^+ = 0 \qquad \text{on } \Sigma, \tag{4.66}$$

where

$$\vec{W}(\vec{X}, t) := \boldsymbol{F}^{-1} \cdot (\vec{v} - \vec{w}). \tag{4.67}$$

In contrast to the area element da, the area element dA may change discontinuously on Σ, i.e., $dA^+ \neq dA^-$, if there is tangential slip along this discontinuity surface, such as along a slipping discontinuity surface (see Fig. 9.1). On the other hand, on a welded discontinuity surface with no tangential slip, $dA^+ = dA^-$, and this factor can be dropped from the jump condition (4.66).

4.3.2 Equation of Motion

A simple form of the Lagrangian equation of motion can be obtained in terms of the first Piola–Kirchhoff stress tensor that is related to the Lagrangian Cauchy stress tensor by (3.22). First, the divergence of the Cauchy stress tensor is expressed using the identity $(1.48)_2$ as

$$\operatorname{div} \boldsymbol{t}(\vec{x}, t) = J^{-1}\operatorname{Div}(J\boldsymbol{F}^{-1} \cdot \boldsymbol{t}(\vec{X}, t)) = J^{-1}\operatorname{Div} \boldsymbol{T}^{(1)}(\vec{X}, t), \tag{4.68}$$

where $\boldsymbol{t}(\vec{x}, t)$ and $\boldsymbol{t}(\vec{X}, t)$ are the Eulerian and Lagrangian Cauchy stress tensors, respectively. Using (4.63), the equation of motion (4.24) is then expressed in the Lagrangian form as

$$\mathrm{Div}\, \boldsymbol{T}^{(1)} + \varrho_0 \vec{F} = \varrho_0 \frac{D\vec{v}}{Dt} \qquad \text{in } V - \Sigma, \tag{4.69}$$

where $\vec{F}(\vec{X}, t)$ is the Lagrangian representation of the body force,

$$\vec{F}(\vec{X}, t) := \vec{f}(\vec{\chi}(\vec{X}, t), t). \tag{4.70}$$

Note that the divergence of the Eulerian Cauchy stress tensor $\mathrm{div}\, \boldsymbol{t}(\vec{x}, t)$ in (4.24) is transformed into the divergence of the first Piola–Kirchhoff stress tensor $\mathrm{Div}\, \boldsymbol{T}^{(1)}(\vec{X}, t)$ in (4.69). In fact, the original definition (3.21) of $\boldsymbol{T}^{(1)}$ was motivated by this transformation.

As for the conservation of mass, the jump condition (4.25) can be expressed in the Lagrangian form. Using (1.81), $(3.22)_2$ and (4.67) leads to

$$\left[\varrho_0(\vec{N} dA \cdot \vec{W})\vec{v} - \vec{N} dA \cdot \boldsymbol{T}^{(1)}\right]_-^+ = \vec{0} \qquad \text{on } \Sigma. \tag{4.71}$$

Note that the Lagrangian equation of motion and the jump conditions can also be expressed in terms of the second Piola–Kirchhoff stress tensor by the transformation $\boldsymbol{T}^{(1)} = \boldsymbol{T}^{(2)} \cdot \boldsymbol{F}^T$.

4.3.3 Symmetries of the Piola–Kirchhoff Stress Tensors

Substituting $(3.22)_2$ and $(3.27)_2$ into the symmetry condition (4.32) of the Cauchy stress tensor $\boldsymbol{t}$ results in

$$\left(\boldsymbol{T}^{(1)}\right)^T = \boldsymbol{F} \cdot \boldsymbol{T}^{(1)} \cdot \boldsymbol{F}^{-T}, \qquad \left(\boldsymbol{T}^{(2)}\right)^T = \boldsymbol{T}^{(2)}. \tag{4.72}$$

Hence, $\boldsymbol{T}^{(2)}$ is symmetric whenever $\boldsymbol{t}$ is symmetric (i.e., in the non-polar case), but $\boldsymbol{T}^{(1)}$ is generally not symmetric.

4.3.4 Energy Equation

Let $\vec{Q}$ be defined as the heat flux with respect to the area element $d\vec{A}$ in the reference configuration,

$$\vec{q} \cdot d\vec{a} =: \vec{Q} \cdot d\vec{A}, \tag{4.73}$$

where $\vec{q}$ is the heat flux with respect to the area element $d\vec{a}$ in the present configuration. Using Nanson's formula (1.81), the heat fluxes in the present and reference configurations are related by

$$\vec{q} = J^{-1}\boldsymbol{F} \cdot \vec{Q}, \qquad \vec{Q} = J\boldsymbol{F}^{-1} \cdot \vec{q}. \tag{4.74}$$

The heat flux in the reference configuration and the Piola–Kirchhoff stress tensors will be employed to transform the energy equation from the Eulerian form to the Lagrangian form. First, the divergence of heat flux vector $\vec{q}$ is expressed by using the identity $(1.48)_2$ as

$$\operatorname{div} \vec{q}(\vec{x}, t) = J^{-1}\operatorname{Div}(J\boldsymbol{F}^{-1} \cdot \vec{q}(\vec{X}, t)) = J^{-1}\operatorname{Div} \vec{Q}(\vec{X}, t), \tag{4.75}$$

where $\vec{q}(\vec{x}, t)$ and $\vec{q}(\vec{X}, t)$ are the Eulerian and Lagrangian representations of heat flux vector. Next, the stress power in the present configuration, $\boldsymbol{t} : \boldsymbol{d}$, is expressed in terms of quantities referred to the reference configuration. Using two tensor identities

$$(\boldsymbol{A} \cdot \boldsymbol{B}) : \boldsymbol{C} = \boldsymbol{A} : (\boldsymbol{B} \cdot \boldsymbol{C}), \qquad (\boldsymbol{A} \cdot \boldsymbol{B}) : (\boldsymbol{C} \cdot \boldsymbol{D}) = (\boldsymbol{D} \cdot \boldsymbol{A}) : (\boldsymbol{B} \cdot \boldsymbol{C}), \tag{4.76}$$

valid for the second-order tensors $\boldsymbol{A}$, $\boldsymbol{B}$, $\boldsymbol{C}$ and $\boldsymbol{D}$, and $(2.12)_2$ and $(3.22)_2$, we have

$$\begin{aligned} \boldsymbol{t} : \boldsymbol{d} = \boldsymbol{t} : \boldsymbol{l} &= J^{-1}(\boldsymbol{F} \cdot \boldsymbol{T}^{(1)}) : (\dot{\boldsymbol{F}} \cdot \boldsymbol{F}^{-1}) = J^{-1}(\boldsymbol{F}^{-1} \cdot \boldsymbol{F}) : (\boldsymbol{T}^{(1)} \cdot \dot{\boldsymbol{F}}) \\ &= J^{-1}\boldsymbol{I} : (\boldsymbol{T}^{(1)} \cdot \dot{\boldsymbol{F}}) = J^{-1}(\boldsymbol{I} \cdot \boldsymbol{T}^{(1)}) : \dot{\boldsymbol{F}}, \end{aligned}$$

or,

$$\boldsymbol{t} : \boldsymbol{d} = J^{-1}\boldsymbol{T}^{(1)} : \dot{\boldsymbol{F}}, \tag{4.77}$$

where $\dot{\boldsymbol{F}}$ is the material time derivative of the deformation gradient tensor. Introducing the second Piola–Kirchhoff stress tensor $\boldsymbol{T}^{(2)}$ instead of $\boldsymbol{T}^{(1)}$, the stress power can alternatively be expressed as

$$\begin{aligned} \boldsymbol{t} : \boldsymbol{d} &= J^{-1}(\boldsymbol{T}^{(2)} \cdot \boldsymbol{F}^T) : \dot{\boldsymbol{F}} = J^{-1}\boldsymbol{T}^{(2)} : (\boldsymbol{F}^T \cdot \dot{\boldsymbol{F}}) = J^{-1}\boldsymbol{T}^{(2)} : (\boldsymbol{F}^T \cdot \boldsymbol{l} \cdot \boldsymbol{F}) \\ &= J^{-1}\boldsymbol{T}^{(2)} : (\boldsymbol{F}^T \cdot \boldsymbol{d} \cdot \boldsymbol{F}), \end{aligned}$$

or, using (2.27) gives

$$\boldsymbol{t} : \boldsymbol{d} = J^{-1}\boldsymbol{T}^{(2)} : \dot{\boldsymbol{E}}, \tag{4.78}$$

where $\dot{\boldsymbol{E}}$ is the material time derivative of the Lagrangian strain tensor $\boldsymbol{E}$.

We are now ready to express the conservation of energy (4.38) in the Lagrangian form. Using (4.75) and (4.78), along with (4.63), yields

$$\varrho_0 \frac{D\varepsilon}{Dt} = \boldsymbol{T}^{(2)} : \dot{\boldsymbol{E}} - \operatorname{Div} \vec{Q} + \varrho_0 h \qquad \text{in } V - \Sigma. \tag{4.79}$$

In view of (4.77) and (4.78), the dissipative term $\boldsymbol{T}^{(2)} : \dot{\boldsymbol{E}}$ is also equal to $\boldsymbol{T}^{(1)} : \dot{\boldsymbol{F}}$.

The Lagrangian form of the energy jump condition (4.37) is

$$\left[\varrho_0(\vec{N}dA \cdot \vec{W})(\varepsilon + \frac{1}{2}\varrho\vec{v} \cdot \vec{v}) - \vec{N}dA \cdot \boldsymbol{T}^{(1)} \cdot \vec{v} + \vec{N}dA \cdot \vec{Q}\right]_-^+ = 0 \qquad \text{on } \Sigma, \tag{4.80}$$

which is derived by substituting (4.65) and (4.67) into (4.37) and considering (4.73).

4.3.5 Entropy Inequality

The same methodology can be applied to the entropy inequality to express it in the Lagrangian form. Let $\vec{S}$ be defined as the entropy flux with respect to the area element $d\vec{A}$ in the reference configuration,

$$\vec{s} \cdot d\vec{a} =: \vec{S} \cdot d\vec{A}, \tag{4.81}$$

where $\vec{s}$ is the entropy flux with respect to the area element $d\vec{a}$ in the present configuration. Using Nanson's formula (1.81), the Lagrangian and Eulerian entropy fluxes are related by

$$\vec{s} = J^{-1}\boldsymbol{F} \cdot \vec{S}, \qquad \vec{S} = J\boldsymbol{F}^{-1} \cdot \vec{s}. \tag{4.82}$$

As in the preceding section, the entropy inequality (4.39) and the entropy jump condition (4.40) can be expressed in the Lagrangian form as

$$\varrho_0 \frac{D\eta}{Dt} + \operatorname{Div} \vec{S} - \varrho_0 z \geq 0 \qquad \text{in } V - \Sigma, \tag{4.83}$$

$$\left[\varrho_0\eta(\vec{N}dA \cdot \vec{W}) + \vec{N}dA \cdot \vec{S}\right]_-^+ \geq 0 \qquad \text{on } \Sigma. \tag{4.84}$$

Chapter 5
Moving Reference Frames

The frame-indifference principle, one of the fundamental concepts of continuum mechanics, means that a physical process remains unchanged when it is observed by different observers. For example, in Chap. 6, we will require that the form of the constitutive equations be invariant under changes of observer. In this chapter, we study a special class of coordinate transformations that is associated with a change of observer. We then consider the effects of this transformation on various kinematic quantities.

5.1 Observer Transformation

The notion of a frame helps us to formulate the principle of frame-indifference mathematically. A *reference frame* (or, simply, a *frame*) represents an observer who measures positions in space (e.g. with a ruler) and instants of time (with a clock).[1] An *event* in the physical world is perceived by an observer as occurring at a particular point $\vec{x}$ in a three-dimensional Euclidean space and at a particular time t.

Let an event at point P be recorded by two observers in two different frames, one fixed (unstarred) and the other in motion (starred).[2] Each frame has a reference point, the so-called *origin*, from which an observer measures distances or defines position vectors in space. Figure 5.1 shows two such frames and the relationship between the position vectors of the point P measured in both frames. Let $\vec{x}$ be the

[1] A frame should not be confused with a coordinate system as they are not the same. A frame can only be used to observe motion. But to describe the motion mathematically and perform operations on the vectors, we need a coordinate system. An observer in a frame is free to choose any coordinate system that may be convenient for describing observations made from that frame.

[2] Both are later associated with the present configuration of a body and are, therefore, called the spatial frames.

Z. Martinec, *Principles of Continuum Mechanics*, Nečas Center Series,
https://doi.org/10.1007/978-3-030-05390-1_5

Fig. 5.1 Unstarred (fixed) and starred (moving) frames employed to record an event at point P

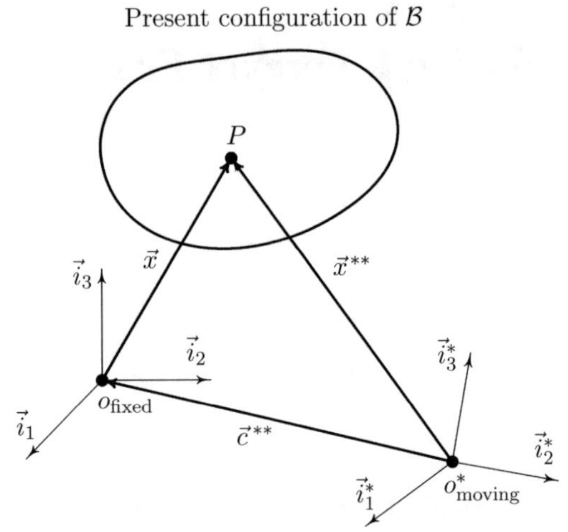

position vector of P relative to the fixed frame and $\vec{x}^{**}$ the position vector of the same point in the moving frame. They are connected by the relationship

$$\vec{x}^{**} = \vec{x} + \vec{c}^{**}, \tag{5.1}$$

where $\vec{c}^{**}$ is the displacement between the unstarred and starred reference points (origins). This relation is independent of the choice of coordinate system. However, the point P may refer its position vector not only to the origin, but also to a coordinate system that is attached to the origin. The coordinate system can be chosen arbitrarily or matched to a realistic situation.

In Fig. 5.1, these coordinate systems are Cartesian. The component form of (5.1) is

$$x^*_{k^*}\vec{i}^*_{k^*} = x_k\vec{i}_k + c^*_{k^*}\vec{i}^*_{k^*}. \tag{5.2}$$

The Cartesian base vectors $\vec{i}_k$ and $\vec{i}^*_{k^*}$ associated with the unstarred and starred spatial frames are related by $(1.13)_1$

$$\vec{i}_k = \delta_{kk^*}\vec{i}^*_{k^*}, \tag{5.3}$$

where δ_{kk^*} are the shifter symbols between the two spatial frames. Substituting for $\vec{i}_k$ from (5.3) into (5.2) and comparing the components at $\vec{i}^*_{k^*}$ gives

$$x^*_{k^*} = O_{k^*k}x_k + c^*_{k^*}, \tag{5.4}$$

where

$$O_{k^*k} := \delta_{kk^*}. \tag{5.5}$$

Equation (5.4) expresses the fact that the position of each point in space can be represented by the components $x^*_{k^*}$ in the moving coordinate frame as well as by the components x_k in the fixed coordinate frame.

The component form (5.4) can be rewritten in tensor notation as follows. Multiplying (5.4) by $\vec{i}_{k^*}$ and defining two vectors[3]

$$\vec{x}^{\,*} := x^*_{k^*}\vec{i}_{k^*}, \qquad \vec{c}^{\,*} := c^*_{k^*}\vec{i}_{k^*}, \tag{5.6}$$

we obtain

$$\vec{x}^{\,*} = \delta_{kk^*}x_k\vec{i}_{k^*} + \vec{c}^{\,*}. \tag{5.7}$$

Using $x_k = \vec{i}_k \cdot \vec{x}$, the identity $\vec{i}_{k^*}(\vec{i}_k \cdot \vec{x}) = (\vec{i}_{k^*} \otimes \vec{i}_k) \cdot \vec{x}$ and introducing the tensor $\boldsymbol{O}$,

$$\boldsymbol{O} := O_{k^*k}(\vec{i}_{k^*} \otimes \vec{i}_k), \tag{5.8}$$

(5.7) may be written in tensor notation as

$$\vec{x}^{\,*} = \boldsymbol{O}(t) \cdot \vec{x} + \vec{c}^{\,*}(t). \tag{5.9}$$

Alternatively, (5.9) can be viewed as a matrix form of (5.4), where $\boldsymbol{O}$ represents a matrix composed of the elements O_{k^*k}, and $\vec{x}^{\,*}$, $\vec{x}$ and $\vec{c}^{\,*}$ represent vector triplets[4] composed of the elements $x^*_{k^*}$, x_k and $c^*_{k^*}$, respectively, and the dot represents the multiplication of a matrix with a vector triplet. Whether (5.9) is understood in tensor or matrix sense, it is important to keep in mind that such an equation is a symbolic notation for the relationships between vector or tensor components in different coordinate systems.

The orthogonality property (1.16) of the shifter symbols implies that $\boldsymbol{O}(t)$ is an orthogonal tensor,

$$\boldsymbol{O}(t) \cdot \boldsymbol{O}^T(t) = \boldsymbol{O}^T(t) \cdot \boldsymbol{O}(t) = \boldsymbol{I}, \tag{5.10}$$

[3]Note that we distinguish between three different vectors, $\vec{x} = x_k\vec{i}_k$, $\vec{x}^{\,*} = x^*_k\vec{i}_k$ and $\vec{x}^{\,**} = x^*_k\vec{i}^{\,*}_k$. Vector notation becomes ambiguous if the vectors $\vec{x}^{\,*}$ and $\vec{x}^{\,**}$ are denoted by the same symbol $\vec{x}^{\,*}$; compare (5.1) and (5.9) in this case.

[4]A *vector* is a geometric object that has, by definition, an invariant property under an arbitrary change of coordinate system. Even though the components of a vector change when a coordinate system changes, they must change in a very specific way. A *vector triplet* is a triplet of numbers representing a vector in a particular coordinate system.

where $\boldsymbol{I}$ is the identity tensor. With this orthogonality, (5.9) expresses a rigid (i.e., independent of location) motion of the starred spatial frame with respect to the unstarred spatial frame, where $\vec{c}^{\,*}(t)$ corresponds to the translation and $\boldsymbol{O}(t)$ to the rotation of the starred frame. As indicated, both $\vec{c}^{\,*}(t)$ and $\boldsymbol{O}(t)$ are assumed to be time-dependent, which expresses the fact that the observers can move relative to each other over time. The transformation (5.9) is often referred to as the *observer transformation* $(\vec{x}, t) \to (\vec{x}^{\,*}, t)$, where $(\vec{x}, t)$ and $(\vec{x}^{\,*}, t)$ are the position and time of the same event observed by the two observers, respectively.[5]

Let us consider three separate events at points P_1, P_2 and P_3 recorded by two observers in the respective unstarred and starred frames. By the observer transformation (5.9), we have

$$\vec{x}_2^{\,*} - \vec{x}_1^{\,*} = \boldsymbol{O}(t) \cdot (\vec{x}_2 - \vec{x}_1).$$

A similar relationship holds for the difference vector $\vec{x}_3^{\,*} - \vec{x}_1^{\,*}$. The square of the distance between two points is

$$\begin{aligned}(\vec{x}_2^{\,*} - \vec{x}_1^{\,*}) \cdot (\vec{x}_2^{\,*} - \vec{x}_1^{\,*}) &= \left[\boldsymbol{O}(t) \cdot (\vec{x}_2 - \vec{x}_1)\right] \cdot \boldsymbol{O}(t) \cdot (\vec{x}_2 - \vec{x}_1) \\ &= (\vec{x}_2 - \vec{x}_1) \cdot \boldsymbol{O}^T(t) \cdot \boldsymbol{O}(t) \cdot (\vec{x}_2 - \vec{x}_1) \\ &= (\vec{x}_2 - \vec{x}_1) \cdot (\vec{x}_2 - \vec{x}_1).\end{aligned}$$

Likewise, the scalar product of the two difference vectors is

$$(\vec{x}_3^{\,*} - \vec{x}_1^{\,*}) \cdot (\vec{x}_2^{\,*} - \vec{x}_1^{\,*}) = (\vec{x}_3 - \vec{x}_1) \cdot (\vec{x}_2 - \vec{x}_1).$$

These equations show that the observer transformation preserves the distances between points and the angles between vectors,[6] hence, it preserves the Euclidean geometry. Thus, the observer transformation (5.9) is also called the *Euclidean transformation*. Although all orthogonal tensors preserve length and angles, we consider $\boldsymbol{O}(t)$ in (5.9) to be a proper orthogonal tensor, that is, $\det \boldsymbol{O}(t) = +1$, so that orientation is also preserved.

[5]The most general change of frame $(\vec{x}, t) \to (\vec{x}^{\,*}, t^*)$ is, in addition, characterised by a shift in time

$$t^* = t - a,$$

where a is a particular time. The shift-in-time transformation is trivial and does not affect the derived relationships. For example, the consequence of frame-indifference with respect to the shift of time is that the constitutive functionals do not depend explicitly on the current time. We will consider this as a starting assumption in defining a general form (6.5) of the constitutive equations.

[6]The unstarred and starred observers are also called *equivalent observers* due to these properties.

5.2 Frame-Indifference

Let us suppose that different observers are examining a deforming body by making measurements. This will involve measurements of spatial quantities associated with the present configuration, for example, the velocity or acceleration, and quantities associated with the reference configuration, for example, the Lagrangian strain tensor in that configuration. It will also involve two-point tensors such as the rotation tensor or the deformation gradient tensor, which are associated with both the present and reference configurations.

We assume that all observers observe the reference configuration to be the same, that is, they record the same set of particles in the reference configuration.[7] The observers then move relative to each other and their measurements of the quantity associated with the present configuration will generally differ. A quantity is called *frame-indifferent* (or, often, *objective*) if its measurements remain unchanged when made by equivalent observers.

Mathematically, a scalar-, vector- and tensor-valued quantity ϕ is frame-indifferent if it is invariant under all observer transformations (5.9), that is, if $\phi^{**} = \phi$, where ϕ is a quantity described in the unstarred frame and ϕ^{**} is the same quantity described in the starred frame. For instance, tensor $\boldsymbol{a}$ is frame-indifferent if its components transform under the observer transformation (5.9) according to the rule

$$a^*_{k^*l^*} = O_{k^*k}(t) a_{kl} O_{l^*l}(t), \tag{5.11}$$

where a_{kl} and $a^*_{k^*l^*}$ are components of $\boldsymbol{a}$ relative to the unstarred and starred frames, respectively. To see this, let us rewrite the transformation relation (5.3) for the base vectors in terms of the components of the tensor $\boldsymbol{O}$. In view of (5.5), (5.3) becomes

$$\vec{i}_k = O_{k^*k}(t)\vec{i}^{\,*}_{k^*}. \tag{5.12}$$

Then,

$$\boldsymbol{a} = a_{kl}(\vec{i}_k \otimes \vec{i}_l) = a_{kl} O_{k^*k} O_{l^*l}(\vec{i}^{\,*}_{k^*} \otimes \vec{i}^{\,*}_{l^*}) \stackrel{!}{=} \boldsymbol{a}^{**} = a^*_{k^*l^*}(\vec{i}^{\,*}_{k^*} \otimes \vec{i}^{\,*}_{l^*}),$$

which yields (5.11). Introducing tensor $\boldsymbol{a}^*$,

$$\boldsymbol{a}^* := a^*_{k^*l^*}(\vec{i}_{k^*} \otimes \vec{i}_{l^*}), \tag{5.13}$$

[7]This does not affect the generality of what follows; the notion of a frame-indifferent quantity is independent of the chosen reference configuration.

the component form (5.11) may be written symbolically as[8]

$$\boldsymbol{a}^* = \boldsymbol{O}(t) \cdot \boldsymbol{a} \cdot \boldsymbol{O}^T(t). \tag{5.14}$$

In an analogous manner, a scalar- and vector-valued physical quantities, λ and $\vec{u}$, respectively, are frame-indifferent if they transform under the rigid motion of the observer frame according to

$$\lambda^* = \lambda, \tag{5.15}$$

$$u^*_{k^*} = O_{k^*k}(t)u_k, \qquad \text{or, symbolically,} \qquad \vec{u}^{\,*} = \boldsymbol{O}(t) \cdot \vec{u}. \tag{5.16}$$

5.3 Frame-Indifference of Some Kinematic Quantities

Let us now examine the frame-indifference property of various kinematic quantities. Whether a particular kinematic quantity is frame-indifferent or not is derived from its definition and properties.

We begin with the motion function (1.31). Suppose that the motion is observed by an unstarred observer as $x_k = \chi_k(\vec{X}, t)$ and by a starred observer as $x^*_{k^*} = \chi^*_{k^*}(\vec{X}, t)$. The observer transformation (5.9) gives

$$\vec{\chi}^{\,*}(\vec{X}, t) = \boldsymbol{O}(t) \cdot \vec{\chi}(\vec{X}, t) + \vec{c}^{\,*}(t), \tag{5.17}$$

where $\vec{\chi}(\vec{X}, t) := \chi_k(\vec{X}, t)\,\vec{i}_k$ and $\vec{\chi}^*(\vec{X}, t) := \chi^*_{k^*}(\vec{X}, t)\,\vec{i}_{k^*}$. Comparing $(5.16)_2$ and (5.17) shows that motion is not a frame-indifferent vector.

We continue with the Eulerian velocity $\vec{v}$ and Eulerian acceleration $\vec{a}$. Differentiation of (5.17) with respect to time yields the relationships between the velocities and accelerations in the starred and unstarred frames

$$\vec{v}^{\,*}(\vec{x}^{\,*}, t) = \boldsymbol{O}(t) \cdot \vec{v}(\vec{x}, t) + \dot{\boldsymbol{O}}(t) \cdot \vec{x} + \dot{\vec{c}}^{\,*}(t), \tag{5.18}$$

$$\vec{a}^{\,*}(\vec{x}^{\,*}, t) = \boldsymbol{O}(t) \cdot \vec{a}(\vec{x}, t) + 2\,\dot{\boldsymbol{O}}(t) \cdot \vec{v}(\vec{x}, t) + \ddot{\boldsymbol{O}}(t) \cdot \vec{x} + \ddot{\vec{c}}^{\,*}(t). \tag{5.19}$$

By analogy with (2.26), we introduce the *angular velocity tensor* $\boldsymbol{\Omega}$ that represents the spin of the starred frame with respect to the unstarred frame,

$$\boldsymbol{\Omega}(t) := \dot{\boldsymbol{O}}(t) \cdot \boldsymbol{O}^T(t). \tag{5.20}$$

[8]Note again that we distinguish between three different tensors $\boldsymbol{a}$, $\boldsymbol{a}^*$ and $\boldsymbol{a}^{**}$.

The relationship

$$\mathbf{0} = (\boldsymbol{O} \cdot \boldsymbol{O}^T)^{\cdot} = \dot{\boldsymbol{O}} \cdot \boldsymbol{O}^T + \boldsymbol{O} \cdot \dot{\boldsymbol{O}}^T = \dot{\boldsymbol{O}} \cdot \boldsymbol{O}^T + (\dot{\boldsymbol{O}} \cdot \boldsymbol{O}^T)^T = \boldsymbol{\Omega} + \boldsymbol{\Omega}^T$$

shows that $\boldsymbol{\Omega}$ is a skew-symmetric tensor. Moreover,

$$\begin{aligned}\dot{\boldsymbol{\Omega}} &= (\dot{\boldsymbol{O}} \cdot \boldsymbol{O}^T)^{\cdot} = \ddot{\boldsymbol{O}} \cdot \boldsymbol{O}^T + \dot{\boldsymbol{O}} \cdot \dot{\boldsymbol{O}}^T = \ddot{\boldsymbol{O}} \cdot \boldsymbol{O}^T + \dot{\boldsymbol{O}} \cdot (\boldsymbol{O}^T \cdot \boldsymbol{O}) \cdot \dot{\boldsymbol{O}}^T \\ &= \ddot{\boldsymbol{O}} \cdot \boldsymbol{O}^T + \boldsymbol{\Omega} \cdot \boldsymbol{\Omega}^T = \ddot{\boldsymbol{O}} \cdot \boldsymbol{O}^T - \boldsymbol{\Omega} \cdot \boldsymbol{\Omega},\end{aligned}$$

which yields

$$\ddot{\boldsymbol{O}} \cdot \boldsymbol{O}^T = \dot{\boldsymbol{\Omega}} + \boldsymbol{\Omega} \cdot \boldsymbol{\Omega}. \tag{5.21}$$

Using (5.9), (5.20) and (5.21), the transformation formulae (5.18) and (5.19) for the velocity and acceleration become

$$\vec{v}^*(\vec{x}^*, t) = \boldsymbol{O}(t) \cdot \vec{v}(\vec{x}, t) + \boldsymbol{\Omega}(t) \cdot \left[\vec{x}^* - \vec{c}^*(t)\right] + \dot{\vec{c}}^*(t), \tag{5.22}$$

$$\begin{aligned}\vec{a}^*(\vec{x}^*, t) = {} & \boldsymbol{O}(t) \cdot \vec{a}(\vec{x}, t) + 2\,\boldsymbol{\Omega}(t) \cdot \left[\vec{v}^*(\vec{x}^*, t) - \dot{\vec{c}}^*(t)\right] \\ & - \boldsymbol{\Omega}(t) \cdot \boldsymbol{\Omega}(t) \cdot \left[\vec{x}^* - \vec{c}^*(t)\right] + \dot{\boldsymbol{\Omega}}(t) \cdot \left[\vec{x}^* - \vec{c}^*(t)\right] + \ddot{\vec{c}}^*(t).\end{aligned} \tag{5.23}$$

Inspection of these equations reveals that both velocity and acceleration are not frame-indifferent vectors. The terms causing the failure of frame-indifference have the following descriptions:

$\boldsymbol{\Omega} \cdot (\vec{x}^* - \vec{c}^*)$	*angular velocity,*
$\dot{\vec{c}}^*$	*translational velocity,*
$2\boldsymbol{\Omega} \cdot (\vec{v}^* - \dot{\vec{c}}^*)$	*Coriolis acceleration,*
$-\boldsymbol{\Omega} \cdot \boldsymbol{\Omega} \cdot (\vec{x}^* - \vec{c}^*)$	*centrifugal acceleration,*
$\dot{\boldsymbol{\Omega}} \cdot (\vec{x}^* - \vec{c}^*)$	*Euler acceleration,*
$\ddot{\vec{c}}^*$	*translational acceleration.*

Among all Euclidean transformations, we can choose those that transform the acceleration in a frame-indifferent manner. In such a case, we require $\vec{a}^* = \boldsymbol{O} \cdot \vec{a}$, which is satisfied when $\boldsymbol{\Omega} = \mathbf{0}$ and $\ddot{\vec{c}}^* = \vec{0}$, giving

$$\vec{c}^*(t) = \vec{V}^* t + \vec{c}_0^*, \qquad \boldsymbol{O}(t) = \boldsymbol{O}, \tag{5.24}$$

where $\vec{V}^*$, $\vec{c}_0^*$ and $\boldsymbol{O}$ are time-independent. The change of frame defined by such constants is

$$\vec{x}^* = \boldsymbol{O} \cdot \vec{x} + \vec{V}^* t + \vec{c}_0^*, \tag{5.25}$$

which is called the *Galilean transformation*. It means that the starred frame moves with a constant velocity in a straight line (a uniform and rectilinear motion) with respect to the unstarred frame. The acceleration is frame-indifferent with respect to the Galilean transformation, whereas the velocity is not.

The notion of an inertial frame is closely related to the Galilean transformation. An *inertial frame* is a frame in which Newton's first law applies. That is, a body with zero net force acting upon is not accelerating in an inertial frame, but is at rest or in uniform motion in a straight line, unless acted upon by an external force. Hence, a frame related to an inertial frame by the Galilean transformation is an inertial frame.

In contrast to the velocity field, which is not frame-indifferent, the divergence of the velocity field is a frame-indifferent scalar,

$$\operatorname{div}^* \vec{v}^* = \operatorname{div} \vec{v}. \tag{5.26}$$

To show this, we apply the divergence operator to (5.22) and successively write

$$\begin{aligned}
\operatorname{div}^* \vec{v}^* &= \frac{\partial v^*_{k^*}}{\partial x^*_{k^*}} = \frac{\partial}{\partial x^*_{k^*}} \left[O_{k^*k} v_k + \Omega_{k^*k}(x^*_k - c^*_k) + \dot{c}^*_k \right] = O_{k^*k} \frac{\partial v_k}{\partial x^*_{k^*}} + \Omega_{k^*k} \frac{\partial x^*_k}{\partial x^*_{k^*}} \\
&= O_{k^*k} \frac{\partial v_k}{\partial x^*_{k^*}} + \Omega_{kk} = O_{k^*k} \frac{\partial v_k}{\partial x^*_{k^*}} = O_{k^*k} \frac{\partial v_k}{\partial x_l} \frac{\partial x_l}{\partial x^*_{k^*}} = O_{k^*k} \frac{\partial v_k}{\partial x_l} O_{k^*l} \\
&= \delta_{kl} \frac{\partial v_k}{\partial x_l} = \frac{\partial v_k}{\partial x_k} = \operatorname{div} \vec{v}.
\end{aligned}$$

To study the effect of an observer transformation on the basic conservation equations derived in Chap. 4, let us now prove that (1) the spatial gradient of a frame-indifferent scalar is a frame-indifferent vector, (2) the spatial divergence of a frame-indifferent vector is a frame-indifferent scalar, (3) the spatial divergence of a frame-indifferent tensor is a frame-indifferent vector, and (4) the double-dot product of two frame-indifferent tensors is a frame-indifferent scalar. The proofs are as follows:

1. Let λ be a frame-indifferent scalar. Equations (5.9) and (5.15) give

$$(\operatorname{grad}^* \lambda^*)_{k*} = \frac{\partial \lambda^*}{\partial x^*_{k^*}} = \frac{\partial \lambda}{\partial x_k} \frac{\partial x_k}{\partial x^*_{k^*}} = O_{k^*k} \frac{\partial \lambda}{\partial x_k} = O_{k^*k} \, (\operatorname{grad} \lambda)_k .$$

Multiplying by $\vec{i}_{k^*}$ and introducing a new vector, $\operatorname{grad} \lambda^* := \dfrac{\partial \lambda^*}{\partial x^*_{k^*}} \vec{i}_{k^*}$ yields

$$\operatorname{grad} \lambda^* = \boldsymbol{O}(t) \cdot \operatorname{grad} \lambda. \tag{5.27}$$

2. Let $\vec{u}$ be a frame-indifferent vector. Equations (5.9) and (5.16) give

$$\operatorname{div}^* \vec{u}^* = \frac{\partial u^*_{k^*}}{\partial x^*_{k^*}} = \frac{\partial (O_{k^*l} u_l)}{\partial x_k} \frac{\partial x_k}{\partial x^*_{k^*}} = O_{k^*l} \frac{\partial u_l}{\partial x_k} O_{k^*k} = \delta_{kl} \frac{\partial u_l}{\partial x_k} = \frac{\partial u_k}{\partial x_k}$$

such that

$$\operatorname{div}^* \vec{u}^* = \operatorname{div} \vec{u}. \tag{5.28}$$

3. The third statement is verified by using (5.9)–(5.11):

$$(\operatorname{div}^* \boldsymbol{a}^*)_{l^*} = \frac{\partial a^*_{k^*l^*}}{\partial x^*_{k^*}} = \frac{\partial (O_{k^*k} a_{kl} O_{l^*l})}{\partial x_m} \frac{\partial x_m}{\partial x^*_{k^*}} = O_{k^*k} O_{l^*l} \frac{\partial a_{kl}}{\partial x_m} O_{k^*m}$$
$$= \delta_{km} O_{l^*l} \frac{\partial a_{kl}}{\partial x_m} = O_{l^*l} (\operatorname{div} \boldsymbol{a})_l .$$

Multiplying by $\vec{i}_{l^*}$ and introducing a new vector $\operatorname{div} \boldsymbol{a}^* := \dfrac{\partial a^*_{k^*l^*}}{\partial x^*_{k^*}} \vec{i}_{l^*}$ yields

$$\operatorname{div} \boldsymbol{a}^* = \boldsymbol{O}(t) \cdot \operatorname{div} \boldsymbol{a}. \tag{5.29}$$

4. Let $\boldsymbol{A}$ and $\boldsymbol{B}$ be frame-indifferent tensors. Then, using (4.76), (5.10) and (5.14), we find that

$$\boldsymbol{A}^* : \boldsymbol{B}^* = (\boldsymbol{O} \cdot \boldsymbol{A} \cdot \boldsymbol{O}^T) : (\boldsymbol{O} \cdot \boldsymbol{B} \cdot \boldsymbol{O}^T) = (\boldsymbol{O}^T \cdot \boldsymbol{O}) : (\boldsymbol{A} \cdot \boldsymbol{O}^T \cdot \boldsymbol{O} \cdot \boldsymbol{B})$$
$$= \boldsymbol{I} : (\boldsymbol{A} \cdot \boldsymbol{B}) = (\boldsymbol{I} \cdot \boldsymbol{A}) : \boldsymbol{B}.$$

Hence,

$$\boldsymbol{A}^* : \boldsymbol{B}^* = \boldsymbol{A} : \boldsymbol{B}, \tag{5.30}$$

which completes the proofs of the statements.

5.4 Observer Transformation of the Deformation Gradient

The transformation rule for the deformation gradient is

$$\boldsymbol{F}^*(\vec{X}, t) = \boldsymbol{O}(t) \cdot \boldsymbol{F}(\vec{X}, t), \tag{5.31}$$

where $\boldsymbol{F}^*(\vec{X}, t) := \chi^*_{k^*,K} \left(\vec{i}_{k^*} \otimes \vec{I}_K \right)$. To show this, we express the deformation gradient in the starred frame according to $(1.35)_1$ and substitute from (5.17)

$$F^*_{k^*K} = \frac{\partial \chi^*_{k^*}}{\partial X_K} = \frac{\partial}{\partial X_K} \left(O_{k^*k} \, \chi_k + c^*_{k^*} \right) = O_{k^*k} \frac{\partial \chi_k}{\partial X_K} = O_{k^*k} F_{kK},$$

which, on multiplication by the dyad $\vec{i}_{k^*} \otimes \vec{I}_K$, gives (5.31). Thus, the two-point deformation gradient tensor $\boldsymbol{F}$ is not a frame-indifferent tensor. However, the three columns of $\boldsymbol{F}$ (for $K = 1, 2, 3$) are frame-indifferent vectors.

In the proof of (5.31), we assume that the reference configuration is not affected by the change of observer frame or, in other words, observers observe the reference configuration to be the same, $\vec{X}^* = \vec{X}$. This follows the original formulation of the principle of material frame-indifference by Truesdell and Noll (1965). However, this assumption is not made in some literature, e.g. Liu and Sampaio (2014). Instead, the reference configuration is assumed to be affected by a change of the observer considering the reference configuration to be a placement of a body, as any other configuration of the body.[9]

Let us verify that the Jacobian J, the Green deformation tensor $\boldsymbol{C}$ and the right-stretch tensor $\boldsymbol{U}$ are all frame-indifferent scalars, while the rotation tensor $\boldsymbol{R}$ is a frame-indifferent vector, and the left-stretch tensor $\boldsymbol{V}$, the Finger deformation tensor $\boldsymbol{b}$ and strain-rate tensor $\boldsymbol{d}$ are all frame-indifferent tensors:

$$J^* = J, \qquad \boldsymbol{C}^* = \boldsymbol{C}, \qquad \boldsymbol{U}^* = \boldsymbol{U}, \tag{5.34}$$

$$\boldsymbol{R}^* = \boldsymbol{O}(t) \cdot \boldsymbol{R}, \tag{5.35}$$

$$\boldsymbol{V}^* = \boldsymbol{O}(t) \cdot \boldsymbol{V} \cdot \boldsymbol{O}^T(t),\ \boldsymbol{b}^* = \boldsymbol{O}(t) \cdot \boldsymbol{b} \cdot \boldsymbol{O}^T(t),\ \boldsymbol{d}^* = \boldsymbol{O}(t) \cdot \boldsymbol{d} \cdot \boldsymbol{O}^T(t). \tag{5.36}$$

On the contrary, the velocity gradient $\boldsymbol{l}$ and the spin tensor $\boldsymbol{w}$ are not frame-indifferent tensors:

$$\boldsymbol{l}^* = \boldsymbol{O}(t) \cdot \boldsymbol{l} \cdot \boldsymbol{O}^T(t) + \boldsymbol{\Omega}(t), \qquad \boldsymbol{w}^* = \boldsymbol{O}(t) \cdot \boldsymbol{w} \cdot \boldsymbol{O}^T(t) + \boldsymbol{\Omega}(t). \tag{5.37}$$

---

[9] To present this alternative concept, let t_0 be the instant when the reference configuration is chosen. The observer transformation (5.9) for the reference configuration is

$$\vec{X}^* = \boldsymbol{O}(t_0) \cdot \vec{X} + \vec{c}^{\,*}(t_0), \tag{5.32}$$

where $\boldsymbol{O}(t_0)$ and $\vec{c}^{\,*}(t_0)$ describe a rigid motion of the observer frame at the instant t_0, and $\vec{X} = X_K \vec{I}_K$, $\vec{X}^* = X^*_{K^*} \vec{I}_{K^*}$, where X_K and $X^*_{K^*}$ are the coordinates of a point associated with an event in the reference configuration, which is recorded in the fixed and moving frames, respectively. Then, the deformation gradient in the two starred frames, one related to the present configuration and the other to the reference configuration, can be expressed as

$$F^*_{k^*K^*} = \frac{\partial \chi^*_{k^*}}{\partial X^*_{K^*}} = \frac{\partial \chi^*_{k^*}}{\partial X_K}\frac{\partial X_K}{\partial X^*_{K^*}} = O_{k^*k}\frac{\partial \chi_k}{\partial X_K}[\boldsymbol{O}(t_0)]_{K^*K} = O_{k^*k} F_{kK} [\boldsymbol{O}(t_0)]_{K^*K},$$

or, multiplying both sides of the last equation by the dyad $\vec{i}_{k^*} \otimes \vec{I}_{K^*}$ gives

$$\boldsymbol{F}^*(\vec{X}, t) = \boldsymbol{O}(t) \cdot \boldsymbol{F}(\vec{X}, t) \cdot \boldsymbol{O}^T(t_0). \tag{5.33}$$

Note that $\boldsymbol{F}^*(\vec{X}, t)$ is the same as in (5.31). The transformation rule (5.33) differs from (5.31), the standard formula obtained under the assumption that the reference configuration is not affected by the change of observer frame, so that $\boldsymbol{O}(t_0)$ reduces to the identity transformation.

The proofs are straightforward by using (1.45), (1.51), (1.52), (1.57), (1.62), (2.13), (2.20), (5.10) and (5.31):

$$
\begin{aligned}
J^* &= \det \boldsymbol{F}^* = \det(\boldsymbol{O}\cdot\boldsymbol{F}) = \det\boldsymbol{O}\det\boldsymbol{F} = \det\boldsymbol{F} = J,\\
\boldsymbol{C}^* &= (\boldsymbol{F}^*)^T\cdot\boldsymbol{F}^* = \boldsymbol{F}^T\cdot\boldsymbol{O}^T\cdot\boldsymbol{O}\cdot\boldsymbol{F} = \boldsymbol{F}^T\cdot\boldsymbol{F} = \boldsymbol{C},\\
\boldsymbol{U}^* &= \sqrt{\boldsymbol{C}^*} = \sqrt{\boldsymbol{C}} = \boldsymbol{U},\\
\boldsymbol{R}^* &= \boldsymbol{F}^*\cdot(\boldsymbol{U}^*)^{-1} = \boldsymbol{O}\cdot\boldsymbol{F}\cdot\boldsymbol{U}^{-1} = \boldsymbol{O}\cdot\boldsymbol{R},\\
\boldsymbol{V}^* &= \boldsymbol{R}^*\cdot\boldsymbol{U}^*\cdot(\boldsymbol{R}^*)^T = \boldsymbol{O}\cdot\boldsymbol{R}\cdot\boldsymbol{U}\cdot\boldsymbol{R}^T\cdot\boldsymbol{O}^T = \boldsymbol{O}\cdot\boldsymbol{V}\cdot\boldsymbol{O}^T,\\
\boldsymbol{b}^* &= \boldsymbol{F}^*\cdot(\boldsymbol{F}^*)^T = \boldsymbol{O}\cdot\boldsymbol{F}\cdot\boldsymbol{F}^T\cdot\boldsymbol{O}^T = \boldsymbol{O}\cdot\boldsymbol{b}\cdot\boldsymbol{O}^T,\\
\boldsymbol{l}^* &= (\boldsymbol{F}^*)^{\cdot}\cdot(\boldsymbol{F}^*)^{-1} = (\boldsymbol{O}\cdot\dot{\boldsymbol{F}}+\dot{\boldsymbol{O}}\cdot\boldsymbol{F})(\boldsymbol{O}\cdot\boldsymbol{F})^{-1} = \boldsymbol{O}\cdot\dot{\boldsymbol{F}}\cdot\boldsymbol{F}^{-1}\cdot\boldsymbol{O}^T+\dot{\boldsymbol{O}}\cdot\boldsymbol{O}^T\\
&= \boldsymbol{O}\cdot\boldsymbol{l}\cdot\boldsymbol{O}^T+\boldsymbol{\Omega},\\
\boldsymbol{d}^* &= \frac{1}{2}(\boldsymbol{l}^*+\boldsymbol{l}^{*T}) = \frac{1}{2}(\boldsymbol{O}\cdot\boldsymbol{l}\cdot\boldsymbol{O}^T+\boldsymbol{O}\cdot\boldsymbol{l}^T\cdot\boldsymbol{O}^T+\boldsymbol{\Omega}+\boldsymbol{\Omega}^T)\\
&= \boldsymbol{O}\cdot\frac{1}{2}(\boldsymbol{l}+\boldsymbol{l}^T)\cdot\boldsymbol{O}^T = \boldsymbol{O}\cdot\boldsymbol{d}\cdot\boldsymbol{O}^T,\\
\boldsymbol{w}^* &= \frac{1}{2}(\boldsymbol{l}^*-\boldsymbol{l}^{*T}) = \frac{1}{2}(\boldsymbol{O}\cdot\boldsymbol{l}\cdot\boldsymbol{O}^T-\boldsymbol{O}\cdot\boldsymbol{l}^T\cdot\boldsymbol{O}^T+\boldsymbol{\Omega}-\boldsymbol{\Omega}^T)\\
&= \boldsymbol{O}\cdot\frac{1}{2}(\boldsymbol{l}-\boldsymbol{l}^T)\cdot\boldsymbol{O}^T+\boldsymbol{\Omega} = \boldsymbol{O}\cdot\boldsymbol{w}\cdot\boldsymbol{O}^T+\boldsymbol{\Omega}.
\end{aligned}
$$

5.5 Frame-Indifferent Time Derivatives

Let us now deal with the material time derivative of a frame-indifferent scalar, a frame-indifferent vector and a frame-indifferent tensor. For a frame-indifferent scalar λ, for which $\lambda^{**} = \lambda^* = \lambda$, it trivially holds that

$$\dot{\lambda}^* = \dot{\lambda}, \tag{5.38}$$

that is, the material time derivative of a frame-indifferent scalar is a frame-indifferent scalar.

For a frame-indifferent vector $\vec{u}$, for which $\vec{u}^* = \boldsymbol{O}(t)\cdot\vec{u}$, the material time derivative is

$$\dot{\vec{u}}^* = \boldsymbol{O}\cdot\dot{\vec{u}}+\dot{\boldsymbol{O}}\cdot\vec{u} = \boldsymbol{O}\cdot\dot{\vec{u}}+\dot{\boldsymbol{O}}\cdot\boldsymbol{O}^T\cdot\vec{u}^*,$$

or, using (5.20) gives

$$\dot{\vec{u}}^* = \boldsymbol{O} \cdot \dot{\vec{u}} + \boldsymbol{\Omega} \cdot \vec{u}^*. \tag{5.39}$$

Obviously, the magnitudes of $\dot{\vec{u}}^*$ and $\dot{\vec{u}}$ are not equal due to the presence of the term $\boldsymbol{\Omega} \cdot \vec{u}^*$, which violates the frame-indifference criterion. Hence, the material time derivative of a frame-indifferent vector is not a frame-indifferent vector. There are a few possibilities where one can define the time derivative of a frame-indifferent vector to be frame-indifferent, and to obey the rules of the ordinary time derivative.

For example, the *Jaumann–Zaremba* or *corotational* time derivative of a frame-indifferent vector $\vec{u}$, defined by

$$\frac{D_{\text{Jau}}\vec{u}}{Dt} := \dot{\vec{u}} - \boldsymbol{w} \cdot \vec{u}, \tag{5.40}$$

where $\boldsymbol{w}$ is the spin tensor of motion, is a frame-indifferent vector. The proof is straightforward by using (5.20), $(5.37)_2$ and (5.39):

$$\begin{aligned}
\frac{D_{\text{Jau}}\vec{u}^*}{Dt} &= \dot{\vec{u}}^* - \boldsymbol{w}^* \cdot \vec{u}^* \\
&= \boldsymbol{O} \cdot \dot{\vec{u}} + \boldsymbol{\Omega} \cdot \vec{u}^* - (\boldsymbol{O} \cdot \boldsymbol{w} \cdot \boldsymbol{O}^T + \boldsymbol{\Omega}) \cdot \boldsymbol{O} \cdot \vec{u} \\
&= \boldsymbol{O} \cdot \dot{\vec{u}} + \boldsymbol{\Omega} \cdot \boldsymbol{O} \cdot \vec{u} - \boldsymbol{O} \cdot \boldsymbol{w} \cdot \boldsymbol{O}^T \cdot \boldsymbol{O} \cdot \vec{u} - \boldsymbol{\Omega} \cdot \boldsymbol{O} \cdot \vec{u} = \boldsymbol{O} \cdot \dot{\vec{u}} - \boldsymbol{O} \cdot \boldsymbol{w} \cdot \vec{u},
\end{aligned}$$

which gives

$$\frac{D_{\text{Jau}}\vec{u}^*}{Dt} = \boldsymbol{O}(t) \cdot \frac{D_{\text{Jau}}\vec{u}}{Dt}. \tag{5.41}$$

Similarly, the material time derivative of a frame-indifferent tensor $\boldsymbol{a}$, for which $\boldsymbol{a}^* = \boldsymbol{O}(t) \cdot \boldsymbol{a} \cdot \boldsymbol{O}^T(t)$, is

$$\begin{aligned}
\dot{\boldsymbol{a}}^* &= \boldsymbol{O} \cdot \dot{\boldsymbol{a}} \cdot \boldsymbol{O}^T + \dot{\boldsymbol{O}} \cdot \boldsymbol{a} \cdot \boldsymbol{O}^T + \boldsymbol{O} \cdot \boldsymbol{a} \cdot \dot{\boldsymbol{O}}^T \\
&= \boldsymbol{O} \cdot \dot{\boldsymbol{a}} \cdot \boldsymbol{O}^T + \dot{\boldsymbol{O}} \cdot \boldsymbol{O}^T \cdot \boldsymbol{a}^* \cdot \boldsymbol{O} \cdot \boldsymbol{O}^T + \boldsymbol{O} \cdot \boldsymbol{O}^T \cdot \boldsymbol{a}^* \cdot \boldsymbol{O} \cdot \dot{\boldsymbol{O}}^T,
\end{aligned}$$

or, using (5.20) gives

$$\dot{\boldsymbol{a}}^* = \boldsymbol{O} \cdot \dot{\boldsymbol{a}} \cdot \boldsymbol{O}^T + \boldsymbol{\Omega} \cdot \boldsymbol{a}^* - \boldsymbol{a}^* \cdot \boldsymbol{\Omega}, \tag{5.42}$$

which means that the material time derivative of a frame-indifferent tensor is not frame-indifferent.

The *Jaumann–Zaremba* or *corotational* time derivative of a frame-indifferent tensor $\boldsymbol{a}$, defined by

$$\frac{D_{\text{Jau}}\boldsymbol{a}}{Dt} := \dot{\vec{a}} - \boldsymbol{w} \cdot \boldsymbol{a} + \boldsymbol{a} \cdot \boldsymbol{w}, \tag{5.43}$$

is a frame-indifferent tensor. The proof is again straightforward by using (5.20), $(5.37)_2$ and (5.42):

$$\begin{aligned}
\frac{D_{\text{Jau}}\boldsymbol{a}^*}{Dt} &= \dot{\boldsymbol{a}}^* - \boldsymbol{w}^* \cdot \boldsymbol{a}^* + \boldsymbol{a}^* \cdot \boldsymbol{w}^* \\
&= \boldsymbol{O} \cdot \dot{\boldsymbol{a}} \cdot \boldsymbol{O}^T + \boldsymbol{\Omega} \cdot \boldsymbol{a}^* - \boldsymbol{a}^* \cdot \boldsymbol{\Omega} \\
&\quad -(\boldsymbol{O} \cdot \boldsymbol{w} \cdot \boldsymbol{O}^T + \boldsymbol{\Omega}) \cdot \boldsymbol{O} \cdot \boldsymbol{a} \cdot \boldsymbol{O}^T + \boldsymbol{O} \cdot \boldsymbol{a} \cdot \boldsymbol{O}^T \cdot (\boldsymbol{O} \cdot \boldsymbol{w} \cdot \boldsymbol{O}^T + \boldsymbol{\Omega}) \\
&= \boldsymbol{O} \cdot \dot{\boldsymbol{a}} \cdot \boldsymbol{O}^T + \boldsymbol{\Omega} \cdot \boldsymbol{O} \cdot \boldsymbol{a} \cdot \boldsymbol{O}^T - \boldsymbol{O} \cdot \boldsymbol{a} \cdot \boldsymbol{O}^T \cdot \boldsymbol{\Omega} \\
&\quad -\boldsymbol{O} \cdot \boldsymbol{w} \cdot \boldsymbol{a} \cdot \boldsymbol{O}^T - \boldsymbol{\Omega} \cdot \boldsymbol{O} \cdot \boldsymbol{a} \cdot \boldsymbol{O}^T + \boldsymbol{O} \cdot \boldsymbol{a} \cdot \boldsymbol{w} \cdot \boldsymbol{O}^T + \boldsymbol{O} \cdot \boldsymbol{a} \cdot \boldsymbol{O}^T \cdot \boldsymbol{\Omega} \\
&= \boldsymbol{O} \cdot \dot{\boldsymbol{a}} \cdot \boldsymbol{O}^T - \boldsymbol{O} \cdot \boldsymbol{w} \cdot \boldsymbol{a} \cdot \boldsymbol{O}^T + \boldsymbol{O} \cdot \boldsymbol{a} \cdot \boldsymbol{w} \cdot \boldsymbol{O}^T,
\end{aligned}$$

which gives

$$\frac{D_{\text{Jau}}\boldsymbol{a}^*}{Dt} = \boldsymbol{O}(t) \cdot \frac{D_{\text{Jau}}\boldsymbol{a}}{Dt} \cdot \boldsymbol{O}^T(t). \tag{5.44}$$

The Oldroyd derivative is another possibility to introduce the frame-indifferent time derivative of vectors and tensors. Let $\vec{u}$ and $\boldsymbol{a}$ be a frame-indifferent vector and tensor, respectively. The *Oldroyd derivatives* of $\vec{u}$ and $\boldsymbol{a}$ are defined by

$$\frac{D_{\text{Old}}\vec{u}}{Dt} := \dot{\vec{u}} - \boldsymbol{l} \cdot \vec{u}, \tag{5.45}$$

$$\frac{D_{\text{Old}}\boldsymbol{a}}{Dt} := \dot{\boldsymbol{a}} - \boldsymbol{l} \cdot \boldsymbol{a} - \boldsymbol{a} \cdot \boldsymbol{l}^T, \tag{5.46}$$

where $\boldsymbol{l}$ is the velocity gradient defined by (2.13). The frame-indifference of the Oldroyd derivative of a frame-indifferent vector follows from (5.20), $(5.37)_1$ and (5.39):

$$\begin{aligned}
\frac{D_{\text{Old}}\vec{u}^*}{Dt} &= \dot{\vec{u}}^* - \boldsymbol{l}^* \cdot \vec{u}^* \\
&= \boldsymbol{O} \cdot \dot{\vec{u}} + \boldsymbol{\Omega} \cdot \vec{u}^* - (\boldsymbol{O} \cdot \boldsymbol{l} \cdot \boldsymbol{O}^T + \boldsymbol{\Omega}) \cdot \boldsymbol{O} \cdot \vec{u} \\
&= \boldsymbol{O} \cdot \dot{\vec{u}} + \boldsymbol{\Omega} \cdot \boldsymbol{O} \cdot \vec{u} - \boldsymbol{O} \cdot \boldsymbol{l} \cdot \boldsymbol{O}^T \cdot \boldsymbol{O} \cdot \vec{u} - \boldsymbol{\Omega} \cdot \boldsymbol{O} \cdot \vec{u} = \boldsymbol{O} \cdot \dot{\vec{u}} - \boldsymbol{O} \cdot \boldsymbol{l} \cdot \vec{u},
\end{aligned}$$

which gives

$$\frac{D_{\text{Old}}\vec{u}^*}{Dt} = \boldsymbol{O}(t) \cdot \frac{D_{\text{Old}}\vec{u}}{Dt}. \tag{5.47}$$

The proof of (5.46) follows again from (5.20), (5.37)$_1$ and (5.42):

$$\begin{aligned}
\frac{D_{\text{Old}}\boldsymbol{a}^*}{Dt} &= \dot{\boldsymbol{a}}^* - \boldsymbol{l}^* \cdot \boldsymbol{a}^* + \boldsymbol{a}^* \cdot \boldsymbol{l}^{*T} \\
&= \boldsymbol{O} \cdot \dot{\boldsymbol{a}} \cdot \boldsymbol{O}^T + \boldsymbol{\Omega} \cdot \boldsymbol{a}^* - \boldsymbol{a}^* \cdot \boldsymbol{\Omega} \\
&\quad -(\boldsymbol{O} \cdot \boldsymbol{l} \cdot \boldsymbol{O}^T + \boldsymbol{\Omega}) \cdot \boldsymbol{O} \cdot \boldsymbol{a} \cdot \boldsymbol{O}^T - \boldsymbol{O} \cdot \boldsymbol{a} \cdot \boldsymbol{O}^T \cdot (\boldsymbol{O} \cdot \boldsymbol{l}^T \cdot \boldsymbol{O}^T + \boldsymbol{\Omega}^T) \\
&= \boldsymbol{O} \cdot \dot{\boldsymbol{a}} \cdot \boldsymbol{O}^T + \boldsymbol{\Omega} \cdot \boldsymbol{O} \cdot \boldsymbol{a} \cdot \boldsymbol{O}^T - \boldsymbol{O} \cdot \boldsymbol{a} \cdot \boldsymbol{O}^T \cdot \boldsymbol{\Omega} \\
&\quad -\boldsymbol{O} \cdot \boldsymbol{l} \cdot \boldsymbol{a} \cdot \boldsymbol{O}^T - \boldsymbol{\Omega} \cdot \boldsymbol{O} \cdot \boldsymbol{a} \cdot \boldsymbol{O}^T - \boldsymbol{O} \cdot \boldsymbol{a} \cdot \boldsymbol{l}^T \cdot \boldsymbol{O}^T - \boldsymbol{O} \cdot \boldsymbol{a} \cdot \boldsymbol{O}^T \cdot \boldsymbol{\Omega}^T \\
&= \boldsymbol{O} \cdot \dot{\boldsymbol{a}} \cdot \boldsymbol{O}^T - \boldsymbol{O} \cdot \boldsymbol{l} \cdot \boldsymbol{a} \cdot \boldsymbol{O}^T - \boldsymbol{O} \cdot \boldsymbol{a} \cdot \boldsymbol{l}^T \cdot \boldsymbol{O}^T,
\end{aligned}$$

which gives

$$\frac{D_{\text{Old}}\boldsymbol{a}^*}{Dt} = \boldsymbol{O}(t) \cdot \frac{D_{\text{Old}}\boldsymbol{a}}{Dt} \cdot \boldsymbol{O}^T(t). \tag{5.48}$$

5.6 The Postulate of Frame-Indifference

In the preceding sections, we studied the effect of the observer transformation on velocity, acceleration and other kinematic quantities. We will now specify the behaviour the fields that represent mass, force, internal energy and heating. We assume that the mass density ϱ, the stress vector $\vec{t}_{(\vec{n})}$, the normal heat flux $\vec{q} \cdot \vec{n}$, the internal energy ε and temperature θ, being associated with the intrinsic properties of a material body, are observed the same by equivalent observers. They are thus postulated to be frame-indifferent, an assumption that may be viewed as a part of the principle of frame-indifference. Note that this postulate concerns only frame-indifference properties of the primitive quantities related to the intrinsic properties of a material body, so that they are universal for any deformable body and therefore are not related to the material properties of the body.

Considering this postulate, let us show that the Cauchy stress tensor $\boldsymbol{t}$ and the heat flux vector $\vec{q}$ are frame-indifferent. The assumption of the frame-indifference of $\vec{t}_{(\vec{n})}$ means that

$$\vec{t}^*_{(\vec{n}^*)} = \boldsymbol{O} \cdot \vec{t}_{(\vec{n})}, \tag{5.49}$$

or, using the Cauchy stress formula (3.14) gives

$$\vec{n}^* \cdot \boldsymbol{t}^* = \boldsymbol{O} \cdot (\vec{n} \cdot \boldsymbol{t}), \tag{5.50}$$

where $\boldsymbol{t}^* := t^*_{k^*l^*}(\vec{i}_{k^*} \otimes \vec{i}_{l^*})$. Employing the frame-indifference property of the unit normal vector[10] $\vec{n}$, i.e., $\vec{n}^* = \boldsymbol{O} \cdot \vec{n}$, leads to

$$(\boldsymbol{O} \cdot \vec{n}) \cdot \boldsymbol{t}^* = \boldsymbol{O} \cdot (\vec{n} \cdot \boldsymbol{t}), \tag{5.51}$$

or, equivalently,

$$\vec{n} \cdot (\boldsymbol{O}^T \cdot \boldsymbol{t}^* - \boldsymbol{t} \cdot \boldsymbol{O}^T) = \vec{0}, \tag{5.52}$$

which must hold for all surfaces passing through a material point. Hence, $\boldsymbol{O}^T \cdot \boldsymbol{t}^* - \boldsymbol{t} \cdot \boldsymbol{O}^T = \boldsymbol{0}$. With the orthogonality property (5.10) of $\boldsymbol{O}$, we finally obtain

$$\boldsymbol{t}^* = \boldsymbol{O}(t) \cdot \boldsymbol{t} \cdot \boldsymbol{O}^T(t), \tag{5.53}$$

which shows that the Cauchy stress tensor is a frame-indifferent tensor. Similarly, the assumption of the frame-indifference of $\vec{q} \cdot \vec{n}$ implies that the heat flux vector $\vec{q}$ is a frame-indifferent vector such that

$$\vec{q}^* = \boldsymbol{O}(t) \cdot \vec{q}. \tag{5.54}$$

The assumption of the frame-indifference of temperature and the assumption that the reference configuration is not affected by a change of the observer imply that the Lagrangian temperature gradient vector $\text{Grad}\,\theta$ employed in formulating the constitutive equations in Chap. 6 is a frame-indifferent scalar,

$$\text{Grad}^*\,\theta^* = \text{Grad}\,\theta. \tag{5.55}$$

The proof is immediate,

$$\text{Grad}^*\,\theta^* = \text{Grad}\,\theta^* = \text{Grad}\,\theta.$$

Similarly, the frame-indifference property (5.53) of the Cauchy stress tensor implies the second Piola–Kirchhoff stress tensor $\boldsymbol{T}^{(2)}$ is a frame-indifferent scalar,

$$\boldsymbol{T}^{(2)*} = \boldsymbol{T}^{(2)}. \tag{5.56}$$

To show this, we use $(3.27)_1$, (5.31), $(5.34)_1$ and (5.53), and successively write

$$\begin{aligned} \boldsymbol{T}^{(2)*} &= J^*(\boldsymbol{F}^*)^{-1} \cdot \boldsymbol{t}^* \cdot (\boldsymbol{F}^*)^{-T} = J\boldsymbol{F}^{-1} \cdot \boldsymbol{O}^T \cdot \boldsymbol{O} \cdot \boldsymbol{t} \cdot \boldsymbol{O}^T \cdot \boldsymbol{O} \cdot \boldsymbol{F}^{-T} \\ &= J\boldsymbol{F}^{-1} \cdot \boldsymbol{t} \cdot \boldsymbol{F}^{-T} = \boldsymbol{T}^{(2)}. \end{aligned}$$

[10]The unit normal vector to the surface is defined by (2.47) as the gradient of an objective scalar function. In view of (5.27), $\vec{n}$ is then a frame-indifferent vector.

5.7 Observer Transformation of Basic Field Equations

The basic field equations of mass, linear momentum, angular momentum and energy are formulated in Chap. 4 with respect to an inertial frame. We will now determine how these equations change under the observer transformation.

We begin with the continuity equation (4.18). The assumption of the frame-indifference of the mass density ϱ together with (5.38) implies that the material time derivative of ϱ is frame indifferent,

$$\varrho^* = \varrho, \qquad \dot{\varrho}^* = \dot{\varrho}. \tag{5.57}$$

Recall from (5.26) that the divergence of the velocity field is a frame-indifferent scalar. Using this result along with (5.57), (4.18) transforms from the unstarred frame to the starred frame as

$$\frac{D\varrho^*}{Dt} + \varrho^* \operatorname{div}^* \vec{v}^* = 0, \tag{5.58}$$

which has exactly the same form as in the unstarred frame. Hence, the continuity equation is invariant under observer transformations.

We continue with the equation of motion (4.24). Equation (5.23) shows that the acceleration is not, in general, frame-indifferent, but transforms as

$$\vec{a}^* = \boldsymbol{O} \cdot \vec{a} + \vec{i}^*, \tag{5.59}$$

where $\vec{a}$ and $\vec{a}^*$ are the accelerations of a material point observed in the unstarred and starred frames, respectively, and

$$\vec{i}^* := 2\,\boldsymbol{\Omega} \cdot (\vec{v}^* - \dot{\vec{c}}^*) - \boldsymbol{\Omega} \cdot \boldsymbol{\Omega} \cdot (\vec{x}^* - \vec{c}^*) + \dot{\boldsymbol{\Omega}} \cdot (\vec{x}^* - \vec{c}^*) + \ddot{\vec{c}}^*. \tag{5.60}$$

Multiplying (4.24) from the left by $\boldsymbol{O}$ gives

$$\boldsymbol{O} \cdot \operatorname{div} \boldsymbol{t} + \varrho\, \boldsymbol{O} \cdot \vec{f} = \varrho\, \boldsymbol{O} \cdot \vec{a}. \tag{5.61}$$

Recall from (5.29) and (5.53) that the Cauchy stress tensor is a frame-indifferent tensor and its spatial divergence is a frame-indifferent vector. Using these results, the assumption of the frame-indifference of the mass density and (5.59), (5.61) takes the form

$$\operatorname{div} \boldsymbol{t}^* + \varrho^* (\boldsymbol{O} \cdot \vec{f} + \vec{i}^*) = \varrho^* \vec{a}^*. \tag{5.62}$$

In addition, assuming the body force $\vec{f}$ is a frame-indifferent vector leads to

$$\operatorname{div} \boldsymbol{t}^* + \varrho^* (\vec{f}^* + \vec{i}^*) = \varrho^* \vec{a}^*. \tag{5.63}$$

Comparison of this equation with (4.24) reveals that the equation of motion in the starred frame has the same form as in the unstarred frame, except that the body force $\vec{f}$ is replaced by $\vec{f}^* + \vec{i}^*$. The vector $\vec{i}^*$, consisting of the centrifugal, Coriolis, Euler and translation accelerations, is an acceleration field due to the motion of the spatial frame relative to an inertial frame. It vanishes in an inertial frame. In a non-inertial frame, for example, a rotating and/or accelerating frame, $\vec{i}^*$ is perceived as an *inertial* or *apparent force* (per unit of mass) contributing to the body force. We conclude that the equation of motion is frame dependent through the inertial force $\vec{i}^*$.

We finally examine the effect of an observer transformation on the energy equation (4.38). The assumption of the frame-indifference of the internal energy density ε together with (5.38) implies that the material time derivative of ε is frame indifferent,

$$\varepsilon^* = \varepsilon, \qquad \dot{\overline{\varepsilon^*}} = \dot{\varepsilon}. \tag{5.64}$$

Recall from $(5.36)_3$ and (5.53) that the strain-rate tensor and the Cauchy stress tensor are frame-indifferent tensors. In view of (5.30), their stress power is a frame-indifferent scalar, $\boldsymbol{t}^* : \boldsymbol{d}^* = \boldsymbol{t} : \boldsymbol{d}$. Moreover, in view of (5.54) and (5.28), the heat flux vector is a frame-indifferent vector and its spatial divergence is a frame-indifferent scalar. Using these results and the postulate of the frame-indifference of the mass density, (4.38) transforms from the unstarred frame to the starred frame as

$$\varrho^* \frac{D\varepsilon^*}{Dt} = \boldsymbol{t}^* : \boldsymbol{d}^* - \operatorname{div}^* \vec{q}^* + \varrho^* h^*, \tag{5.65}$$

where the heat supply h was assumed to be a frame-indifferent scalar field. The last equation has exactly the same form as in the unstarred frame. Hence, the energy equation is invariant under observer transformations.

Chapter 6
Constitutive Equations

The equations listed in the preceding chapters apply to any material body that deforms under the action of external forces. Mathematically, they do not, by themselves, have a unique solution because the deformations are not related to internal contact forces. The governing equations are completed by constitutive equations that provide the missing connection. Physically, the constitutive equations characterise the particular material and its reaction to applied external forces. They therefore describe the macroscopic behaviour resulting from the internal *constitution* of the material.

In this chapter, we introduce the general ideas and procedures underlying the formulation of constitutive equations, and apply them to set up the constitutive equation for a simple material. A general form of this equation is then specified for fluids and solids. Examples of various forms of idealised material responses are given at the end of this chapter.

6.1 The Need for Constitutive Equations

The basic principles of continuum mechanics, discussed in Chap. 4, namely, conservation of mass, linear momentum, angular momentum and energy resulted in the fundamental equations:

$$\frac{\partial \varrho}{\partial t} + \operatorname{div}(\varrho \vec{v}) = 0, \tag{6.1}$$

$$\operatorname{div} \boldsymbol{t} + \varrho \vec{f} = \varrho \frac{D\vec{v}}{Dt}, \tag{6.2}$$

Z. Martinec, *Principles of Continuum Mechanics*, Nečas Center Series,
https://doi.org/10.1007/978-3-030-05390-1_6

$$t = t^T, \tag{6.3}$$

$$\varrho \frac{D\varepsilon}{Dt} = t : d - \operatorname{div} \vec{q} + \varrho h. \tag{6.4}$$

In total, they constitute five independent equations (one for mass, three for linear momentum and one for energy) for 15 unknown field variables:

- mass density ϱ,
- velocity $\vec{v}$,
- Cauchy's stress tensor $\boldsymbol{t}$,
- internal energy ε,
- heat flux $\vec{q}$,
- temperature θ,

provided that the temperature θ enters (6.1)–(6.4) implicitly, and the body forces $\vec{f}$ and the distribution of internal heat supplies h are given. Clearly, these equations are not sufficient to determine these unknowns, except for some trivial situations, for example, rigid-body motions in the absence of heat conduction. Hence, 10 additional equations must be supplied to make the problem well-posed.

In the derivation of (6.1)–(6.4), no differentiation has been made between various types of materials. It is therefore not surprising that these equations are not sufficient to fully explain the motions of materials with various types of physical properties. The character of the material is brought into the formulation through the so-called *constitutive equations*, which specify the mechanical and thermal properties of particular materials based on their internal constitution. Mathematically, the usefulness of these constitutive equations is to describe the relationships between the kinematic, mechanical and thermal field variables, and to formulate well-posed continuum mechanics problems. Physically, constitutive equations define various idealised materials that serve as models for the behaviour of real materials. However, it is not possible to write one equation capable of representing a given material over its entire range of applications, since many materials behave quite differently under changing levels of loading, such as an elastic-plastic response with increasing stress. Thus, in this sense, it is perhaps better to think of constitutive equations as being representative of a particular *behaviour* rather than of a particular *material.*

6.2 Formulation of Thermomechanical Constitutive Equations

In this section, we deal with the constitutive equations for *thermomechanical materials*. Chemical changes and electromagnetic effects are excluded because a large class of materials do not undergo chemical transition or produce appreciable electromagnetic effects when deformed. However, deformation and motion generally produce heat. Conversely, materials subjected to thermal changes deform.

The effects of thermal changes on the material behaviour depends on the range and severity of such changes.

The thermomechanical constitutive equations are relationships between a set of thermomechanical variables. They may be expressed as an implicit tensor-valued functional $\mathcal{R}$ of the 15 unknown field variables,

$$\mathop{\mathcal{R}}_{\substack{\vec{X}' \in \mathcal{B} \\ \tau \leq t}} \left[\varrho(\vec{X}', \tau), \vec{\chi}(\vec{X}', \tau), \theta(\vec{X}', \tau), \boldsymbol{t}(\vec{X}', \tau), \vec{q}(\vec{X}', \tau), \varepsilon(\vec{X}', \tau), \vec{X} \right] = 0, \tag{6.5}$$

where τ represents all past times and t is the present time. The constraints $\vec{X}' \in \mathcal{B}$ and $\tau \leq t$ express the *principle of determinism*, postulating that the present state of the thermomechanical variables at a material point $\vec{X}$ of the body $\mathcal{B}$ at time t is uniquely determined by the past history of the motion and temperature of all material points of $\mathcal{B}$. The principle of determinism is a principle of exclusions, meaning it excludes the dependence of the material behaviour on any point outside the body and any future events.

We will restrict the functional in (6.5) to be of a type that does not change with time, that is, it does not depend on the present time t explicitly, but only implicitly via thermomechanical variables. Such a functional is invariant with respect to translation in time.

The *constitutive functional* $\mathcal{R}$ describes the material properties of a given material particle $\mathcal{X}$ with the position $\vec{X}$.[1] The functional form may, in general, be different for different particles and $\mathcal{R}$ may thus vary with changing position within the body $\mathcal{B}$. Such a material is called *inhomogeneous*. If the functional $\mathcal{R}$ is independent of $\vec{X}$, the material is termed *homogeneous*.

For a simple material (see Sect. 6.3), the implicit constitutive equation (6.5) is assumed to be solved uniquely for the present values of thermomechanical variables. In this case, (6.5) is replaced by a set of explicit functional equations

$$\begin{aligned}
\boldsymbol{t}(\vec{X}, t) &= \mathop{\mathcal{F}}_{\substack{\vec{X}' \in \mathcal{B} \\ \tau \leq t}} \left[\varrho(\vec{X}', \tau), \vec{\chi}(\vec{X}', \tau), \theta(\vec{X}', \tau), \vec{X} \right], \\
\vec{q}(\vec{X}, t) &= \mathop{\mathcal{Q}}_{\substack{\vec{X}' \in \mathcal{B} \\ \tau \leq t}} \left[\varrho(\vec{X}', \tau), \vec{\chi}(\vec{X}', \tau), \theta(\vec{X}', \tau), \vec{X} \right], \\
\varepsilon(\vec{X}, t) &= \mathop{\mathcal{E}}_{\substack{\vec{X}' \in \mathcal{B} \\ \tau \leq t}} \left[\varrho(\vec{X}', \tau), \vec{\chi}(\vec{X}', \tau), \theta(\vec{X}', \tau), \vec{X} \right],
\end{aligned} \tag{6.6}$$

[1] This is a significant advantage of the Lagrangian description because material properties are always associated with a material particle $\mathcal{X}$ with the position $\vec{X}$, while in the Eulerian description, various material particles may pass through a given present position $\vec{x}$.

where $\mathcal{F}$, $\mathcal{Q}$ and $\mathcal{E}$ are, respectively, tensor-valued, vector-valued and scalar-valued functionals. Note that all of these functionals are assumed to depend on the same set of *constitutive variables* $\varrho(\vec{X}', \tau)$, $\vec{\chi}(\vec{X}', \tau)$, $\theta(\vec{X}', \tau)$ and $\vec{X}$. This is known as the *principle of equipresence*.

However, the implicit functional equation (6.5) does not necessarily determine $\boldsymbol{t}$, $\vec{q}$ and ε at $(\vec{X}, t)$ explicitly. For instance, the stress at $(\vec{X}, t)$ may depend not only on the motion and temperature at all other points of the body, but also on the histories of the stress, heat flux and internal energy. Various approximations to (6.5) exist in which the dependence on $\boldsymbol{t}(\vec{X}', \tau)$ is replaced by the history of various orders of stress rates, heat rates, etc. For example, the constitutive equation (6.5) may be written explicitly for the stress rates $\dot{\boldsymbol{t}}$ at $(\vec{X}, t)$,

$$\dot{\boldsymbol{t}}\,(\vec{X}, t) = \underset{\substack{\vec{X}' \in \mathcal{B} \\ \tau \le t}}{\mathcal{F}} \left[\boldsymbol{t}(\vec{X}, t), \varrho(\vec{X}', \tau), \vec{\chi}(\vec{X}', \tau), \theta(\vec{X}', \tau), \vec{X}\right]. \tag{6.7}$$

More generally,

$$\boldsymbol{t}^{(p)}(\vec{X}, t) = \underset{\substack{\vec{X}' \in \mathcal{B} \\ \tau \le t}}{\mathcal{F}} \left[\boldsymbol{t}^{(p-1)}(\vec{X}, t), \boldsymbol{t}^{(p-2)}(\vec{X}, t), \dots, \boldsymbol{t}(\vec{X}, t), \varrho(\vec{X}', \tau), \right.$$
$$\left. \vec{\chi}(\vec{X}', \tau), \theta(\vec{X}', \tau), \vec{X}\right], \tag{6.8}$$

which involves the stress rates up to the pth order at $(\vec{X}, t)$. This type of generalisation is, for instance, needed to interpret creep data.

The following text considers the explicit constitutive equations (6.6) only. Together with the 5 basic conservation equations (6.1)–(6.4), they form 15 equations for 15 unknowns. Since $\boldsymbol{t}$, $\vec{q}$, and ε in (6.6) are expressed explicitly, it is, in principle, possible to eliminate these variables in (6.1)–(6.4). We then obtain five field equations of thermodynamics for five unknown field variables: mass density ϱ, velocity $\vec{v}$ and temperature θ. Any solution of the field equations of thermodynamics for a given material is called a *thermodynamic process*.

We now deduce the consequences of additional restrictions on the functionals $\mathcal{F}$, $\mathcal{Q}$ and $\mathcal{E}$. Since the procedure is similar for all functionals, for the sake of brevity, we carry out the analysis only for the stress functional $\mathcal{F}$. The results for $\mathcal{Q}$ and $\mathcal{E}$ are then expressed by analogy.

6.3 Simple Materials

The constitutive functionals $\mathcal{F}$, $\mathcal{Q}$ and $\mathcal{E}$ are subject to another fundamental principle: the *principle of local action*, which postulates that the motion and temperature at material points distant from $\vec{X}$ do not appreciably affect the stress, heat flux and internal energy at point $\vec{X}$. Suppose that the functions $\varrho(\vec{X}', \tau)$, $\vec{\chi}(\vec{X}', \tau)$ and $\theta(\vec{X}', \tau)$ admit the Taylor series expansion about $\vec{X}$ for all $\tau < t$. According to the principle of local action, the global history functions can be approximated in a small neighbourhood of $\vec{X}$ by the Taylor series up to a certain order. In particular, when only a linear approximation is concerned, we have

$$\begin{aligned} \varrho(\vec{X}', \tau) &\approx \varrho(\vec{X}, \tau) + \operatorname{Grad} \varrho(\vec{X}, \tau) \cdot d\vec{X}, \\ \vec{\chi}(\vec{X}', \tau) &\approx \vec{\chi}(\vec{X}, \tau) + \boldsymbol{F}(\vec{X}, \tau) \cdot d\vec{X}, \\ \theta(\vec{X}', \tau) &\approx \theta(\vec{X}, \tau) + \operatorname{Grad} \theta(\vec{X}, \tau) \cdot d\vec{X}, \end{aligned} \tag{6.9}$$

where $\boldsymbol{F}(\vec{X}, \tau)$ is the deformation gradient tensor at $\vec{X}$ and time τ, and $d\vec{X} = \vec{X}' - \vec{X}$. Since the relative motion and temperature history of an infinitesimal neighbourhood of $\vec{X}$ is completely determined by the history of mass density, deformation and temperature gradients at $\vec{X}$, the stress $\boldsymbol{t}(\vec{X}, t)$ must be determined by the history of $\operatorname{Grad} \varrho(\vec{X}, \tau)$, $\boldsymbol{F}(\vec{X}, \tau)$ and $\operatorname{Grad} \theta(\vec{X}, \tau)$ for $\tau \leq t$. Such a material was called a *simple material* by Noll (1958).[2] In other words, the behaviour of the material point $\vec{X}$ within a simple material is not affected by the histories of distant points from $\vec{X}$. To any desired degree of accuracy, the whole configuration of a sufficiently small neighbourhood of the material point $\vec{X}$ is determined by the history of $\operatorname{Grad} \varrho(\vec{X}, \tau)$, $\boldsymbol{F}(\vec{X}, \tau)$ and $\operatorname{Grad} \theta(\vec{X}, \tau)$, and we may say that the stress $\boldsymbol{t}(\vec{X}, t)$, which was assumed to be determined by the local configuration, is completely determined by $\operatorname{Grad} \varrho(\vec{X}, \tau)$, $\boldsymbol{F}(\vec{X}, \tau)$ and $\operatorname{Grad} \theta(\vec{X}, \tau)$. That is, the general constitutive equation $(6.6)_1$ reduces to the form

$$\begin{aligned} \boldsymbol{t}(\vec{X}, t) = \mathop{\mathcal{F}}_{\tau \leq t} \Big[& \varrho(\vec{X}, \tau), \operatorname{Grad} \varrho(\vec{X}, \tau), \vec{\chi}(\vec{X}, \tau), \boldsymbol{F}(\vec{X}, \tau), \\ & \theta(\vec{X}, \tau), \operatorname{Grad} \theta(\vec{X}, \tau), \vec{X} \Big]. \end{aligned} \tag{6.10}$$

Note that although the constitutive functional for a simple material depends only on local values at $\vec{X}$, it is still general enough to define a material with memory of past local deformation and temperature. The class of simple materials includes most materials of practical interest such as elastic solids, Kelvin–Voigt viscoelastic solids, Navier-Stokes fluids and non-Newtonian fluids.

Given the deformation gradient $\boldsymbol{F}$, the mass density in the reference configuration is expressed through the continuity equation (4.63) as

$$\varrho(\vec{X}, \tau) = \frac{\varrho_0(\vec{X})}{\det \boldsymbol{F}(\vec{X}, \tau)}. \tag{6.11}$$

---

[2] Note that if higher-order gradients in (6.9) are retained, we obtain non-simple materials of various classes. For example, we obtain a polar material of *couple stress* by including the second-order gradients into argument of $\mathcal{F}$.

We can thus drop $\varrho(\vec{X}, \tau)$ from the argument of the functional $\mathcal{F}$ since $\varrho(\vec{X}, \tau)$ is expressible in terms of $\boldsymbol{F}(\vec{X}, \tau)$ (the factor $\varrho_0(\vec{X})$ is a fixed expression—not dependent on time—for a given reference configuration). Moreover, by applying the gradient operator to (6.11), the term $\mathrm{Grad}\,\varrho(\vec{X}, \tau)$ can be expressed in terms of $\mathrm{Grad}\,\boldsymbol{F}(\vec{X}, \tau)$. For a simple material, however, $\mathrm{Grad}\,\boldsymbol{F}$ is small compared to $\boldsymbol{F}$ and can be neglected. In summary, the principle of local action applied to a simple material leads to

$$\boldsymbol{t}(\vec{X}, t) = \mathop{\mathcal{F}}_{\tau \le t} \left[\vec{\chi}(\vec{X}, \tau), \boldsymbol{F}(\vec{X}, \tau), \theta(\vec{X}, \tau), \vec{G}_\theta(\vec{X}, \tau), \vec{X} \right], \tag{6.12}$$

where $\vec{G}_\theta$ stands for the temperature gradient, $\vec{G}_\theta := \mathrm{Grad}\,\theta$.

6.4 The Principle of Material Frame-Indifference

The constitutive functionals are subject to yet another fundamental principle. Since these functionals characterise the intrinsic properties of a material body, they must be indifferent to changes of the observer frame. This principle, called the *principle of material frame-indifference*, states that a constitutive equation must be form-invariant under rigid motions of the observer frame or, in other words, the material properties cannot depend on the motion of an observer.

To express this principle mathematically, let us write the constitutive equation (6.12) in the unstarred and starred frames,

$$\begin{aligned} \boldsymbol{t}(\vec{X}, t) &= \mathop{\mathcal{F}}_{\tau \le t} \left[\vec{\chi}(\vec{X}, \tau), \boldsymbol{F}(\vec{X}, \tau), \theta(\vec{X}, \tau), \vec{G}_\theta(\vec{X}, \tau), \vec{X} \right], \\ \boldsymbol{t}^*(\vec{X}, t) &= \mathop{\mathcal{F}^*}_{\tau \le t} \left[\vec{\chi}^*(\vec{X}, \tau), \boldsymbol{F}^*(\vec{X}, \tau), \theta^*(\vec{X}, \tau), \vec{G}^*_\theta(\vec{X}, \tau), \vec{X} \right]. \end{aligned} \tag{6.13}$$

The postulate of frame-indifference assumes that, among other quantities, temperature is frame-indifferent such that $\theta^* = \theta$. Moreover, the motion function $\vec{\chi}$, deformation gradient $\boldsymbol{F}$ and temperature gradient $\vec{G}_\theta$ transform according to (5.17), (5.31) and (5.55), respectively. In view of these equations, substituting (6.13) into the frame-indifference condition (5.53) for the Cauchy stress tensor leads to

$$\begin{aligned} &\mathop{\mathcal{F}^*}_{\tau \le t} \left[\boldsymbol{O}(\tau) \cdot \vec{\chi}(\vec{X}, \tau) + \vec{c}^{\,*}(\tau), \boldsymbol{O}(\tau) \cdot \boldsymbol{F}(\vec{X}, \tau), \theta(\vec{X}, \tau), \vec{G}_\theta(\vec{X}, \tau), \vec{X} \right] \\ &= \boldsymbol{O}(t) \cdot \mathop{\mathcal{F}}_{\tau \le t} \left[\vec{\chi}(\vec{X}, \tau), \boldsymbol{F}(\vec{X}, \tau), \theta(\vec{X}, \tau), \vec{G}_\theta(\vec{X}, \tau), \vec{X} \right] \cdot \boldsymbol{O}^T(t), \end{aligned} \tag{6.14}$$

which relates the constitutive functionals for two different observers. It shows that different observers cannot independently propose their own constitutive functionals.

Instead, the condition (6.14) determines the constitutive functional $\mathcal{F}^*$ once the constitutive functional $\mathcal{F}$ is given or vice versa.

In general, the unstarred and starred functionals $\mathcal{F}$ and $\mathcal{F}^*$ may differ, but the principle of material frame-indifference postulates that the form of the constitutive functional $\mathcal{F}$ must be the same under any rigid motions of the observer frame. Mathematically,

$$\mathcal{F}^*[\,\bullet\,] = \mathcal{F}[\,\bullet\,], \tag{6.15}$$

where the (different) arguments are those of (6.14). Hence, (6.14) simplifies to

$$\begin{aligned} &\mathop{\mathcal{F}}_{\tau\le t}\left[\boldsymbol{O}(\tau)\cdot\vec{\chi}(\vec{X},\tau)+\vec{c}^{\,*}(\tau),\, \boldsymbol{O}(\tau)\cdot\boldsymbol{F}(\vec{X},\tau), \theta(\vec{X},\tau), \vec{G}_\theta(\vec{X},\tau), \vec{X}\right] \\ &\quad = \boldsymbol{O}(t)\cdot \mathop{\mathcal{F}}_{\tau\le t}\left[\vec{\chi}(\vec{X},\tau),\, \boldsymbol{F}(\vec{X},\tau), \theta(\vec{X},\tau), \vec{G}_\theta(\vec{X},\tau), \vec{X}\right]\cdot\boldsymbol{O}^T(t). \end{aligned} \tag{6.16}$$

This relation must hold for any arbitrary orthogonal tensor function $\boldsymbol{O}(t)$ and any arbitrary vector function $\vec{c}^{\,*}(t)$. Since the condition (6.16) involves only a single observer frame, it imposes a restriction on the constitutive functional $\mathcal{F}$.

Let us consider a special case of a rigid translation of the observer frame such that the origin of the observer frame moves with the material point $\vec{X}$,

$$\boldsymbol{O}(\tau) = \boldsymbol{I}\,, \qquad\qquad \vec{c}^{\,*}(\tau) = -\vec{\chi}(\vec{X},\tau). \tag{6.17}$$

This means that the reference observer frame is translated, so that the material point $\vec{X}$ at any time τ remains at the origin of this frame. From (5.17) it follows that $\vec{\chi}^*(\vec{X},\tau) = \vec{0}$ and (6.16) becomes

$$\begin{aligned} &\mathop{\mathcal{F}}_{\tau\le t}\left[\vec{0},\, \boldsymbol{F}(\vec{X},\tau), \theta(\vec{X},\tau), \vec{G}_\theta(\vec{X},\tau), \vec{X}\right] \\ &\qquad = \mathop{\mathcal{F}}_{\tau\le t}\left[\vec{\chi}(\vec{X},\tau),\, \boldsymbol{F}(\vec{X},\tau), \theta(\vec{X},\tau), \vec{G}_\theta(\vec{X},\tau), \vec{X}\right], \end{aligned} \tag{6.18}$$

which must hold for all deformation and temperature histories. Thus, the stress at the material point $\vec{X}$ and time t cannot depend **explicitly** on the history of the motion of this point. It also means that the velocity, acceleration and all other higher-order time derivatives of motion have no influence on the material laws. Consequently, the general constitutive equation (6.12) for a simple material reduces to the form

$$\boldsymbol{t}(\vec{X},t) = \mathop{\mathcal{F}}_{\tau\le t}\left[\boldsymbol{F}(\vec{X},\tau), \theta(\vec{X},\tau), \vec{G}_\theta(\vec{X},\tau), \vec{X}\right], \tag{6.19}$$

and the material frame-indifference restriction (6.16) simplifies to

$$\boldsymbol{O}(t)\;\cdot \mathop{\mathcal{F}}_{\tau\le t}\left[\boldsymbol{F}(\vec{X},\tau), \theta(\vec{X},\tau), \vec{G}_\theta(\vec{X},\tau), \vec{X}\right]\cdot\boldsymbol{O}^T(t)$$

$$= \underset{\tau \le t}{\mathcal{F}} \left[\boldsymbol{O}(\tau) \cdot \boldsymbol{F}(\vec{X}, \tau), \theta(\vec{X}, \tau), \vec{G}_\theta(\vec{X}, \tau), \vec{X} \right], \tag{6.20}$$

which must hold for all orthogonal tensors $\boldsymbol{O}(t)$, deformation gradients $\boldsymbol{F}(\vec{X}, t)$, temperatures $\theta(\vec{X}, t)$ and temperature gradients $\vec{G}_\theta(\vec{X}, t)$. The condition (6.20) is the most well-known result in constitutive theories of continuum mechanics.

6.5 Reduction by Polar Decomposition

The condition (6.20) will now be used to reduce the constitutive equation (6.19). Let us recall the polar decomposition (1.50) of the deformation gradient $\boldsymbol{F}(\vec{X}, t) = \boldsymbol{R}(\vec{X}, t) \cdot \boldsymbol{U}(\vec{X}, t)$ into the rotation tensor $\boldsymbol{R}$ and the right-stretch tensor $\boldsymbol{U} = \sqrt{\boldsymbol{C}} = \sqrt{\boldsymbol{F}^T \cdot \boldsymbol{F}}$. We now make a special choice of the orthogonal tensor $\boldsymbol{O}(t)$ in (6.20); for any fixed reference position $\vec{X}$, we set $\boldsymbol{O}(t) = \boldsymbol{R}^T(\vec{X}, t)$ for all t. This choice of $\boldsymbol{O}(t)$ yields

$$\begin{aligned} \boldsymbol{R}^T(\vec{X}, t) \cdot \underset{\tau \le t}{\mathcal{F}} \left[\boldsymbol{F}(\vec{X}, \tau), \theta(\vec{X}, \tau), \vec{G}_\theta(\vec{X}, \tau), \vec{X} \right] \cdot \boldsymbol{R}(\vec{X}, t) \\ = \underset{\tau \le t}{\mathcal{F}} \left[\boldsymbol{R}^T(\vec{X}, \tau) \cdot \boldsymbol{F}(\vec{X}, \tau), \theta(\vec{X}, \tau), \vec{G}_\theta(\vec{X}, \tau), \vec{X} \right], \end{aligned} \tag{6.21}$$

which, with $\boldsymbol{U} = \boldsymbol{R}^T \cdot \boldsymbol{F}$, reduces to

$$\begin{aligned} &\underset{\tau \le t}{\mathcal{F}} \left[\boldsymbol{F}(\vec{X}, \tau), \theta(\vec{X}, \tau), \vec{G}_\theta(\vec{X}, \tau), \vec{X} \right] \\ &\quad = \boldsymbol{R}(\vec{X}, t) \cdot \underset{\tau \le t}{\mathcal{F}} \left[\boldsymbol{U}(\vec{X}, \tau), \theta(\vec{X}, \tau), \vec{G}_\theta(\vec{X}, \tau), \vec{X} \right] \cdot \boldsymbol{R}^T(\vec{X}, t). \end{aligned} \tag{6.22}$$

This reduced form has been obtained for a special choice of $\boldsymbol{O}(t)$ in the principle of material frame-indifference (6.20), which means that (6.22) is a necessary condition for satisfying the principle of material frame-indifference. We now prove that (6.22) is also a sufficient condition for satisfying this principle. Suppose that $\mathcal{F}$ is of the form (6.22) and consider an arbitrary orthogonal tensor history $\boldsymbol{O}(t)$. Since the polar decomposition of $\boldsymbol{O} \cdot \boldsymbol{F}$ is $(\boldsymbol{O} \cdot \boldsymbol{R}) \cdot \boldsymbol{U}$, (6.22) for $\boldsymbol{O} \cdot \boldsymbol{F}$ takes the form

$$\begin{aligned} &\underset{\tau \le t}{\mathcal{F}} \left[\boldsymbol{O}(\tau) \cdot \boldsymbol{F}(\vec{X}, \tau), \theta(\vec{X}, \tau), \vec{G}_\theta(\vec{X}, \tau), \vec{X} \right] \\ &= \boldsymbol{O}(t) \cdot \boldsymbol{R}(\vec{X}, t) \cdot \underset{\tau \le t}{\mathcal{F}} \left[\boldsymbol{U}(\vec{X}, \tau), \theta(\vec{X}, \tau), \vec{G}_\theta(\vec{X}, \tau), \vec{X} \right] \cdot \left[\boldsymbol{O}(t) \cdot \boldsymbol{R}(\vec{X}, t) \right]^T, \end{aligned}$$

which reduces to (6.20) in view of (6.22). Therefore, the reduced form (6.22) is necessary and sufficient to satisfy the principle of material frame-indifference.

We have proved that the constitutive equation for a simple material may be expressed in the form

$$\boldsymbol{t}(\vec{X},t) = \boldsymbol{R}(\vec{X},t)\cdot \underset{\tau\leq t}{\mathcal{F}}\left[\boldsymbol{U}(\vec{X},\tau),\theta(\vec{X},\tau),\vec{G}_\theta(\vec{X},\tau),\vec{X}\right]\cdot \boldsymbol{R}^T(\vec{X},t). \tag{6.23}$$

A constitutive equation of this kind, in which the functionals are not subject to any further restriction, is called a *reduced form*. The result (6.23) shows that while the stretch history of a simple material may affect its present stress, past rotations have no effect at all. The present rotation enters (6.23) explicitly.

There are many other reduced forms of the constitutive equation for a simple material. Replacing the stretch $\boldsymbol{U}$ by the Green deformation tensor $\boldsymbol{C}$, where $\boldsymbol{U} = \sqrt{\boldsymbol{C}}$, and denoting the functional $\mathcal{F}(\sqrt{\boldsymbol{C}},\cdots)$ as $\mathcal{F}_1(\boldsymbol{C},\cdots)$, we obtain

$$\boldsymbol{t}(\vec{X},t) = \boldsymbol{R}(\vec{X},t)\cdot \underset{\tau\leq t}{\mathcal{F}_1}\left[\boldsymbol{C}(\vec{X},\tau),\theta(\vec{X},\tau),\vec{G}_\theta(\vec{X},\tau),\vec{X}\right]\cdot \boldsymbol{R}^T(\vec{X},t). \tag{6.24}$$

Similarly, expressing the rotation $\boldsymbol{R}$ through the deformation gradient, $\boldsymbol{R} = \boldsymbol{F}\cdot\boldsymbol{U}^{-1}$ and replacing the stretch $\boldsymbol{U}$ by Green's deformation tensor $\boldsymbol{C}$, Eq. (6.24) can be put into another reduced form

$$\boldsymbol{t}(\vec{X},t) = \boldsymbol{F}(\vec{X},t)\cdot \underset{\tau\leq t}{\mathcal{F}_2}\left[\boldsymbol{C}(\vec{X},\tau),\theta(\vec{X},\tau),\vec{G}_\theta(\vec{X},\tau),\vec{X}\right]\cdot \boldsymbol{F}^T(\vec{X},t) \tag{6.25}$$

after introducing a new functional $\mathcal{F}_2$ of the deformation history $\boldsymbol{C}$. Note that $\mathcal{F}_1$ and $\mathcal{F}_2$ are materially frame-indifferent functionals.

The principle of material frame-indifference applied to the constitutive equations (6.6) for the heat flux and internal energy in an analogous manner results in

$$\begin{aligned} \vec{q}(\vec{X},t) &= \boldsymbol{R}(\vec{X},t)\cdot \underset{\tau\leq t}{\mathcal{Q}}\left[\boldsymbol{C}(\vec{X},\tau),\theta(\vec{X},\tau),\vec{G}_\theta(\vec{X},\tau),\vec{X}\right], \\ \varepsilon(\vec{X},t) &= \underset{\tau\leq t}{\mathcal{E}}\left[\boldsymbol{C}(\vec{X},\tau),\theta(\vec{X},\tau),\vec{G}_\theta(\vec{X},\tau),\vec{X}\right]. \end{aligned} \tag{6.26}$$

Another useful reduced form may be obtained if the second Piola–Kirchhoff stress tensor $\boldsymbol{T}^{(2)}$, defined by $(3.27)_1$ as $\boldsymbol{T}^{(2)} = (\det\boldsymbol{F})\,\boldsymbol{F}^{-1}\cdot\boldsymbol{t}\cdot\boldsymbol{F}^{-T}$, is used in the constitutive equation instead of the Cauchy stress tensor $\boldsymbol{t}$. Writing

$$J = \det\boldsymbol{F} = \det(\boldsymbol{R}\cdot\boldsymbol{U}) = \det\boldsymbol{U} = \sqrt{\det\boldsymbol{C}} \tag{6.27}$$

and defining a new functional $\mathcal{G}$ of the deformation history $\boldsymbol{C}$, the constitutive equation (6.25) can be rewritten for $\boldsymbol{T}^{(2)}$ as

$$\boldsymbol{T}^{(2)}(\vec{X},t) = \underset{\tau\leq t}{\mathcal{G}}\left[\boldsymbol{C}(\vec{X},\tau),\theta(\vec{X},\tau),\vec{G}_\theta(\vec{X},\tau),\vec{X}\right]. \tag{6.28}$$

According to this result, the tensor $\boldsymbol{T}^{(2)}$ depends only on the Green deformation tensor and not on the rotation. Similarly, the constitutive functional for the heat flux vector in the reference configuration, defined by $(4.74)_2$ as $\vec{Q} = J\boldsymbol{F}^{-1} \cdot \vec{q}$, has the form

$$\vec{Q}(\vec{X}, t) = \underset{\tau \le t}{\mathcal{Q}_1} \left[\boldsymbol{C}(\vec{X}, \tau), \theta(\vec{X}, \tau), \vec{G}_\theta(\vec{X}, \tau), \vec{X} \right] \tag{6.29}$$

after introducing a new functional $\mathcal{Q}_1$ of the deformation history $\boldsymbol{C}$.

6.6 Kinematic Constraints

A condition of *kinematic* or *internal constraint* is a geometric restriction on the set of all possible motions of a material body. A condition of constraint can be defined as a restriction on the deformation gradients such that

$$\lambda(\boldsymbol{F}(t)) = 0, \tag{6.30}$$

where we drop the position $\vec{X}$ to shorten the notation in this section, e.g. $\boldsymbol{F}(t) \equiv \boldsymbol{F}(\vec{X}, t)$. The requirement that kinematic constraints be frame-indifferent implies that the constraint condition (6.30) has to be replaced by

$$\lambda(\boldsymbol{C}(t)) = 0. \tag{6.31}$$

Within the context of kinematic constraints, the principle of determinism must be modified; it now postulates that only a part of the stress tensor is related to the history of the deformation. Thus, the stress tensor (6.19) obtains an additional term

$$\boldsymbol{t}(t) = \boldsymbol{\pi}(t) + \underset{\tau \le t}{\mathcal{F}} \left[\boldsymbol{F}(\tau), \theta(\tau), \vec{G}_\theta(\tau) \right]. \tag{6.32}$$

The tensor $\boldsymbol{t} - \boldsymbol{\pi}$ is called the *determinate stress* because it is uniquely determined by the motion. The tensor $\boldsymbol{\pi}$ is called the *indeterminate stress* and represents the *reaction stress* produced by the kinematic constraint (6.31), since it is not determined by the motion. By analogy with analytical mechanics, it is assumed that the reaction stress does not perform any work, i.e., the stress power vanishes for all motions compatible with the constraint condition (6.31) such that

$$\boldsymbol{\pi}(t) : \boldsymbol{d}(t) = 0, \tag{6.33}$$

where $\boldsymbol{d}$ is the strain-rate tensor defined by $(2.20)_1$.

To derive the Lagrangian form of (6.33), we first modify the constitutive equation (6.28) for $\boldsymbol{T}^{(2)}$ as

$$\boldsymbol{T}^{(2)}(t) = \boldsymbol{\Pi}(t) + \underset{\tau \le t}{\mathcal{G}} \Big[\boldsymbol{C}(\tau), \theta(\tau), \vec{G}_\theta(\tau)\Big]. \tag{6.34}$$

Taking into account (2.27) and $(3.27)_2$, the double-dot product $\boldsymbol{\pi} : \boldsymbol{d}$ can be expressed in terms of the Lagrangian reaction stress $\boldsymbol{\Pi}$ as

$$\boldsymbol{\pi} : \boldsymbol{d} = \frac{1}{J}(\boldsymbol{F} \cdot \boldsymbol{\Pi} \cdot \boldsymbol{F}^T) : \frac{1}{2}(\boldsymbol{F}^{-T} \cdot \dot{\boldsymbol{C}} \cdot \boldsymbol{F}^{-1}) = \frac{1}{2J}(\boldsymbol{\Pi} : \dot{\boldsymbol{C}}),$$

where $(4.76)_2$ has been applied in the last equality. Hence, the Lagrangian form of the constraint (6.33) is

$$\boldsymbol{\Pi}(t) : \dot{\boldsymbol{C}}(t) = 0, \tag{6.35}$$

which means that the reaction stress $\boldsymbol{\Pi}$ has no power to work for all motions compatible with the constraint (6.31). This compatibility can be evaluated in a more specific form. Differentiating (6.31) with respect to time gives

$$\frac{d\lambda(\boldsymbol{C}(t))}{dC_{KL}} \dot{C}_{KL} = 0, \qquad \text{or} \qquad \frac{d\lambda(\boldsymbol{C}(t))}{d\boldsymbol{C}} : \dot{\boldsymbol{C}} = 0, \tag{6.36}$$

which shows that the normal $d\lambda/d\boldsymbol{C}$ to the surface $\lambda(\boldsymbol{C}) = \text{const.}$ is orthogonal to all strain rates $\dot{\boldsymbol{C}}$ that are allowed by the constraint (6.31). Equation (6.35) suggests that the same orthogonality holds for the reaction stress $\boldsymbol{\Pi}$, whence it follows that $\boldsymbol{\Pi}$ must be parallel to $d\lambda/d\boldsymbol{C}$,

$$\boldsymbol{\Pi}(t) = \alpha(t) \frac{d\lambda(\boldsymbol{C}(t))}{d\boldsymbol{C}}. \tag{6.37}$$

Here, the factor α is left undetermined and must be regarded as an independent field variable in the principle of conservation of linear momentum.

The same arguments apply if several constraints are specified simultaneously. For

$$\lambda_i(\boldsymbol{C}(t)) = 0, \quad i = 1, 2, \ldots, \tag{6.38}$$

we have

$$\boldsymbol{T}^{(2)}(t) = \sum_i \boldsymbol{\Pi}_i(t) + \underset{\tau \le t}{\mathcal{G}} [\boldsymbol{C}(\tau), \theta(\tau), \text{Grad}\,\theta(\tau)] \tag{6.39}$$

with

$$\boldsymbol{\Pi}_i(t) = \alpha_i(t) \frac{d\lambda_i(\boldsymbol{C}(t))}{d\boldsymbol{C}}. \tag{6.40}$$

As an example, we consider the *volume-preserving* or *isochoric motion* for which $J = 1$. Since $J = \varrho_0/\varrho$ by (4.63), the volume-preserving motion is identical to the *density-preserving motion* for which the mass density of a particle remains unchanged during motion. The volume- or density-preserving constraints are traditionally regarded as the *incompressibility constraint*. However, incompressibility may also mean that the equation of state for mass density, that is, the equation expressing mass density as a function of temperature θ and thermodynamic pressure p, $\varrho = \hat{\varrho}(\theta, p)$, is independent of pressure, that is, $\varrho = \hat{\varrho}(\theta)$. In view of (1.45) and (1.57), the *volume-preserving* constraint becomes

$$J = \sqrt{\det \boldsymbol{C}} = 1, \tag{6.41}$$

which places a restriction on $\boldsymbol{C}$, namely, that the components of $\boldsymbol{C}$ are not all independent. In this special case, the constraint (6.31) becomes

$$\lambda(\boldsymbol{C}) = \det \boldsymbol{C} - 1. \tag{6.42}$$

Using Jacobi's identity $(1.46)_1$, i.e.,

$$\frac{d}{d\boldsymbol{A}}(\det \boldsymbol{A}) = (\det \boldsymbol{A})\boldsymbol{A}^{-T}, \tag{6.43}$$

valid for any invertible second-order tensor $\boldsymbol{A}$, we obtain

$$\frac{d}{d\boldsymbol{C}}(\det \boldsymbol{C} - 1) = (\det \boldsymbol{C})\boldsymbol{C}^{-1} = \boldsymbol{C}^{-1}. \tag{6.44}$$

Equation (6.37) then gives

$$\boldsymbol{\Pi}(t) = \alpha(t)\, \boldsymbol{C}^{-1}(t). \tag{6.45}$$

Writing $-\pi$ in place of α, (6.34) becomes

$$\boldsymbol{T}^{(2)}(t) = -\pi(t)\boldsymbol{C}^{-1}(t) + \underset{\tau \le t}{\mathcal{G}} \left[\boldsymbol{C}(\tau), \theta(\tau), \vec{G}_\theta(\tau)\right], \tag{6.46}$$

where π is called the *constraint pressure*. Correspondingly, the Cauchy stress tensor is expressed as

$$\boldsymbol{t}(t) = -\pi(t)\boldsymbol{I} + \boldsymbol{R}(t) \cdot \underset{\tau \le t}{\mathcal{F}_1} \left[\boldsymbol{C}(\tau), \theta(\tau), \vec{G}_\theta(\tau)\right] \cdot \boldsymbol{R}^T(t), \tag{6.47}$$

which shows that the stress for an incompressible simple material is determined by the constitutive equation only to within the constraint pressure π.

6.7 Material Symmetry

In this section, we will confine ourselves to a homogeneous material body. We say that a material body is *homogeneous* if there is at least one reference configuration in which a constitutive functional has the same form for all particles of the body, i.e., the explicit dependence of the constitutive functional on the position $\vec{X}$ disappears. Such a reference configuration is termed a *homogeneous configuration*. For instance, (6.19) for a homogeneous body reduces to

$$\boldsymbol{t}(\vec{X}, t) = \underset{\tau \leq t}{\mathcal{F}_R} \left[\boldsymbol{F}(\vec{X}, \tau), \theta(\vec{X}, \tau), \vec{G}_\theta(\vec{X}, \tau) \right]. \tag{6.48}$$

Note that the functional $\mathcal{F}$ is now labelled by the subscript R since $\mathcal{F}$ depends on the choice of reference configuration.

Material symmetry, if it exists, can be characterised by invariance properties of the constitutive equations with respect to a change of reference configuration. To make this statement more precise and to exploit its consequences, we consider two different reference configurations, denoted by R and $\hat{R}$ (see Fig. 6.1), that are related by the one-to-one mapping

$$\vec{\hat{X}} = \vec{\Lambda}(\vec{X}) \iff \vec{X} = \vec{\Lambda}^{-1}(\vec{\hat{X}}), \tag{6.49}$$

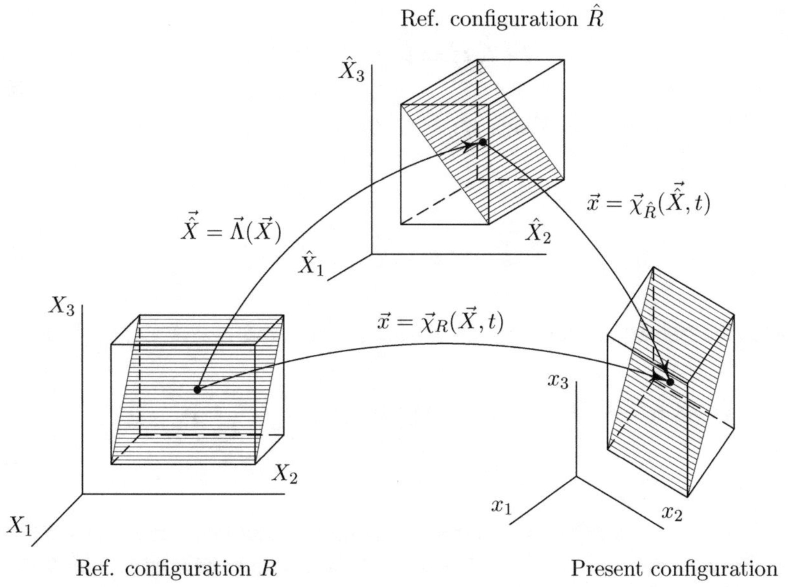

Fig. 6.1 Motion of a body relative to two reference configurations

where $\vec{X}$ and $\vec{\hat{X}}$ are the positions of the particle $\mathcal{X}$ in the reference configurations R and $\hat{R}$, respectively, $\vec{X} = X_K \vec{I}_K$ and $\vec{\hat{X}} = \hat{X}_{\hat{K}} \vec{\hat{I}}_{\hat{K}}$. Then, the motion of the particle in the present configuration $\vec{x} = \vec{\chi}_R(\vec{X}, t)$ takes the form $\vec{x} = \vec{\chi}_{\hat{R}}(\vec{\hat{X}}, t)$, giving the identity

$$\vec{\chi}_R(\vec{X}, t) = \vec{\chi}_{\hat{R}}(\vec{\Lambda}(\vec{X}), t), \tag{6.50}$$

which holds for all $\vec{X}$ and t. Differentiating the motion function $\vec{\chi}_R(\vec{X}, t)$ and temperature $\theta(\vec{X}, t)$ with respect to $\vec{X}$ and using the chain rule enabled by (6.50) leads to the transformation rules for the deformation and temperature gradients,

$$F_{kK} = \frac{\partial(\vec{\chi}_R)_k}{\partial X_K} = \frac{\partial(\vec{\chi}_{\hat{R}})_k}{\partial \Lambda_{\hat{L}}} \frac{\partial \Lambda_{\hat{L}}}{\partial X_K} = \hat{F}_{k\hat{L}} P_{\hat{L}K},$$

$$(\vec{G}_\theta)_K = \frac{\partial \theta}{\partial X_K} = \frac{\partial \hat{\theta}}{\partial \Lambda_{\hat{L}}} \frac{\partial \Lambda_{\hat{L}}}{\partial X_K} = (\vec{\hat{G}}_\theta)_{\hat{L}} P_{\hat{L}K},$$

where

$$\hat{\theta}(\vec{\hat{X}}, t) := \theta(\vec{X}(\vec{\hat{X}}), t). \tag{6.51}$$

In tensor notation, we have

$$\begin{aligned} \boldsymbol{F}(\vec{X}, t) &= \hat{\boldsymbol{F}}(\vec{\hat{X}}, t) \cdot \boldsymbol{P}(\vec{X}), \\ \vec{G}_\theta(\vec{X}, t) &= \vec{\hat{G}}_{\hat{\theta}}(\vec{\hat{X}}, t) \cdot \boldsymbol{P}(\vec{X}), \end{aligned} \tag{6.52}$$

where

$$\hat{\boldsymbol{F}}(\vec{\hat{X}}, t) := (\text{Grad}\, \vec{\chi}_{\hat{R}}\, (\vec{\hat{X}}, t))^T \tag{6.53}$$

denotes the transposed deformation gradient of the motion $\vec{x} = \vec{\chi}_{\hat{R}}(\vec{\hat{X}}, t)$, and

$$\boldsymbol{P}(\vec{X}) := (\text{Grad}\, \vec{\Lambda}(\vec{X}))^T = P_{\hat{K}L}(\vec{X})(\vec{\hat{I}}_{\hat{K}} \otimes \vec{I}_L), \qquad P_{\hat{K}L}(\vec{X}) = \frac{\partial \Lambda_{\hat{K}}}{\partial X_L}, \tag{6.54}$$

is the transposed gradient of the transformation $\vec{\Lambda}$ that maps the configuration R onto the configuration $\hat{R}$. The deformation and temperature gradients therefore depend on the choice of reference configuration. Similarly, the Green deformation tensor $\boldsymbol{C}$, where $\boldsymbol{C} = \boldsymbol{F}^T \cdot \boldsymbol{F}$, transforms according to the rule

$$\boldsymbol{C}(\vec{X}, t) = \boldsymbol{P}^T(\vec{X}) \cdot \hat{\boldsymbol{C}}(\vec{\hat{X}}, t) \cdot \boldsymbol{P}(\vec{X}). \tag{6.55}$$

The constitutive equation (6.48) for a homogeneous body in the reference configuration $\hat{R}$ is

$$\boldsymbol{t}(\vec{\hat{X}}, t) = \underset{\tau \leq t}{\mathcal{F}_{\hat{R}}} \left[\hat{\boldsymbol{F}}(\vec{\hat{X}}, \tau), \hat{\theta}(\vec{\hat{X}}, \tau), \vec{G}_{\hat{\theta}}(\vec{\hat{X}}, \tau) \right]. \tag{6.56}$$

Assuming that two different deformation and temperature histories are related by the same mapping as the associated reference configurations, the functionals $\mathcal{F}_R$ and $\mathcal{F}_{\hat{R}}$ are related by the identity

$$\underset{\tau \leq t}{\mathcal{F}_R} \left[\boldsymbol{F}(\vec{X}, \tau), \theta(\vec{X}, \tau), \vec{G}_\theta(\vec{X}, \tau) \right] = \underset{\tau \leq t}{\mathcal{F}_{\hat{R}}} \left[\hat{\boldsymbol{F}}(\vec{\hat{X}}, \tau), \hat{\theta}(\vec{\hat{X}}, \tau), \vec{G}_{\hat{\theta}}(\vec{\hat{X}}, \tau) \right], \tag{6.57}$$

where (6.52) is implicitly considered. Thus, the arguments in configuration R can be expressed in terms of those in configuration $\hat{R}$ such that

$$\begin{aligned} &\underset{\tau \leq t}{\mathcal{F}_R} \left[\hat{\boldsymbol{F}}(\vec{\hat{X}}, \tau) \cdot \boldsymbol{P}(\vec{X}), \hat{\theta}(\vec{\hat{X}}, \tau), \vec{G}_{\hat{\theta}}(\vec{\hat{X}}, \tau) \cdot \boldsymbol{P}(\vec{X}) \right] \\ &\quad = \underset{\tau \leq t}{\mathcal{F}_{\hat{R}}} \left[\hat{\boldsymbol{F}}(\vec{\hat{X}}, \tau), \hat{\theta}(\vec{\hat{X}}, \tau), \vec{G}_{\hat{\theta}}(\vec{\hat{X}}, \tau) \right]. \end{aligned} \tag{6.58}$$

Since (6.58) holds for any deformation and temperature history, we obtain the identity

$$\begin{aligned} &\underset{\tau \leq t}{\mathcal{F}_R} \left[\boldsymbol{F}(\vec{X}, \tau) \cdot \boldsymbol{P}(\vec{X}), \theta(\vec{X}, \tau), \vec{G}_\theta(\vec{X}, \tau) \cdot \boldsymbol{P}(\vec{X}) \right] \\ &\quad = \underset{\tau \leq t}{\mathcal{F}_{\hat{R}}} \left[\boldsymbol{F}(\vec{X}, \tau), \theta(\vec{X}, \tau), \vec{G}_\theta(\vec{X}, \tau) \right], \end{aligned} \tag{6.59}$$

which can alternatively be arranged by expressing the arguments in configuration $\hat{R}$ in terms of those in configuration R,

$$\begin{aligned} &\underset{\tau \leq t}{\mathcal{F}_R} \left[\boldsymbol{F}(\vec{X}, \tau), \theta(\vec{X}, \tau), \vec{G}_\theta(\vec{X}, \tau) \right] \\ &\quad = \underset{\tau \leq t}{\mathcal{F}_{\hat{R}}} \left[\boldsymbol{F}(\vec{X}, \tau) \cdot \boldsymbol{P}^{-1}(\vec{X}), \theta(\vec{X}, \tau), \vec{G}_\theta(\vec{X}, \tau) \cdot \boldsymbol{P}^{-1}(\vec{X}) \right]. \end{aligned} \tag{6.60}$$

To introduce the concept of material symmetry, consider a particle that was in configuration R at time $t = -\infty$, and subsequently experienced certain histories of the deformation gradient $\boldsymbol{F}$, the temperature θ and the temperature gradient $\vec{G}_\theta$. At the present time t, the resulting stress is given by (6.48),

$$\boldsymbol{t}(\vec{X}, t) = \underset{\tau \leq t}{\mathcal{F}_R} \left[\boldsymbol{F}(\vec{X}, \tau), \theta(\vec{X}, \tau), \vec{G}_\theta(\vec{X}, \tau) \right]. \tag{6.61}$$

Let the same particle, but in the configuration $\hat{R}$, experience the same histories of deformation gradient, temperature and temperature gradient. The resulting stress at the present time t is

$$\hat{\boldsymbol{t}}(\vec{X}, t) = \underset{\tau \le t}{\mathcal{F}_{\hat{R}}} \left[\boldsymbol{F}(\vec{X}, \tau), \theta(\vec{X}, \tau), \vec{G}_\theta(\vec{X}, \tau) \right]. \tag{6.62}$$

Since $\mathcal{F}_R$ and $\mathcal{F}_{\hat{R}}$ are not, in general, the same functionals, it follows that $\boldsymbol{t} \neq \hat{\boldsymbol{t}}$. However, it may happen that these values coincide, which expresses a certain symmetry of material. The condition for this case is

$$\underset{\tau \le t}{\mathcal{F}_R} \left[\boldsymbol{F}(\vec{X}, \tau), \theta(\vec{X}, \tau), \vec{G}_\theta(\vec{X}, \tau) \right] = \underset{\tau \le t}{\mathcal{F}_{\hat{R}}} \left[\boldsymbol{F}(\vec{X}, \tau), \theta(\vec{X}, \tau), \vec{G}_\theta(\vec{X}, \tau) \right]. \tag{6.63}$$

Expressing the right-hand side according to the identity (6.59) gives

$$\begin{aligned} &\underset{\tau \le t}{\mathcal{F}_R} \left[\boldsymbol{F}(\vec{X}, \tau), \theta(\vec{X}, \tau), \vec{G}_\theta(\vec{X}, \tau) \right] \\ &\qquad = \underset{\tau \le t}{\mathcal{F}_R} \left[\boldsymbol{F}(\vec{X}, \tau) \cdot \boldsymbol{P}(\vec{X}), \theta(\vec{X}, \tau), \vec{G}_\theta(\vec{X}, \tau) \cdot \boldsymbol{P}(\vec{X}) \right]. \end{aligned} \tag{6.64}$$

If this relation holds for all deformation gradient, temperature and temperature gradient histories, we say that the material at particle $\mathcal{X}$ is *symmetric* with respect to the transformation $\boldsymbol{P} : R \to \hat{R}$.

Any such $\boldsymbol{P}$ corresponds to a static local deformation (at a given particle $\mathcal{X}$) from R to another reference configuration $\hat{R}$ such that any deformation and temperature history leads to the same stress in both R and $\hat{R}$ at $\mathcal{X}$. Hence, $\boldsymbol{P}$ corresponds to a change of reference configuration that cannot be detected by any experiment at a given particle $\mathcal{X}$. The tensor $\boldsymbol{P}$, for which (6.64) applies, is called the *symmetry* transformation.

To find the property of $\boldsymbol{P}$, we replace $\boldsymbol{F}$ in (6.64) with $\boldsymbol{F} \cdot \boldsymbol{P}$ and $\vec{G}_\theta(\tau)$ with $\vec{G}_\theta(\tau) \cdot \boldsymbol{P}$ such that

$$\underset{\tau \le t}{\mathcal{F}_R} \left[\boldsymbol{F}(\tau) \cdot \boldsymbol{P} \cdot \boldsymbol{P}, \theta(\tau), \vec{G}_\theta(\tau) \cdot \boldsymbol{P} \cdot \boldsymbol{P} \right] = \underset{\tau \le t}{\mathcal{F}_R} \left[\boldsymbol{F}(\tau) \cdot \boldsymbol{P}, \theta(\tau), \vec{G}_\theta(\tau) \cdot \boldsymbol{P} \right],$$

where the position $\vec{X}$ has been dropped from the notation. The right-hand side can be arranged by (6.64) to give

$$\underset{\tau \le t}{\mathcal{F}_R} \left[\boldsymbol{F}(\tau) \cdot \boldsymbol{P} \cdot \boldsymbol{P}, \theta(\tau), \vec{G}_\theta(\tau) \cdot \boldsymbol{P} \cdot \boldsymbol{P} \right] = \underset{\tau \le t}{\mathcal{F}_R} \left[\boldsymbol{F}(\tau), \theta(\tau), \vec{G}_\theta(\tau) \right].$$

Continuing recursively n times leads to

$$\underset{\tau\le t}{\mathcal{F}_R}\left[\boldsymbol{F}(\tau)\cdot\boldsymbol{P}^n,\theta(\tau),\vec{G}_\theta(\tau)\cdot\boldsymbol{P}^n\right]=\underset{\tau\le t}{\mathcal{F}_R}\left[\boldsymbol{F}(\tau),\theta(\tau),\vec{G}_\theta(\tau)\right].$$

If $|\det \boldsymbol{P}| < 1$, then $\lim_{n\to\infty}\det(\boldsymbol{F}\cdot\boldsymbol{P}^n) = 0$. In the opposite case where $|\det \boldsymbol{P}| > 1$, $\lim_{n\to\infty}\det(\boldsymbol{F}\cdot\boldsymbol{P}^n) = \infty$. In both cases, the regularity of the tensor $\boldsymbol{F}$ is violated. Hence, it must hold that

$$\det \boldsymbol{P} = \pm 1. \tag{6.65}$$

This *volume-preserving transformation* is described as being *unimodular*.

The properties of material symmetry can be represented in terms of symmetry groups. The set of all unimodular transformations that leave the constitutive equation invariant with respect to R, that is, for which (6.64) holds, forms a group.[3] This group is called the *symmetry group* $g_R(\mathcal{F})$ of a material characterised by the constitutive functional $\mathcal{F}$ at particle $\mathcal{X}$ whose position in the reference configuration R is $\vec{X}$,

$$\begin{aligned}g_R(\mathcal{F}):=\Big\{&\boldsymbol{H}(\vec{X})|\quad \det\boldsymbol{H}(\vec{X})=\pm1\quad\text{and}\\ &\underset{\tau\le t}{\mathcal{F}_R}\left[\boldsymbol{F}(\vec{X},\tau)\cdot\boldsymbol{H}(\vec{X}),\theta(\vec{X},\tau),\vec{G}_\theta(\vec{X},\tau)\cdot\boldsymbol{H}(\vec{X})\right]\\ &=\underset{\tau\le t}{\mathcal{F}_R}\left[\boldsymbol{F}(\vec{X},\tau),\theta(\vec{X},\tau),\vec{G}_\theta(\vec{X},\tau)\right]\text{ for }\forall\boldsymbol{F}(\vec{X},\tau),\ \forall\theta(\vec{X},\tau),\ \forall\vec{G}_\theta(\vec{X},\tau)\Big\}.\end{aligned} \tag{6.66}$$

The operation defined on $g_R(\mathcal{F})$ is the scalar product of tensors, and the identity element is the identity tensor.

To show that $g_R(\mathcal{F})$ is a group, let $\boldsymbol{H}_1$ and $\boldsymbol{H}_2$ be two elements of $g_R(\mathcal{F})$. Then, we can replace $\boldsymbol{F}$ in (6.66) with $\boldsymbol{F}\cdot\boldsymbol{H}_1$, $\vec{G}_\theta$ with $\vec{G}_\theta\cdot\boldsymbol{H}_1$, and take $\boldsymbol{H}=\boldsymbol{H}_2$ to give

$$\begin{aligned}&\underset{\tau\le t}{\mathcal{F}_R}\left[\boldsymbol{F}(\tau)\cdot\boldsymbol{H}_1\cdot\boldsymbol{H}_2,\theta(\tau),\vec{G}_\theta(\tau)\cdot\boldsymbol{H}_1\cdot\boldsymbol{H}_2\right]\\ &\quad=\underset{\tau\le t}{\mathcal{F}_R}\left[\boldsymbol{F}(\tau)\cdot\boldsymbol{H}_1,\theta(\tau),\vec{G}_\theta(\tau)\cdot\boldsymbol{H}_1\right],\end{aligned}$$

where the position $\vec{X}$ has been dropped $\vec{X}$ from the notation. Since $\boldsymbol{H}_1\in g_R(\mathcal{F})$,

$$\underset{\tau\le t}{\mathcal{F}_R}\left[\boldsymbol{F}(\tau)\cdot\boldsymbol{H}_1\cdot\boldsymbol{H}_2,\theta(\tau),\vec{G}_\theta(\tau)\cdot\boldsymbol{H}_1\cdot\boldsymbol{H}_2\right]=\underset{\tau\le t}{\mathcal{F}_R}\left[\boldsymbol{F}(\tau),\theta(\tau),\vec{G}_\theta(\tau)\right],$$

[3] A *group* is a set of abstract elements $\boldsymbol{A}, \boldsymbol{B}, \boldsymbol{C}, \cdots$, with a defined operation $\boldsymbol{A}\cdot\boldsymbol{B}$ (as, for instance, '*multiplication*') such that: (1) the product $\boldsymbol{A}\cdot\boldsymbol{B}$ is defined for all elements $\boldsymbol{A}$ and $\boldsymbol{B}$ of the set, (2) this product $\boldsymbol{A}\cdot\boldsymbol{B}$ is itself an element of the set for all $\boldsymbol{A}$ and $\boldsymbol{B}$ (the set is *closed* under the operation), (3) the set contains an identity element $\boldsymbol{I}$ such that $\boldsymbol{I}\cdot\boldsymbol{A}=\boldsymbol{A}=\boldsymbol{A}\cdot\boldsymbol{I}$ for all $\boldsymbol{A}$, (4) every element has an inverse $\boldsymbol{A}^{-1}$ in the set such that $\boldsymbol{A}\cdot\boldsymbol{A}^{-1}=\boldsymbol{A}^{-1}\cdot\boldsymbol{A}=\boldsymbol{I}$.

which shows that the scalar product $\boldsymbol{H}_1 \cdot \boldsymbol{H}_2 \in g_R(\mathcal{F})$. Furthermore, the identity tensor $\boldsymbol{H} = \boldsymbol{I}$ clearly satisfies (6.66). Finally, if $\boldsymbol{H} \in g_R(\mathcal{F})$ and $\boldsymbol{F}$ is invertible, then $\boldsymbol{F} \cdot \boldsymbol{H}^{-1}$ is an invertible tensor. Hence, (6.66) must hold with $\boldsymbol{F}$ replaced by $\boldsymbol{F} \cdot \boldsymbol{H}^{-1}$ (and $\vec{G}_\theta$ with $\vec{G}_\theta \cdot \boldsymbol{H}^{-1}$),

$$\begin{aligned} &\mathop{\mathcal{F}_R}_{\tau \le t} \left[\boldsymbol{F}(\tau) \cdot \boldsymbol{H}^{-1} \cdot \boldsymbol{H}, \theta(\tau), \vec{G}_\theta(\tau) \cdot \boldsymbol{H}^{-1} \cdot \boldsymbol{H}\right] \\ &\qquad = \mathop{\mathcal{F}_R}_{\tau \le t} \left[\boldsymbol{F}(\tau) \cdot \boldsymbol{H}^{-1}, \theta(\tau), \vec{G}_\theta(\tau) \cdot \boldsymbol{H}^{-1}\right]. \end{aligned}$$

Therefore,

$$\mathop{\mathcal{F}_R}_{\tau \le t} \left[\boldsymbol{F}(\tau), \theta(\tau), \vec{G}_\theta(\tau)\right] = \mathop{\mathcal{F}_R}_{\tau \le t} \left[\boldsymbol{F}(\tau) \cdot \boldsymbol{H}^{-1}, \theta(\tau), \vec{G}_\theta(\tau) \cdot \boldsymbol{H}^{-1}\right], \tag{6.67}$$

which shows that $\boldsymbol{H}^{-1} \in g_R(\mathcal{F})$, thus proving that $g_R(\mathcal{F})$ has the structure of a group. Note that (6.67) is an equivalent condition for material symmetry.

6.8 Material Symmetry of Reduced-Form Constitutive Functionals

We now find a symmetry criterion and define a symmetry group for functionals whose forms are reduced by applying the polar decomposition of the deformation gradient. As an example, let us consider the constitutive equation (6.25) for the Cauchy stress tensor. For a homogeneous body, this constitutive equation in the reference configurations R and $\hat{R}$ has the form

$$\begin{aligned} \boldsymbol{t}(\vec{X}, t) &= \boldsymbol{F}(\vec{X}, t) \cdot \mathop{\mathcal{F}_{2,R}}_{\tau \le t} \left[\boldsymbol{C}(\vec{X}, \tau), \theta(\vec{X}, \tau), \vec{G}_\theta(\vec{X}, \tau)\right] \cdot \boldsymbol{F}^T(\vec{X}, t), \\ \boldsymbol{t}(\vec{\hat{X}}, t) &= \hat{\boldsymbol{F}}(\vec{\hat{X}}, t) \cdot \mathop{\mathcal{F}_{2,\hat{R}}}_{\tau \le t} \left[\hat{\boldsymbol{C}}(\vec{\hat{X}}, \tau), \hat{\theta}(\vec{\hat{X}}, \tau), \vec{\hat{G}}_{\hat{\theta}}(\vec{\hat{X}}, \tau)\right] \cdot \hat{\boldsymbol{F}}^T(\vec{\hat{X}}, t), \end{aligned} \tag{6.68}$$

where the subscripts R and $\hat{R}$ refer to the underlying reference configuration. Using the transformation relations (6.51) and (6.52) for the temperature, deformation and temperature gradients, and (6.55) for the Green deformation tensor, the functional $\mathcal{F}_{2,R}$ satisfies an identity analogous to (6.60):

$$\begin{aligned} &\boldsymbol{P}(\vec{X}) \cdot \mathop{\mathcal{F}_{2,R}}_{\tau \le t} \left[\boldsymbol{C}(\vec{X}, \tau), \theta(\vec{X}, \tau), \vec{G}_\theta(\vec{X}, \tau)\right] \cdot \boldsymbol{P}^T(\vec{X}) \\ &= \mathop{\mathcal{F}_{2,\hat{R}}}_{\tau \le t} \left[\boldsymbol{P}^{-T}(\vec{X}) \cdot \boldsymbol{C}(\vec{X}, \tau) \cdot \boldsymbol{P}^{-1}(\vec{X}), \theta(\vec{X}, \tau), \vec{G}_\theta(\vec{X}, \tau) \cdot \boldsymbol{P}^{-1}(\vec{X})\right]. \end{aligned} \tag{6.69}$$

Since $\mathcal{F}_{2,R}$ and $\mathcal{F}_{2,\hat{R}}$ are not, in general, the same functionals, the functional $\mathcal{F}_{2,\hat{R}}$ on the right cannot be replaced by the functional $\mathcal{F}_{2,R}$. However, it may happen that $\mathcal{F}_{2,R} = \mathcal{F}_{2,\hat{R}}$, which expresses a certain symmetry of the material. The condition for this case is

$$\begin{aligned} &\boldsymbol{P}(\vec{X}) \cdot \underset{\tau \le t}{\mathcal{F}_{2,R}} \left[\boldsymbol{C}(\vec{X},\tau), \theta(\vec{X},\tau), \vec{G}_\theta(\vec{X},\tau) \right] \cdot \boldsymbol{P}^T(\vec{X}) \\ &= \underset{\tau \le t}{\mathcal{F}_{2,R}} \left[\boldsymbol{P}^{-T}(\vec{X}) \cdot \boldsymbol{C}(\vec{X},\tau) \cdot \boldsymbol{P}^{-1}(\vec{X}), \theta(\vec{X},\tau), \vec{G}_\theta(\vec{X},\tau) \cdot \boldsymbol{P}^{-1}(\vec{X}) \right]. \end{aligned} \tag{6.70}$$

We say that the material at particle $\mathcal{X}$ (whose position in the reference configuration is $\vec{X}$) is *symmetric with respect to the transformation* $\boldsymbol{P} : R \to \hat{R}$ if (6.70) is valid for all deformation and temperature histories. The symmetry group $g_R(\mathcal{F}_2)$ of a material characterised by the constitutive functional $\mathcal{F}_2$ is defined by

$$\begin{aligned} g_R(\mathcal{F}_2) := \Big\{ &\boldsymbol{H}(\vec{X}) | \quad \det \boldsymbol{H}(\vec{X}) = \pm 1 \quad \text{and} \\ &\boldsymbol{H}(\vec{X}) \cdot \underset{\tau \le t}{\mathcal{F}_{2,R}} \left[\boldsymbol{C}(\vec{X},\tau), \theta(\vec{X},\tau), \vec{G}_\theta(\vec{X},\tau) \right] \cdot \boldsymbol{H}^T(\vec{X}) \\ &= \underset{\tau \le t}{\mathcal{F}_{2,R}} \left[\boldsymbol{H}^{-T}(\vec{X}) \cdot \boldsymbol{C}(\vec{X},\tau) \cdot \boldsymbol{H}^{-1}(\vec{X}), \theta(\vec{X},\tau), \vec{G}_\theta(\vec{X},\tau) \cdot \boldsymbol{H}^{-1}(\vec{X}) \right] \\ &\text{for } \forall \boldsymbol{C}(\vec{X},\tau),\ \forall \theta(\vec{X},\tau),\ \forall \vec{G}_\theta(\vec{X},\tau) \Big\}, \end{aligned} \tag{6.71}$$

In an analogous manner, it can be shown that the condition of material symmetry for the materially frame-indifferent constitutive functional $\mathcal{G}_R$ in the constitutive equation (6.28) is similar to that of $\mathcal{F}_2$:

$$\begin{aligned} &\boldsymbol{H}(\vec{X}) \cdot \underset{\tau \le t}{\mathcal{G}_R} \left[\boldsymbol{C}(\vec{X},\tau), \theta(\vec{X},\tau), \vec{G}_\theta(\vec{X},\tau) \right] \cdot \boldsymbol{H}^T(\vec{X}) \\ &= \underset{\tau \le t}{\mathcal{G}_R} \left[\boldsymbol{H}^{-T}(\vec{X}) \cdot \boldsymbol{C}(\vec{X},\tau) \cdot \boldsymbol{H}^{-1}(\vec{X}), \theta(\vec{X},\tau), \vec{G}_\theta(\vec{X},\tau) \cdot \boldsymbol{H}^{-1}(\vec{X}) \right]. \end{aligned} \tag{6.72}$$

6.9 Noll's Rule

The symmetry group $g_R(\mathcal{F})$ (or $g_R(\mathcal{F}_2)$) generally depends upon the choice of reference configuration since the symmetries a body with respect to one configuration generally differ from those the body has with respect to another. However, the symmetry groups $g_R(\mathcal{F})$ and $g_{\hat{R}}(\mathcal{F})$ (for the same particle) relative to two different reference configurations R and $\hat{R}$ are related by Noll's rule:

$$g_{\hat{R}}(\mathcal{F}) = \boldsymbol{P} \cdot g_R(\mathcal{F}) \cdot \boldsymbol{P}^{-1}, \tag{6.73}$$

where $\boldsymbol{P}$ is the fixed local deformation tensor for the deformation from R to $\hat{R}$ given by (6.54). Noll's rule means that every tensor $\hat{\boldsymbol{H}} \in g_{\hat{R}}(\mathcal{F})$ must be of the form $\hat{\boldsymbol{H}} = \boldsymbol{P} \cdot \boldsymbol{H} \cdot \boldsymbol{P}^{-1}$ for some tensor $\boldsymbol{H} \in g_R(\mathcal{F})$, and conversely every tensor $\boldsymbol{H} \in g_R(\mathcal{F})$ is of the form $\boldsymbol{H} = \boldsymbol{P}^{-1} \cdot \hat{\boldsymbol{H}} \cdot \boldsymbol{P}$ for some $\hat{\boldsymbol{H}} \in g_{\hat{R}}(\mathcal{F})$. Note that $\boldsymbol{P}$ need not be unimodular because $\boldsymbol{P}$ represents a change of reference configuration and is not necessarily a member of the symmetry group. Noll's rule shows that if $g_R(\mathcal{F})$ is known for one configuration, it is known for all. That is, the symmetries of a material in any one configuration determine its symmetries in every other configuration.

To prove Noll's rule, let $\boldsymbol{H} \in g_R(\mathcal{F})$. Then, (6.66) holds and it is permissible to replace $\boldsymbol{F}$ by $\boldsymbol{F} \cdot \boldsymbol{P}$ and $\vec{G}_\theta$ by $\vec{G}_\theta \cdot \boldsymbol{P}$ such that

$$\mathop{\mathcal{F}_R}_{\tau \le t}\left[\boldsymbol{F}(\tau) \cdot \boldsymbol{P} \cdot \boldsymbol{H}, \theta(\tau), \vec{G}_\theta(\tau) \cdot \boldsymbol{P} \cdot \boldsymbol{H}\right] = \mathop{\mathcal{F}_R}_{\tau \le t}\left[\boldsymbol{F}(\tau) \cdot \boldsymbol{P}, \theta(\tau), \vec{G}_\theta(\tau) \cdot \boldsymbol{P}\right].$$

Applying the identity (6.60) to both sides of this equation gives

$$\mathop{\mathcal{F}_{\hat{R}}}_{\tau \le t}\left[\boldsymbol{F}(\tau) \cdot \boldsymbol{P} \cdot \boldsymbol{H} \cdot \boldsymbol{P}^{-1}, \theta(\tau), \vec{G}_\theta(\tau) \cdot \boldsymbol{P} \cdot \boldsymbol{H} \cdot \boldsymbol{P}^{-1}\right] = \mathop{\mathcal{F}_{\hat{R}}}_{\tau \le t}\left[\boldsymbol{F}(\tau), \theta(\tau), \vec{G}_\theta(\tau)\right],$$

which holds for all deformation gradient, temperature and temperature gradient histories. Thus, if $\boldsymbol{H} \in g_R(\mathcal{F})$, then $\boldsymbol{P} \cdot \boldsymbol{H} \cdot \boldsymbol{P}^{-1} \in g_{\hat{R}}(\mathcal{F})$. The argument can be reversed, which completes the proof of Noll's rule.

By definition, a symmetry group is a subgroup of the group of all unimodular transformations,

$$g_R(\mathcal{F}) \subseteq \mathsf{unim}. \tag{6.74}$$

The 'size' of a specific symmetry group is a measure of the amount of material symmetry. The group of unimodular transformations contains the group of *orthogonal transformations* (for instance, rotations or reflections), and also the group of non-orthogonal volume-preserving transformations (for instance, torsions), that is,

$$\mathsf{orth.} \subset \mathsf{unim.}, \tag{6.75}$$

where $\mathsf{orth.}$ denotes the group of all orthogonal transformations.

The smallest non-trivial symmetry group $\{\boldsymbol{I}, -\boldsymbol{I}\}$ corresponds to materials called *triclinic*. Obviously,

$$\boldsymbol{P} \cdot \{\boldsymbol{I}, -\boldsymbol{I}\} \cdot \boldsymbol{P}^{-1} = \{\boldsymbol{I}, -\boldsymbol{I}\}, \tag{6.76}$$

meaning that no transformation can increase the symmetries of a triclinic material.

The largest symmetry group is the group of all unimodular transformations. Noll's rule shows that if $\boldsymbol{P}$ is an arbitrary invertible tensor, then $\boldsymbol{P} \cdot \boldsymbol{H} \cdot \boldsymbol{P}^{-1}$ is unimodular for all unimodular $\boldsymbol{H}$, and if $\overline{\boldsymbol{H}}$ is any unimodular tensor, the tensor

$P^{-1} \cdot \overline{H} \cdot P$ is a unimodular tensor, and $\overline{H} = P \cdot H \cdot P^{-1}$. Therefore,

$$P \cdot \mathsf{unim.} \cdot P^{-1} = \mathsf{unim.} \tag{6.77}$$

Hence, the maximum symmetry of such a material cannot be destroyed by deformation.

6.10 Classification of Symmetry Properties

6.10.1 Isotropic Materials

We will now classify materials according to their symmetry groups. Firstly, we say that a material is *isotropic* if there exists a reference configuration R in which the symmetry condition (6.66) or (6.71) holds for all orthogonal transformations such that

$$g_R(\mathcal{F}) \supseteq \mathsf{orth}. \tag{6.78}$$

For an isotropic material, (6.74) and (6.78) can be combined to give

$$\mathsf{orth.} \subseteq g_R(\mathcal{F}) \subseteq \mathsf{unim}. \tag{6.79}$$

All groups of isotropic bodies are thus bounded by the orthogonal and unimodular groups. A question then arises as to how many groups bounded by $\mathsf{orth.}$ and $\mathsf{unim.}$ may exist. Group theory states that the orthogonal group is a maximal subgroup of the unimodular group. Consequently, there are only two groups satisfying (6.79), either orthogonal or unimodular; no other group exists between them. Hence, there are two kinds of isotropic materials, either those with $g_R(\mathcal{F}) = \mathsf{orth.}$, or those with $g_R(\mathcal{F}) = \mathsf{unim}$. All other materials are anisotropic and possess a lower degree of symmetry. The concept of symmetry groups can be used to define solids and fluids according to Noll.

6.10.2 Fluids

Fluids have the property that they can be poured from one container to another and, after some time, no evidence of the previous circumstances remains. Such a change in container can be thought of as a change of reference configuration, so that for fluids all reference configurations with the same mass density are indistinguishable. In other words, every configuration, including the present one, can be thought as a reference configuration. According to Noll, a simple material is a *fluid* if its symmetry group has a maximal symmetry, being identical to the set of all unimodular transformations,

$$g_R(\mathcal{F}) = \mathsf{unim.} \tag{6.80}$$

Moreover, Noll's rule implies that the condition (6.80) holds for every configuration if it holds for any one reference configuration R, so that a fluid has maximal symmetry relative to every reference configuration. A fluid is thus a material that does not have a preferred configuration. In terms of symmetry groups, $g_{\hat{R}}(\mathcal{F}) = g_R(\mathcal{F})$ $(= \mathsf{unim.})$. Moreover, a fluid is isotropic due to (6.75).

6.10.3 *Solids*

Unlike fluids, solids typically have the property that any change of shape (as represented by a transformation of the reference configuration) brings the material into a new reference configuration from which its response to deformation and temperature loading histories is different. Hence, according to Noll, a simple material is called a *solid* if there is a reference configuration such that every element of the symmetry group is a rotation,

$$g_R(\mathcal{F}) \subseteq \mathsf{orth.} \tag{6.81}$$

A solid whose symmetry group is equal to the full orthogonal group is termed *isotropic*,

$$g_R(\mathcal{F}) = \mathsf{orth.} \tag{6.82}$$

If the symmetry group of a solid is smaller than the full orthogonal group, the solid is termed *anisotropic*,

$$g_R(\mathcal{F}) \subset \mathsf{orth.} \tag{6.83}$$

There is only a finite number of symmetry groups with this property, and it forms the *32 crystal classes*.

Note that simple fluids and simple solids do not exhaust the possible types of simple materials. There is, for instance, a simple material, termed *liquid crystal*, that is neither a simple fluid nor a simple solid. However, a liquid crystal is a rather specific material, and we will not consider it here.

6.11 Constitutive Equations for Isotropic Materials

The symmetry group $g_R(\mathcal{F})$ of an isotropic material is the group of all orthogonal transformations, $g_R(\mathcal{F}) = \mathsf{orth}$. Let $\boldsymbol{Q}$ be an orthogonal tensor, $\boldsymbol{Q} \in \mathsf{orth}$. The transformation rules for the Finger deformation tensor $\boldsymbol{b}$ and the left-stretch tensor $\boldsymbol{V}$ under the orthogonal transformation of the reference configuration R onto $\hat{R}$ are

$$\hat{\boldsymbol{b}} = \boldsymbol{b}, \qquad \hat{\boldsymbol{V}} = \boldsymbol{V}. \tag{6.84}$$

The first relation follows from the transformation rule $(6.52)_1$ for the deformation gradient $\boldsymbol{F}$,

$$\hat{\boldsymbol{b}} = \hat{\boldsymbol{F}} \cdot \hat{\boldsymbol{F}^T} = (\boldsymbol{F} \cdot \boldsymbol{Q}^{-1}) \cdot (\boldsymbol{Q}^{-T} \cdot \boldsymbol{F}^T) = \boldsymbol{F} \cdot \boldsymbol{Q}^{-1} \cdot \boldsymbol{Q} \cdot \boldsymbol{F}^T = \boldsymbol{F} \cdot \boldsymbol{F}^T = \boldsymbol{b},$$

while the definition $(1.51)_2$ of $\boldsymbol{V}$, i.e., $\boldsymbol{V} = \sqrt{\boldsymbol{b}}$, gives to the second relation. Hence, the tensors $\boldsymbol{b}$ and $\boldsymbol{V}$ are unchanged by an orthogonal transformation of the reference configuration. The polar decomposition of $\boldsymbol{F}$, i.e., $\boldsymbol{F} = \boldsymbol{V} \cdot \boldsymbol{R}$, and the transformation rules $(6.52)_1$ and $(6.84)_2$ give the transformation rule for the rotation tensor $\boldsymbol{R}$,

$$\boldsymbol{R} = \hat{\boldsymbol{R}} \cdot \boldsymbol{Q}, \tag{6.85}$$

which follows from

$$\hat{\boldsymbol{R}} = \hat{\boldsymbol{V}}^{-1} \cdot \hat{\boldsymbol{F}} = \boldsymbol{V}^{-1} \cdot \hat{\boldsymbol{F}} = \boldsymbol{V}^{-1} \cdot \boldsymbol{F} \cdot \boldsymbol{Q}^{-1} = \boldsymbol{V}^{-1} \cdot \boldsymbol{V} \cdot \boldsymbol{R} \cdot \boldsymbol{Q}^{-1} = \boldsymbol{R} \cdot \boldsymbol{Q}^{-1}.$$

Comparing $(6.52)_1$ and (6.85) shows that the transformation rules for the deformation gradient $\boldsymbol{F}$ and the rotation tensor $\boldsymbol{R}$ have the same form, with the respective transformation tensors $\boldsymbol{P}$ and $\boldsymbol{Q}$. In addition, the constitutive equation (6.25) with the functional $\mathcal{F}_2$ multiplied by $\boldsymbol{F}$ and $\boldsymbol{F}^T$ has the same form as the constitutive equation (6.24) with the functional $\mathcal{F}_1$ multiplied by $\boldsymbol{R}$ and $\boldsymbol{R}^T$. This allows us to deduce that the symmetry condition of the functional $\mathcal{F}_1$ from the symmetry condition (6.71) of the functional $\mathcal{F}_2$ is

$$\begin{aligned}
&\boldsymbol{Q}(\vec{X}) \cdot \mathop{\mathcal{F}_1}_{\tau \le t}\left[\boldsymbol{C}(\vec{X}, \tau), \theta(\vec{X}, \tau), \vec{G}_\theta(\vec{X}, \tau)\right] \cdot \boldsymbol{Q}^T(\vec{X}) \\
&= \mathop{\mathcal{F}_1}_{\tau \le t}\left[\boldsymbol{Q}^{-T}(\vec{X}) \cdot \boldsymbol{C}(\vec{X}, \tau) \cdot \boldsymbol{Q}^{-1}(\vec{X}), \theta(\vec{X}, \tau), \vec{G}_\theta(\vec{X}, \tau) \cdot \boldsymbol{Q}^{-1}(\vec{X})\right],
\end{aligned}$$

where the right-hand side can be simplified by using the orthogonality property of the tensor $\boldsymbol{Q}$ and the identity $\boldsymbol{Q} \cdot \vec{G}_\theta = \vec{G}_\theta \cdot \boldsymbol{Q}^T$ such that

$$\begin{aligned}
&\boldsymbol{Q}(\vec{X}) \cdot \mathop{\mathcal{F}_1}_{\tau \le t}\left[\boldsymbol{C}(\vec{X}, \tau), \theta(\vec{X}, \tau), \vec{G}_\theta(\vec{X}, \tau)\right] \cdot \boldsymbol{Q}^T(\vec{X}) \\
&= \mathop{\mathcal{F}_1}_{\tau \le t}\left[\boldsymbol{Q}(\vec{X}) \cdot \boldsymbol{C}(\vec{X}, \tau) \cdot \boldsymbol{Q}^T(\vec{X}), \theta(\vec{X}, \tau), \boldsymbol{Q}(\vec{X}) \cdot \vec{G}_\theta(\vec{X}, \tau)\right].
\end{aligned} \tag{6.86}$$

Similar constraints can be derived for the constitutive functionals for the heat flux $\mathcal{Q}$ and the internal energy $\mathcal{E}$, respectively,

$$\begin{aligned}\boldsymbol{Q}(\vec{X}) \cdot \underset{\tau \leq t}{\mathcal{Q}} & \left[\boldsymbol{C}(\vec{X}, \tau), \theta(\vec{X}, \tau), \vec{G}_\theta(\vec{X}, \tau)\right] \\ &= \underset{\tau \leq t}{\mathcal{Q}} \left[\boldsymbol{Q}(\vec{X}) \cdot \boldsymbol{C}(\vec{X}, \tau) \cdot \boldsymbol{Q}^T(\vec{X}), \theta(\vec{X}, \tau), \boldsymbol{Q}(\vec{X}) \cdot \vec{G}_\theta(\vec{X}, \tau)\right],\end{aligned} \tag{6.87}$$

$$\begin{aligned}\underset{\tau \leq t}{\mathcal{E}} & \left[\boldsymbol{C}(\vec{X}, \tau), \theta(\vec{X}, \tau), \vec{G}_\theta(\vec{X}, \tau)\right] \\ &= \underset{\tau \leq t}{\mathcal{E}} \left[\boldsymbol{Q}(\vec{X}) \cdot \boldsymbol{C}(\vec{X}, \tau) \cdot \boldsymbol{Q}^T(\vec{X}), \theta(\vec{X}, \tau), \boldsymbol{Q}(\vec{X}) \cdot \vec{G}_\theta(\vec{X}, \tau)\right].\end{aligned}$$

Functionals that satisfy constraints (6.86) and (6.87) are called *tensor, vector* and *scalar isotropic functionals* with respect to all orthogonal transformations. All these functionals represent the constitutive equations for an isotropic body.

6.12 The Present Configuration as the Reference

So far, we have employed a reference configuration that is fixed in time, but we can also use a time-varying reference configuration. Thus, one motion may be described in terms of any other. The only suitable time-varying reference configuration is the present configuration. If we take this as the reference, we describe past events as they seem to an observer fixed at the particle $\mathcal{X}$, now at the position $\vec{x}$. The corresponding description is called *relative* and it was introduced in Sect. 1.2.

The relative motion function $\vec{\chi}_t(\vec{x}, \tau)$, defined by (1.8) as

$$\vec{\chi}_t(\vec{x}, \tau) := \vec{\chi}(\vec{\chi}^{\,-1}(\vec{x}, t), \tau), \tag{6.88}$$

describes the deformation of the configuration κ_τ that was occupied by the body at a past time τ relative to the present configuration κ_t. The subscript t at the function $\vec{\chi}$ recalls that the present configuration κ_t is considered to be the reference. The relative description given by the mapping (6.88) is actually a special case of the referential description, differing from the Lagrangian description in that the reference position is now denoted by $\vec{x}$ at time t instead of $\vec{X}$ at time $t = 0$. The variable time τ, being the time when the particle occupied the position $\vec{\xi} = \vec{\chi}_t(\vec{x}, \tau)$, is now considered as an independent variable instead of the time t in the Lagrangian description.

The *relative deformation gradient* $\boldsymbol{F}_t$ is the gradient of the relative motion function:

$$\boldsymbol{F}_t(\vec{x}, \tau) := \left(\operatorname{grad} \vec{\chi}_t(\vec{x}, \tau)\right)^T, \qquad (\boldsymbol{F}_t)_{kl} := \frac{\partial(\vec{\chi}_t)_k}{\partial x_l}. \tag{6.89}$$

It can be expressed in terms of the deformation gradients $\boldsymbol{F}$ and $\boldsymbol{F}^{-1}$. Differentiating (6.88) with respect to X_K and using the chain rule yields

$$\frac{\partial \chi_k(\vec{X},\tau)}{\partial X_K} = \frac{\partial (\vec{\chi}_t)_k(\vec{x},\tau)}{\partial x_l}\frac{\partial \chi_l(\vec{X},t)}{\partial X_K},$$

or

$$(\boldsymbol{F})_{kK}(\vec{X},\tau) = (\boldsymbol{F}_t)_{kl}(\vec{x},\tau)(\boldsymbol{F})_{lK}(\vec{X},t),$$

and

$$\boldsymbol{F}(\vec{X},\tau) = \boldsymbol{F}_t(\vec{x},\tau)\cdot\boldsymbol{F}(\vec{X},t) \qquad \text{or} \qquad \boldsymbol{F}_t(\vec{x},\tau) = \boldsymbol{F}(\vec{X},\tau)\cdot\boldsymbol{F}^{-1}(\vec{x},t) \tag{6.90}$$

in tensor notation. Specifically, when $\tau = t$, this gives $\boldsymbol{F}_t(\vec{x},t) = \boldsymbol{I}$.

The Green deformation tensor $\boldsymbol{C}$ may also be developed in terms of the relative deformation gradient as

$$\boldsymbol{C}(\vec{X},\tau) = \boldsymbol{F}^T(\vec{X},\tau)\cdot\boldsymbol{F}(\vec{X},\tau) = \boldsymbol{F}^T(\vec{X},t)\cdot\boldsymbol{F}_t^T(\vec{x},\tau)\cdot\boldsymbol{F}_t(\vec{x},\tau)\cdot\boldsymbol{F}(\vec{X},t).$$

Defining the *relative Green deformation tensor* by

$$\boldsymbol{c}_t(\vec{x},\tau) := \boldsymbol{F}_t^T(\vec{x},\tau)\cdot\boldsymbol{F}_t(\vec{x},\tau), \tag{6.91}$$

we have

$$\boldsymbol{C}(\vec{X},\tau) = \boldsymbol{F}^T(\vec{X},t)\cdot\boldsymbol{c}_t(\vec{x},\tau)\cdot\boldsymbol{F}(\vec{X},t). \tag{6.92}$$

6.13 Isotropic Constitutive Functionals in the Relative Representation

Now, let the motion be considered in the relative representation (6.88). Thus, the present configuration at the current time t serves as the reference, while the configuration at a past time τ is taken as the present. First, the Lagrangian temperature gradient, $\vec{G}_\theta := \mathrm{Grad}\,\theta$, can be expressed in terms of the Eulerian temperature gradient, $\vec{g}_\theta := \mathrm{grad}\,\theta$, as[4]

$$\vec{G}_\theta(\vec{X},\tau) = \vec{g}_\theta(\vec{x},\tau)\cdot\boldsymbol{F}(\vec{X},t). \tag{6.93}$$

[4]The temperature, as any other variable, has both an Eulerian and a Lagrangian representation. For example, the Lagrangian temperature is defined as $\Theta(\vec{X},t) := \theta(\vec{x}(\vec{X},t),t)$, where $\theta(\vec{x},t)$ is the Eulerian temperature. We make, however, an exception in the notation and use $\theta(\vec{X},t)$ instead of $\Theta(\vec{X},t)$ for the Lagrangian representation of temperature.

The proof is straightforward,

$$\vec{G}_\theta(\vec{X},\tau) = \frac{\partial\theta(\vec{X},\tau)}{\partial X_K}\,\vec{I}_K = \frac{\partial\theta(\vec{x},\tau)}{\partial x_k}\frac{\partial\chi_k(\vec{X},t)}{\partial X_K}\,\vec{I}_K = \left[\vec{g}_\theta(\vec{x},\tau)\right]_k \left[\boldsymbol{F}(\vec{x},t)\right]_{kK}\vec{I}_K$$
$$= \vec{g}_\theta(\vec{x},\tau)\cdot\boldsymbol{F}(\vec{X},t).$$

We consider the reduced-form constitutive equation (6.24) for a homogeneous body,

$$\boldsymbol{t}(\vec{X},t) = \boldsymbol{R}(\vec{X},t)\cdot\underset{\tau\le t}{\mathcal{F}_1}\left[\boldsymbol{C}(\vec{X},\tau),\theta(\vec{X},\tau),\vec{G}_\theta(\vec{X},\tau)\right]\cdot\boldsymbol{R}^T(\vec{X},t). \tag{6.94}$$

Since t is fixed in the relative description of motion, we may choose $\boldsymbol{Q}(\vec{X}) = \boldsymbol{R}(\vec{X},t)$ in (6.86) for the symmetry condition of the functional $\mathcal{F}_1$ such that

$$\begin{aligned}&\boldsymbol{R}(\vec{X},t)\cdot\underset{\tau\le t}{\mathcal{F}_1}\left[\boldsymbol{C}(\vec{X},\tau),\theta(\vec{X},\tau),\vec{G}_\theta(\vec{X},\tau)\right]\cdot\boldsymbol{R}^T(\vec{X},t)\\ &=\underset{\tau\le t}{\mathcal{F}_1}\left[\boldsymbol{R}(\vec{X},t)\cdot\boldsymbol{C}(\vec{X},\tau)\cdot\boldsymbol{R}^T(\vec{X},t),\theta(\vec{X},\tau),\vec{G}_\theta(\vec{X},\tau)\cdot\boldsymbol{R}^T(\vec{X},t)\right],\end{aligned} \tag{6.95}$$

whose left-hand side is equal to the Cauchy stress tensor according to (6.94),

$$\boldsymbol{t}(\vec{X},t) =\underset{\tau\le t}{\mathcal{F}_1}\left[\boldsymbol{R}(\vec{X},t)\cdot\boldsymbol{C}(\vec{X},\tau)\cdot\boldsymbol{R}^T(\vec{X},t),\theta(\vec{X},\tau),\vec{G}_\theta(\vec{X},\tau)\cdot\boldsymbol{R}^T(\vec{X},t)\right]. \tag{6.96}$$

Using (6.92) and (6.93) together with the polar decomposition $\boldsymbol{F} = \boldsymbol{V}\cdot\boldsymbol{R}$, the arguments of $\mathcal{F}_1$ can be arranged as

$$\begin{aligned}\boldsymbol{R}(\vec{X},t)\cdot\boldsymbol{C}(\vec{X},\tau)\cdot\boldsymbol{R}^T(\vec{X},t) &= \boldsymbol{R}(\vec{X},t)\cdot\boldsymbol{F}^T(\vec{X},t)\cdot\boldsymbol{c}_t(\vec{x},\tau)\cdot\boldsymbol{F}(\vec{X},t)\cdot\boldsymbol{R}^T(\vec{X},t)\\ &= \boldsymbol{V}(\vec{x},t)\cdot\boldsymbol{c}_t(\vec{x},\tau)\cdot\boldsymbol{V}(\vec{x},t)\\ &= \sqrt{\boldsymbol{b}(\vec{x},t)}\cdot\boldsymbol{c}_t(\vec{x},\tau)\cdot\sqrt{\boldsymbol{b}(\vec{x},t)}\end{aligned}$$

and

$$\begin{aligned}\vec{G}_\theta(\vec{X},\tau)\cdot\boldsymbol{R}^T(\vec{X},t) &= \vec{g}_\theta(\vec{x},\tau)\cdot\boldsymbol{F}(\vec{X},t)\cdot\boldsymbol{R}^T(\vec{X},t) = \vec{g}_\theta(\vec{x},\tau)\cdot\boldsymbol{V}(\vec{x},t)\\ &= \vec{g}_\theta(\vec{x},\tau)\cdot\sqrt{\boldsymbol{b}(\vec{x},t)},\end{aligned}$$

where $\boldsymbol{b}$ is the Finger deformation tensor, $\boldsymbol{b} = \boldsymbol{F}\cdot\boldsymbol{F}^T = \boldsymbol{V}\cdot\boldsymbol{V}$. The constitutive equation (6.96) now takes the form

$$\boldsymbol{t}(\vec{x},t) = \underset{\tau\le t}{\mathcal{F}_1}\left[\sqrt{\boldsymbol{b}(\vec{x},t)}\cdot\boldsymbol{c}_t(\vec{x},\tau)\cdot\sqrt{\boldsymbol{b}(\vec{x},t)},\theta(\vec{x},\tau),\vec{g}_\theta(\vec{x},\tau)\cdot\sqrt{\boldsymbol{b}(\vec{x},t)}\right]. \tag{6.97}$$

The information contained in the last constitutive equation can alternatively be expressed as

$$\boldsymbol{t}(\vec{x}, t) = \underset{\tau \leq t}{\mathcal{F}_1}\left[\boldsymbol{c}_t(\vec{x}, \tau), \theta(\vec{x}, \tau), \vec{g}_\theta(\vec{x}, \tau); \boldsymbol{b}(\vec{x}, t)\right], \tag{6.98}$$

where the semicolon denotes that the Finger deformation tensor at present time, i.e., $\boldsymbol{b}(\vec{x}, t)$, appears in the functional $\mathcal{F}_1$ as a parameter. This shows that it is not possible to express the effects of past deformation on the stress entirely by measuring deformation with respect to the present configuration, but a fixed reference configuration is, in general, required to allow for the effects of the present deformation, as indicated by $\boldsymbol{b}(\vec{x}, t)$ in (6.98).

Equation (6.98) is a general form of the constitutive equation, expressed in a relative representation, for a simple homogeneous isotropic material. It remains to be shown that this constitutive equation is invariant under an arbitrary orthogonal transformation $\boldsymbol{Q}$ of the reference configuration R, not only for our special choice of $\boldsymbol{Q} = \boldsymbol{R}(\vec{X}, t)$. Since none of the arguments in (6.98) depends on the reference configuration, this requirement is satisfied trivially. For instance, $(6.84)_1$ shows that the Finger deformation tensor does not change by an orthogonal transformation of the reference configuration, i.e., $\hat{\boldsymbol{b}} = \boldsymbol{b}$.

The construction (6.98) does not guarantee that the constitutive functional $\mathcal{F}_1$ is automatically materially frame-indifferent. The frame-indifference (5.53) of the Cauchy stress tensor implies that the functional $\mathcal{F}_1$ must transform under a rigid motion of the spatial observer frame as

$$\underset{\tau \leq t}{\mathcal{F}_1}\left[\boldsymbol{c}_t^*(\tau), \theta^*(\tau), \vec{g}_{\theta*}^*(\tau); \boldsymbol{b}^*(t)\right] = \boldsymbol{O}(t) \cdot \underset{\tau \leq t}{\mathcal{F}_1}\left[\boldsymbol{c}_t(\tau), \theta(\tau), \vec{g}_\theta(\tau); \boldsymbol{b}(t)\right] \cdot \boldsymbol{O}^T(t), \tag{6.99}$$

where the position $\vec{x}$ has been dropped to shorten the notation, e.g. $\boldsymbol{c}_t(\vec{x}, \tau) \equiv \boldsymbol{c}_t(\tau)$ or $\boldsymbol{b}(\vec{x}, t) \equiv \boldsymbol{b}(t)$. The transformation rules for the relative deformation gradient, the relative Green deformation tensor, the Eulerian temperature gradient and the Finger deformation tensor are

$$\begin{aligned} \boldsymbol{F}_t^*(\tau) &= \boldsymbol{O}(\tau) \cdot \boldsymbol{F}_t(\tau) \cdot \boldsymbol{O}^T(t), \\ \boldsymbol{c}_t^*(\tau) &= \boldsymbol{O}(t) \cdot \boldsymbol{c}_t(\tau) \cdot \boldsymbol{O}^T(t), \\ \vec{g}_{\theta*}^*(\tau) &= \boldsymbol{O}(t) \cdot \vec{g}_\theta(\tau), \\ \boldsymbol{b}^*(t) &= \boldsymbol{O}(t) \cdot \boldsymbol{b}(t) \cdot \boldsymbol{O}^T(t). \end{aligned} \tag{6.100}$$

The proofs are straightforward:

$$\begin{aligned} \boldsymbol{F}_t^*(\tau) &= \boldsymbol{F}^*(\tau) \cdot [\boldsymbol{F}^*(t)]^{-1} = \boldsymbol{O}(\tau) \cdot \boldsymbol{F}(\tau) \cdot \boldsymbol{F}^{-1}(t) \cdot \boldsymbol{O}^{-1}(t) \\ &= \boldsymbol{O}(\tau) \cdot \boldsymbol{F}_t(\tau) \cdot \boldsymbol{O}^T(t), \end{aligned}$$

$$\boldsymbol{c}_t^*(\tau) = [\boldsymbol{F}^*(t)]^{-T} \cdot \boldsymbol{C}^*(\tau) \cdot [\boldsymbol{F}^*(t)]^{-1}$$
$$= \boldsymbol{O}(t) \cdot \boldsymbol{F}^{-T}(t) \cdot \boldsymbol{C}(\tau) \cdot \boldsymbol{F}^{-1}(t) \cdot \boldsymbol{O}^{-1}(t) = \boldsymbol{O}(t) \cdot \boldsymbol{c}_t(\tau) \cdot \boldsymbol{O}^T(t),$$
$$\vec{g}_{\theta*}^*(\tau) = [\boldsymbol{F}^*(t)]^{-T} \cdot \vec{G}_{\theta*}^*(\tau) = \boldsymbol{O}(t) \cdot \boldsymbol{F}^{-T}(t) \cdot \vec{G}_\theta(\tau) = \boldsymbol{O}(t) \cdot \vec{g}_\theta(\tau).$$

In view of (6.100) and bearing in mind that the temperature θ was postulated to be invariant under transformations of observer frame, the constraint (6.99) simplifies to

$$\underset{\tau \le t}{\mathcal{F}_1}\left[\boldsymbol{O}(t) \cdot \boldsymbol{c}_t(\tau) \cdot \boldsymbol{O}^T(t), \theta(\tau), \boldsymbol{O}(t) \cdot \vec{g}_\theta(\tau); \boldsymbol{O}(t) \cdot \boldsymbol{b}(t) \cdot \boldsymbol{O}^T(t)\right]$$
$$= \boldsymbol{O}(t) \cdot \underset{\tau \le t}{\mathcal{F}_1}\left[\boldsymbol{c}_t(\tau), \theta(\tau), \vec{g}_\theta(\tau); \boldsymbol{b}(t)\right] \cdot \boldsymbol{O}^T(t). \tag{6.101}$$

This shows that the functional $\mathcal{F}_1$ is materially frame-indifferent if it is a spatially isotropic functional of the variables $\boldsymbol{c}_t(\vec{x}, \tau)$, $\theta(\vec{x}, \tau)$ and $\vec{g}_\theta(\vec{x}, \tau)$, as well as a spatially isotropic function of $\boldsymbol{b}(\vec{x}, t)$.[5]

An analogous arrangement can be carried out for the constitutive equations (6.26) for the heat flux and the internal energy:

$$\vec{q}(\vec{x}, t) = \underset{\tau \le t}{\mathcal{Q}}\left[\boldsymbol{c}_t(\vec{x}, \tau), \theta(\vec{x}, \tau), \vec{g}_\theta(\vec{x}, \tau); \boldsymbol{b}(\vec{x}, t)\right],$$
$$\varepsilon(\vec{x}, t) = \underset{\tau \le t}{\mathcal{E}}\left[\boldsymbol{c}_t(\vec{x}, \tau), \theta(\vec{x}, \tau), \vec{g}_\theta(\vec{x}, \tau); \boldsymbol{b}(\vec{x}, t)\right]. \tag{6.102}$$

Note that the constitutive equations (6.98) and (6.102) in the relative representation will be used to describe material behaviour of a fluid, while for a solid we will employ the constitutive equation (6.24) in the Lagrangian representation.

6.14 The General Constitutive Equation for Fluids

A fluid is characterised through the largest symmetry group, which is identical to the set of all unimodular transformations, $g_R(\mathcal{F}) = \text{unim}$. Mathematically, the condition for the material symmetry of the functional $\mathcal{F}$ in (6.98) under a unimodular transformation $\boldsymbol{H}$ is expressed, by analogy with the symmetry condition (6.64), as

$$\underset{\tau \le t}{\mathcal{F}_1}\left[\boldsymbol{c}_t(\vec{x}, \tau), \theta(\vec{x}, \tau), \vec{g}_\theta(\vec{x}, \tau); \boldsymbol{F}(\vec{X}, t) \cdot \boldsymbol{F}^T(\vec{X}, t)\right]$$
$$= \underset{\tau \le t}{\mathcal{F}_1}\left[\boldsymbol{c}_t(\vec{x}, \tau), \theta(\vec{x}, \tau), \vec{g}_\theta(\vec{x}, \tau); \boldsymbol{F}(\vec{X}, t) \cdot \boldsymbol{H}(\vec{X}) \cdot \boldsymbol{H}^T(\vec{X}) \cdot \boldsymbol{F}^T(\vec{X}, t)\right], \tag{6.103}$$

[5]Note that this is spatial isotropy, not material isotropy as discussed in the preceding sections.

since the Finger deformation tensor $\boldsymbol{b} = \boldsymbol{F} \cdot \boldsymbol{F}^T$ transforms under $\boldsymbol{H}$ as $\boldsymbol{F} \cdot \boldsymbol{H} \cdot \boldsymbol{H}^T \cdot \boldsymbol{F}^T$. This general characterisation of a fluid must also hold for the special unimodular transformation

$$\boldsymbol{H} = \left(\det \boldsymbol{F}\right)^{1/3} \boldsymbol{F}^{-1} = \left(\frac{\varrho_0}{\varrho}\right)^{1/3} \boldsymbol{F}^{-1}, \tag{6.104}$$

where $\varrho(\vec{x}, t)$ and $\varrho_0(\vec{X})$ are the mass density of the fluid in the present and reference configurations, respectively. The transformation (6.104) is indeed unimodular since

$$\det \boldsymbol{H} = \left[(\det \boldsymbol{F})^{1/3} \right]^3 \det \boldsymbol{F}^{-1} = 1.$$

For such $\boldsymbol{H}$, $\boldsymbol{F} \cdot \boldsymbol{H} \cdot \boldsymbol{H}^T \cdot \boldsymbol{F}^T = (\det \boldsymbol{F})^{2/3}\, \boldsymbol{I} = (\varrho_0/\varrho)^{2/3} \boldsymbol{I}$, and the constitutive equation (6.98) can be rewritten as

$$\boldsymbol{t}(\vec{x}, t) = \underset{\tau \le t}{\mathcal{F}_t} \left[\boldsymbol{c}_t(\vec{x}, \tau), \theta(\vec{x}, \tau), \vec{g}_\theta(\vec{x}, \tau); \varrho(\vec{x}, t) \right]. \tag{6.105}$$

This is a general form of the constitutive equation for a simple fluid in the relative representation. It shows that the Cauchy stress tensor $\boldsymbol{t}$ for a fluid depends on the relative deformation history, the temperature and temperature gradient histories, and on the present mass density. Since none of $\boldsymbol{c}_t(\vec{x}, \tau)$, $\theta(\vec{x}, \tau)$, $\vec{g}_\theta(\vec{x}, \tau)$ and $\varrho(\vec{x}, t)$ depends on the reference configuration, the constitutive equation (6.105) is invariant under any transformation of the reference configuration. Hence, it satisfies trivially the requirement that $g_R(\mathcal{F}) = \mathsf{unim}$. It is important to note that the stress in a fluid when it has been at rest for all times, past and present, depends on the density alone.

6.15 The Principle of Bounded Memory

This principle states that deformation and temperature events in the distant past have a very small influence on the present behaviour of a material. In other words, the memory of past motions and temperatures at any material point decays rapidly with time. This principle is the time-domain counterpart of the principle of smooth neighbourhood. No unique mathematical formulation can be made of this principle; the following limited interpretation suffices for our purposes.

To express the boundedness of memory, let $h(\vec{X}, \tau)$ be a thermomechanical history in the argument of a constitutive functional. Suppose that $h(\vec{X}, \tau)$ is an analytical function such that it possesses continuous partial derivatives with respect to τ at $\tau = t$. For small $\tau - t$, it can be approximated by a Taylor series expansion about $\tau = t$ up to order N,

$$h(\vec{X}, \tau) = \sum_{n=0}^{N} \frac{1}{n!} \left. \frac{\partial^n h(\vec{X}, \tau)}{\partial \tau^n} \right|_{\tau = t} (\tau - t)^n. \tag{6.106}$$

If the constitutive functional is sufficiently smooth, so that its dependence on $h(\vec{X}, \tau)$ can be replaced by the list of functions

$$\overset{(n)}{h}(\vec{X}, t) := \left.\frac{\partial^n h(\vec{X}, \tau)}{\partial \tau^n}\right|_{\tau=t} \qquad \text{for } n = 0, 1, \ldots, N, \tag{6.107}$$

that is, $\overset{(n)}{h}$ is the nth material time derivative of h, we say that the material is of the *rate type* of order N with respect to the variable h. If the material is of the rate type in all its variables, the constitutive equation (6.28) for a homogeneous solid can be approximated by

$$\begin{aligned} \boldsymbol{T}^{(2)}(\vec{X}, t) = \hat{\boldsymbol{T}}^{(2)} \Big[& \boldsymbol{C}(\vec{X}, t), \dot{\boldsymbol{C}}(\vec{X}, t), \ldots, \overset{(N_C)}{\boldsymbol{C}}(\vec{X}, t), \theta(\vec{X}, t), \dot{\theta}(\vec{X}, t), \ldots, \overset{(N_\theta)}{\theta}(\vec{X}, t), \\ & \vec{G}_\theta(\vec{X}, t), \dot{\vec{G}}_\theta(\vec{X}, t), \ldots, \overset{(N_G)}{\vec{G}_\theta}(\vec{X}, t) \Big]. \end{aligned} \tag{6.108}$$

We note that $\hat{\boldsymbol{T}}^{(2)}$ is no longer functional. It is a tensor-valued function, called the *constitutive function*, of the arguments listed. They involve time rates of the deformation gradient up to order N_C, and the temperature and its gradient up to order N_θ and N_G, respectively. The orders N_C, N_θ and N_G need not be the same.[6]

The general constitutive equation for a fluid is given by (6.105). We again assume that the relative Green deformation tensor $\boldsymbol{c}_t$ can be approximated by a truncated Taylor series expansion of the form (6.106):

$$\boldsymbol{c}_t(\vec{x}, \tau) = \boldsymbol{a}_0(\vec{x}, t) + \boldsymbol{a}_1(\vec{x}, t)(\tau - t) + \boldsymbol{a}_2(\vec{x}, t)\frac{(\tau - t)^2}{2} + \ldots, \tag{6.109}$$

where the expansion coefficient

$$\boldsymbol{a}_n(\vec{x}, t) := \left.\frac{\partial^n \boldsymbol{c}_t(\vec{x}, \tau)}{\partial \tau^n}\right|_{\tau=t} \tag{6.110}$$

---

[6]The functions in the argument of constitutive functionals may not be so smooth as to admit truncated Taylor series expansion. Nevertheless, the constitutive functionals may be such as to smooth out past discontinuities in these argument functions and/or their derivatives. The principle of bounded memory, in this context also called the *principle of fading memory*, is then a requirement on the smoothness of constitutive functionals.

The principle of fading memory, mathematically formulated by Coleman and Noll (1960), starts with the assumption that the so-called Frechét derivatives of constitutive functionals up to an order n exist and are continuous in the neighbourhood of histories at time t in the Hilbert space normed by an influence function of order greater than $n + \frac{1}{2}$. Then, the constitutive functionals can be approximated by linear functionals for which explicit mathematical representations are known. The most important result of the fading memory theory is the possibility to approximate asymptotically sufficiently slow strain and temperature histories by a Taylor series expansion.

is the Rivlin–Ericksen tensor of order n defined by (2.33). It is a frame-indifferent tensor,

$$\boldsymbol{a}_n^*(\vec{x}, t) = \boldsymbol{O}(t) \cdot \boldsymbol{a}_n(\vec{x}, t) \cdot \boldsymbol{O}^T(t). \tag{6.111}$$

To show this, we use $(6.100)_2$

$$\begin{aligned} a_n^*(\vec{x}, t) &= \left.\frac{\partial^n \boldsymbol{c}_t^*(\vec{x}, \tau)}{\partial \tau^n}\right|_{\tau=t} = \left.\frac{\partial^n}{\partial \tau^n}\Big(\boldsymbol{O}(t) \cdot \boldsymbol{c}_t(\vec{x}, \tau) \cdot \boldsymbol{O}^T(t)\Big)\right|_{\tau=t} \\ &= \boldsymbol{O}(t) \cdot \left.\frac{\partial^n \boldsymbol{c}_t(\vec{x}, \tau)}{\partial \tau^n}\right|_{\tau=t} \cdot \boldsymbol{O}^T(t), \end{aligned}$$

which, in view of (6.110), gives (6.111).

We again assume that all functions in the argument of $\mathcal{F}_t$ in (6.105) can be approximated by a truncated Taylor series expansion. Then, the functional for a class of materials with bounded memory may be represented by functions involving various time derivatives of the argument functions,

$$\begin{aligned} \boldsymbol{t}(\vec{x}, t) = \hat{\boldsymbol{t}} \Big[&\boldsymbol{a}_1(\vec{x}, t), \ldots, \boldsymbol{a}_{N_C}(\vec{x}, t), \theta(\vec{x}, t), \dot{\theta}(\vec{x}, t), \ldots, \overset{(N_\theta)}{\theta}(\vec{x}, t), \\ &\vec{g}_\theta(\vec{x}, t), \dot{\vec{g}}_\theta(\vec{x}, t), \ldots, \overset{(N_g)}{\vec{g}_\theta}(\vec{x}, t); \varrho(\vec{x}, t) \Big], \end{aligned} \tag{6.112}$$

where the dependence on $\boldsymbol{a}_0 = \boldsymbol{I}$ has been omitted due to its redundancy. Note that $\hat{\boldsymbol{t}}$ is again no longer functional, but a tensor-valued function. This is an asymptotic approximation to the general constitutive equation (6.105) for a fluid. Coleman and Noll (1960) showed that this approximation is valid for sufficiently slow deformation processes.

Similar expressions hold for the heat flux vector $\vec{q}$ and the internal energy ε. The number of time derivatives for each of the gradients depends on the strength of the memory of these gradients. According to the principle of equipresence, however, all constitutive functionals should be represented in terms of the same list of independent constitutive variables.

6.16 Representation Theorems for Isotropic Functions

We now seek the representation of isotropic scalar-, vector- and tensor-valued functions.

6.16.1 Representation Theorem for a Scalar-Valued Isotropic Function

A scalar-valued function b depending on a symmetric tensor $\boldsymbol{S}$ and a vector $\vec{v}$ is called *isotropic* if it satisfies the identity

$$b(\boldsymbol{S}, \vec{v}) = b(\boldsymbol{Q} \cdot \boldsymbol{S} \cdot \boldsymbol{Q}^T, \boldsymbol{Q} \cdot \vec{v}) \tag{6.113}$$

for all orthogonal tensors $\boldsymbol{Q}$. This means that b does not depend on components S_{kl} and v_k arbitrarily, but it can only be a function of scalar invariants that are independent of any orthogonal transformation.

Hence, we need to determine the set of all independent invariants of the variables $\boldsymbol{S}$ and $\vec{v}$. There are three independent invariants of tensor $\boldsymbol{S}$,

$$I_S = \operatorname{tr} \boldsymbol{S}, \qquad II_S = \frac{1}{2}\left[(\operatorname{tr} \boldsymbol{S})^2 - \operatorname{tr}(\boldsymbol{S}^2)\right], \qquad III_S = \det \boldsymbol{S}, \tag{6.114}$$

and one invariant of vector $\vec{v}$, the scalar product $\vec{v} \cdot \vec{v}$. Combining $\boldsymbol{S}$ and $\vec{v}$, we can create a number of other invariants,

$$\vec{v} \cdot \boldsymbol{S} \cdot \vec{v}, \quad \vec{v} \cdot \boldsymbol{S}^2 \cdot \vec{v}, \quad \vec{v} \cdot \boldsymbol{S}^3 \cdot \vec{v}, \ldots, \vec{v} \cdot \boldsymbol{S}^n \cdot \vec{v}, \ldots, \tag{6.115}$$

although not all of them are mutually independent. According to the Cayley–Hamilton theorem, which states that a tensor satisfies its own characteristic equation, we have

$$\boldsymbol{S}^3 - I_S\, \boldsymbol{S}^2 + II_S\, \boldsymbol{S} - III_S\, \boldsymbol{I} = \boldsymbol{0}, \tag{6.116}$$

where $\boldsymbol{I}$ and $\boldsymbol{0}$ are the unit and zero tensors of the same order as $\boldsymbol{S}$. Multiplying this equation from the left and the right by vector $\vec{v}$, we obtain

$$\vec{v} \cdot \boldsymbol{S}^3 \cdot \vec{v} - I_S\, (\vec{v} \cdot \boldsymbol{S}^2 \cdot \vec{v}) + II_S\, (\vec{v} \cdot \boldsymbol{S} \cdot \vec{v}) - III_S\, (\vec{v} \cdot \vec{v}) = 0. \tag{6.117}$$

We see that $\vec{v} \cdot \boldsymbol{S}^n \cdot \vec{v}$ for $n \geq 3$ can be expressed in terms of $\vec{v} \cdot \boldsymbol{S}^2 \cdot \vec{v}$, $\vec{v} \cdot \boldsymbol{S} \cdot \vec{v}$ and $\vec{v} \cdot \vec{v}$, but these cannot be used to express the three invariants of $\boldsymbol{S}$. Hence, the set of all independent invariants of the variables $\boldsymbol{S}$ and $\vec{v}$ is

$$I_S, \; II_S, \; III_S, \; \vec{v} \cdot \vec{v}, \; \vec{v} \cdot \boldsymbol{S} \cdot \vec{v}, \; \vec{v} \cdot \boldsymbol{S}^2 \cdot \vec{v}. \tag{6.118}$$

We then conclude that the scalar-valued isotropic function b is represented in the form

$$b(\boldsymbol{S}, \vec{v}) = b(I_S, II_S, III_S, \; \vec{v} \cdot \vec{v}, \; \vec{v} \cdot \boldsymbol{S} \cdot \vec{v}, \; \vec{v} \cdot \boldsymbol{S}^2 \cdot \vec{v}). \tag{6.119}$$

6.16.2 Representation Theorem for a Vector-Valued Isotropic Function

A vector-valued function $\vec{b}$ depending on a symmetric tensor $\boldsymbol{S}$ and a vector $\vec{v}$ is called *isotropic* if it satisfies the identity

$$\boldsymbol{Q} \cdot \vec{b}(\boldsymbol{S}, \vec{v}) = \vec{b}(\boldsymbol{Q} \cdot \boldsymbol{S} \cdot \boldsymbol{Q}^T, \boldsymbol{Q} \cdot \vec{v}) \tag{6.120}$$

for all orthogonal tensors $\boldsymbol{Q}$. For an arbitrary vector $\vec{c}$, let us define a function f as

$$f(\boldsymbol{S}, \vec{v}, \vec{c}) := \vec{c} \cdot \vec{b}(\boldsymbol{S}, \vec{v}). \tag{6.121}$$

This function is a scalar-valued isotropic function of the variables $\boldsymbol{S}$, $\vec{v}$ and $\vec{c}$, which follows from

$$\begin{aligned} f(\boldsymbol{Q} \cdot \boldsymbol{S} \cdot \boldsymbol{Q}^T, \boldsymbol{Q} \cdot \vec{v}, \boldsymbol{Q} \cdot \vec{c}) &= \boldsymbol{Q} \cdot \vec{c} \cdot \vec{b}(\boldsymbol{Q} \cdot \boldsymbol{S} \cdot \boldsymbol{Q}^T, \boldsymbol{Q} \cdot \vec{v}) = \boldsymbol{Q} \cdot \vec{c} \cdot \boldsymbol{Q} \cdot \vec{b}(\boldsymbol{S}, \vec{v}) \\ &= \vec{c} \cdot \boldsymbol{Q}^T \cdot \boldsymbol{Q} \cdot \vec{b}(\boldsymbol{S}, \vec{v}) = f(\boldsymbol{S}, \vec{v}, \vec{c}). \end{aligned}$$

By analogy with the preceding section, the set of all independent invariants of $\boldsymbol{S}$, $\vec{v}$ and $\vec{c}$ is

$$\begin{aligned} I_S,\ II_S,\ III_S,\ &\vec{v} \cdot \vec{v},\ \vec{v} \cdot \boldsymbol{S} \cdot \vec{v},\ \vec{v} \cdot \boldsymbol{S}^2 \cdot \vec{v}, \\ &\vec{c} \cdot \vec{v},\ \vec{c} \cdot \boldsymbol{S} \cdot \vec{v},\ \vec{c} \cdot \boldsymbol{S}^2 \cdot \vec{v}, \\ &\vec{c} \cdot \vec{c},\ \vec{c} \cdot \boldsymbol{S} \cdot \vec{c},\ \vec{c} \cdot \boldsymbol{S}^2 \cdot \vec{c}. \end{aligned} \tag{6.122}$$

If f were an arbitrary function of $\boldsymbol{S}$, $\vec{v}$ and $\vec{c}$, it could be represented in terms of these 12 invariants in an arbitrary manner. However, f is linearly dependent on $\vec{c}$, so it can only be represented in terms of invariants that are linear functions of $\vec{c}$,

$$f = a_0\, \vec{c} \cdot \vec{v} + a_1\, \vec{c} \cdot \boldsymbol{S} \cdot \vec{v} + a_2\, \vec{c} \cdot \boldsymbol{S}^2 \cdot \vec{v}, \tag{6.123}$$

where the coefficients a_i depend on the six invariants that are independent of $\vec{c}$. Since $\vec{c}$ was an arbitrary vector in (6.121)–(6.123), we conclude that the vector-valued isotropic function $\vec{b}$ is represented in the form

$$\vec{b} = a_0\, \vec{v} + a_1\, \boldsymbol{S} \cdot \vec{v} + a_2\, \boldsymbol{S}^2 \cdot \vec{v}, \tag{6.124}$$

where

$$a_i = a_i\, (I_S, II_S, III_S, \vec{v} \cdot \vec{v}, \vec{v} \cdot \boldsymbol{S} \cdot \vec{v}, \vec{v} \cdot \boldsymbol{S}^2 \cdot \vec{v}). \tag{6.125}$$

6.16.3 Representation Theorem for a Symmetric Tensor-Valued Isotropic Function

(i) Symmetric Tensor-Valued Isotropic Function of a Symmetric Tensor and a Vector A symmetric tensor-valued function $\boldsymbol{B}$ depending on a symmetric tensor $\boldsymbol{S}$ and a vector $\vec{v}$ is called *isotropic* if it satisfies the identity

$$\boldsymbol{Q} \cdot \boldsymbol{B}(\boldsymbol{S}, \vec{v}) \cdot \boldsymbol{Q}^T = \boldsymbol{B}(\boldsymbol{Q} \cdot \boldsymbol{S} \cdot \boldsymbol{Q}^T, \boldsymbol{Q} \cdot \vec{v}) \tag{6.126}$$

for all orthogonal tensors $\boldsymbol{Q}$. For a tensor $\boldsymbol{C}$, let us define a function f as

$$f(\boldsymbol{S}, \vec{v}, \boldsymbol{C}) := \mathrm{tr}\big(\boldsymbol{C} \cdot \boldsymbol{B}(\boldsymbol{S}, \vec{v})\big). \tag{6.127}$$

This function is a scalar-valued isotropic function of variables $\boldsymbol{S}$, $\vec{v}$ and $\boldsymbol{C}$, which follows from

$$\begin{aligned} f(\boldsymbol{Q} \cdot \boldsymbol{S} \cdot \boldsymbol{Q}^T, \boldsymbol{Q} \cdot \vec{v}, \boldsymbol{Q} \cdot \boldsymbol{C} \cdot \boldsymbol{Q}^T) &= \mathrm{tr}\big(\boldsymbol{Q} \cdot \boldsymbol{C} \cdot \boldsymbol{Q}^T \cdot \boldsymbol{B}(\boldsymbol{Q} \cdot \boldsymbol{S} \cdot \boldsymbol{Q}^T, \boldsymbol{Q} \cdot \vec{v})\big) \\ &= \mathrm{tr}\big(\boldsymbol{Q} \cdot \boldsymbol{C} \cdot \boldsymbol{Q}^T \cdot \boldsymbol{Q} \cdot \boldsymbol{B}(\boldsymbol{S}, \vec{v}) \cdot \boldsymbol{Q}^T\big) = \mathrm{tr}\big(\boldsymbol{Q} \cdot \boldsymbol{C} \cdot \boldsymbol{B}(\boldsymbol{S}, \vec{v}) \cdot \boldsymbol{Q}^T\big) \\ &= \mathrm{tr}\big(\boldsymbol{C} \cdot \boldsymbol{B}(\boldsymbol{S}, \vec{v}) \cdot \boldsymbol{Q}^T \cdot \boldsymbol{Q}\big) = f(\boldsymbol{S}, \vec{v}, \boldsymbol{C}). \end{aligned}$$

By the same argument as in the previous case, f must be represented as a linear combination of all independent invariants that are linear functions of $\boldsymbol{C}$ only:

$$\mathrm{tr}\,\boldsymbol{C},\ \mathrm{tr}(\boldsymbol{C} \cdot \boldsymbol{S}),\ \mathrm{tr}(\boldsymbol{C} \cdot \boldsymbol{S}^2),\ \vec{v} \cdot \boldsymbol{C} \cdot \vec{v},\ \vec{v} \cdot \boldsymbol{C} \cdot \boldsymbol{S} \cdot \vec{v},\ \vec{v} \cdot \boldsymbol{C} \cdot \boldsymbol{S}^2 \cdot \vec{v}. \tag{6.128}$$

Note that it is possible, but not trivial, to prove that there are no other independent invariants linearly dependent on $\boldsymbol{C}$. Hence,

$$\begin{aligned} f = {} & a_0\,\mathrm{tr}\,\boldsymbol{C} + a_1\,\mathrm{tr}(\boldsymbol{C} \cdot \boldsymbol{S}) + a_2\,\mathrm{tr}(\boldsymbol{C} \cdot \boldsymbol{S}^2) + a_3\,\vec{v} \cdot \boldsymbol{C} \cdot \vec{v} + a_4\,\vec{v} \cdot \boldsymbol{C} \cdot \boldsymbol{S} \cdot \vec{v} \\ & + a_5\,\vec{v} \cdot \boldsymbol{C} \cdot \boldsymbol{S}^2 \cdot \vec{v}, \end{aligned} \tag{6.129}$$

where the coefficients a_i depend on the invariants that are independent of $\boldsymbol{C}$, that is, on the six invariants of $\boldsymbol{S}$ and $\vec{v}$. From the identity

$$\vec{v} \cdot \boldsymbol{C} \cdot \vec{v} = \mathrm{tr}(\boldsymbol{C} \cdot \vec{v} \otimes \vec{v}), \tag{6.130}$$

where $\vec{v} \otimes \vec{v}$ is the dyadic product of vector $\vec{v}$ with itself, we also have

$$\begin{aligned} f = {} & a_0\,\mathrm{tr}\,\boldsymbol{C} + a_1\,\mathrm{tr}(\boldsymbol{C} \cdot \boldsymbol{S}) + a_2\,\mathrm{tr}(\boldsymbol{C} \cdot \boldsymbol{S}^2) + a_3\,\mathrm{tr}(\boldsymbol{C} \cdot \vec{v} \otimes \vec{v}) + a_4\,\mathrm{tr}(\boldsymbol{C} \cdot \boldsymbol{S} \cdot \vec{v} \otimes \vec{v}) \\ & + a_5\,\mathrm{tr}(\boldsymbol{C} \cdot \boldsymbol{S}^2 \cdot \vec{v} \otimes \vec{v}). \end{aligned} \tag{6.131}$$

Comparison of (6.131) with (6.127) results in a representation of a symmetric tensor-valued isotropic function $\boldsymbol{B}$ of the form

$$\boldsymbol{B}(\boldsymbol{S},\vec{v}) = a_0\boldsymbol{I} + a_1\boldsymbol{S} + a_2\boldsymbol{S}^2 + a_3\,\vec{v}\otimes\vec{v} + a_4\,\mathrm{sym}(\boldsymbol{S}\cdot\vec{v}\otimes\vec{v}) + a_5\,\mathrm{sym}(\boldsymbol{S}^2\cdot\vec{v}\otimes\vec{v}), \tag{6.132}$$

where

$$a_i = a_i(I_S, II_S, III_S, \vec{v}\cdot\vec{v}, \vec{v}\cdot\boldsymbol{S}\cdot\vec{v}, \vec{v}\cdot\boldsymbol{S}^2\cdot\vec{v}) \tag{6.133}$$

and the symbol 'sym' stands for the symmetric part of a tensor.

(ii) Symmetric Tensor-Valued Isotropic Function of Two Symmetric Tensors A symmetric tensor-valued function $\boldsymbol{B}$ depending on two symmetric tensors $\boldsymbol{S}$ and $\boldsymbol{T}$ is called *isotropic* if it satisfies the identity

$$\boldsymbol{Q}\cdot\boldsymbol{B}(\boldsymbol{S},\boldsymbol{T})\cdot\boldsymbol{Q}^T = \boldsymbol{B}(\boldsymbol{Q}\cdot\boldsymbol{S}\cdot\boldsymbol{Q}^T, \boldsymbol{Q}\cdot\boldsymbol{T}\cdot\boldsymbol{Q}^T) \tag{6.134}$$

for all orthogonal tensors $\boldsymbol{Q}$. For a tensor $\boldsymbol{C}$, let us define a function f as

$$f(\boldsymbol{S},\boldsymbol{T},\boldsymbol{C}) := \mathrm{tr}\big(\boldsymbol{C}\cdot\boldsymbol{B}(\boldsymbol{S},\boldsymbol{T})\big), \tag{6.135}$$

which is a scalar-valued isotropic function of the variables $\boldsymbol{S}$, $\boldsymbol{T}$ and $\boldsymbol{C}$. The set of all independent invariants of these three symmetric tensors that are linear functions of $\boldsymbol{C}$ is

$$\begin{aligned}&\mathrm{tr}\,\boldsymbol{C},\ \mathrm{tr}(\boldsymbol{C}\cdot\boldsymbol{S}),\ \mathrm{tr}(\boldsymbol{C}\cdot\boldsymbol{S}^2),\ \mathrm{tr}(\boldsymbol{C}\cdot\boldsymbol{T}),\ \mathrm{tr}(\boldsymbol{C}\cdot\boldsymbol{T}^2),\ \mathrm{tr}(\boldsymbol{C}\cdot\boldsymbol{S}\cdot\boldsymbol{T}),\ \mathrm{tr}(\boldsymbol{C}\cdot\boldsymbol{S}\cdot\boldsymbol{T}^2),\\ &\quad\mathrm{tr}(\boldsymbol{C}\cdot\boldsymbol{S}^2\cdot\boldsymbol{T}),\ \mathrm{tr}(\boldsymbol{C}\cdot\boldsymbol{S}^2\cdot\boldsymbol{T}^2).\end{aligned} \tag{6.136}$$

Hence, f is represented in the form

$$\begin{aligned}f &= a_0\,\mathrm{tr}\,\boldsymbol{C} + a_1\,\mathrm{tr}(\boldsymbol{C}\cdot\boldsymbol{S}) + a_2\,\mathrm{tr}(\boldsymbol{C}\cdot\boldsymbol{S}^2) + a_3\,\mathrm{tr}(\boldsymbol{C}\cdot\boldsymbol{T}) + a_4\,\mathrm{tr}(\boldsymbol{C}\cdot\boldsymbol{T}^2)\\ &\quad + a_5\,\mathrm{tr}(\boldsymbol{C}\cdot\boldsymbol{S}\cdot\boldsymbol{T}) + a_6\,\mathrm{tr}(\boldsymbol{C}\cdot\boldsymbol{S}\cdot\boldsymbol{T}^2) + a_7\,\mathrm{tr}(\boldsymbol{C}\cdot\boldsymbol{S}^2\cdot\boldsymbol{T}) + a_8\,\mathrm{tr}(\boldsymbol{C}\cdot\boldsymbol{S}^2\cdot\boldsymbol{T}^2),\end{aligned} \tag{6.137}$$

where the coefficients a_i depend on the invariants that are independent of $\boldsymbol{C}$, that is, on the 10 invariants of $\boldsymbol{S}$ and $\boldsymbol{T}$:

$$I_S, II_S, III_S, I_T, II_T, III_T, \mathrm{tr}(\boldsymbol{S}\cdot\boldsymbol{T}), \mathrm{tr}(\boldsymbol{S}\cdot\boldsymbol{T}^2), \mathrm{tr}(\boldsymbol{S}^2\cdot\boldsymbol{T}), \mathrm{tr}(\boldsymbol{S}^2\cdot\boldsymbol{T}^2). \tag{6.138}$$

Comparison of (6.137) with (6.135) yields a representation of a symmetric tensor-valued isotropic function $\boldsymbol{B}$ of the form

$$\boldsymbol{B}(\boldsymbol{S},\boldsymbol{T}) = a_0\,\boldsymbol{I} + a_1\,\boldsymbol{S} + a_2\,\boldsymbol{S}^2 + a_3\,\boldsymbol{T} + a_4\,\boldsymbol{T}^2 + a_5\,\mathrm{sym}(\boldsymbol{S}\cdot\boldsymbol{T}) + a_6\,\mathrm{sym}(\boldsymbol{S}\cdot\boldsymbol{T}^2)$$

$$+ a_7 \,\mathrm{sym}(\boldsymbol{S}^2 \cdot \boldsymbol{T}) + a_8 \,\mathrm{sym}(\boldsymbol{S}^2 \cdot \boldsymbol{T}^2), \tag{6.139}$$

where

$$a_i = a_i\big(I_S, II_S, III_S, I_T, II_T, III_T, \mathrm{tr}(\boldsymbol{S} \cdot \boldsymbol{T}), \mathrm{tr}(\boldsymbol{S} \cdot \boldsymbol{T}^2), \mathrm{tr}(\boldsymbol{S}^2 \cdot \boldsymbol{T}), \mathrm{tr}(\boldsymbol{S}^2 \cdot \boldsymbol{T}^2)\big). \tag{6.140}$$

6.17 Examples of Isotropic Materials with Bounded Memory

6.17.1 Elastic Solids

A solid is described as being *elastic* if the stress tensor $\boldsymbol{t}$ depends only on the deformation gradient $\boldsymbol{F}$ at the present time, and not on the temperature θ nor on its thermomechanical history, so that the constitutive equation (6.19) for the Lagrangian Cauchy stress tensor reduces to the simple form

$$\boldsymbol{t}(\vec{X}, t) = \hat{\boldsymbol{t}}\big[\boldsymbol{F}(\vec{X}, t)\big]. \tag{6.141}$$

Hence, the stress in an elastic material at each particle is uniquely determined by the present deformation from a fixed reference configuration. The material frame-indifference (6.24) requires that the dependence on $\boldsymbol{F}$ is not arbitrary, but has the form

$$\boldsymbol{t}(\vec{X}, t) = \boldsymbol{R}(\vec{X}, t) \cdot \hat{\boldsymbol{t}}\big[\boldsymbol{C}(\vec{X}, t)\big] \cdot \boldsymbol{R}^T(\vec{X}, t). \tag{6.142}$$

For isotropic materials, $\hat{\boldsymbol{t}}$ must be an isotropic function of $\boldsymbol{C}$. Using the representation theorem (6.132) for a tensor-valued isotropic function gives

$$\boldsymbol{t}(\vec{X}, t) = \boldsymbol{R}(\vec{X}, t) \cdot \big[a_0 \boldsymbol{I} + a_1 \boldsymbol{C}(\vec{X}, t) + a_2 \boldsymbol{C}^2(\vec{X}, t)\big] \cdot \boldsymbol{R}^T(\vec{X}, t), \tag{6.143}$$

where the expansion coefficients a_i are scalar functions of the three invariants of $\boldsymbol{C}$,

$$a_i = a_i(I_C, II_C, III_C). \tag{6.144}$$

Note that a_i also depends on $\vec{X}$ and t via the invariants of $\boldsymbol{C}$. With the help of the Finger deformation tensor $\boldsymbol{b}$,

$$\boldsymbol{b} = \boldsymbol{F} \cdot \boldsymbol{F}^T = \boldsymbol{V}^2 = \boldsymbol{V}^T \cdot \boldsymbol{V} = \boldsymbol{R} \cdot \boldsymbol{F}^T \cdot \boldsymbol{F} \cdot \boldsymbol{R}^T = \boldsymbol{R} \cdot \boldsymbol{C} \cdot \boldsymbol{R}^T,$$

its second power,

$$\boldsymbol{b}^2 = \boldsymbol{R} \cdot \boldsymbol{C} \cdot \boldsymbol{R}^T \cdot \boldsymbol{R} \cdot \boldsymbol{C} \cdot \boldsymbol{R}^T = \boldsymbol{R} \cdot \boldsymbol{C}^2 \cdot \boldsymbol{R}^T,$$

and the fact that $\boldsymbol{C}$ and $\boldsymbol{b}$ have the same invariants because $\boldsymbol{U}$ and $\boldsymbol{V}$ have the same eigenvalues, the constitutive equation (6.143) can be expressed in terms of the Eulerian Cauchy stress tensor as

$$\boldsymbol{t}(\vec{x},t) = a_0\boldsymbol{I} + a_1\boldsymbol{b}(\vec{x},t) + a_2\boldsymbol{b}^2(\vec{x},t), \tag{6.145}$$

where

$$a_i = a_i(I_b, II_b, III_b). \tag{6.146}$$

Note that the constitutive equation (6.145) does not imply that $\boldsymbol{t}$ can only be a quadratic equation in the components of $\boldsymbol{b}$ because the scalar functions a_i may be non-linear in the components of $\boldsymbol{b}$.

If the Lagrangian description is considered, the second Piola–Kirchhoff stress tensor $\boldsymbol{T}^{(2)}$ is employed. The constitutive equation (6.108) for an elastic material is reduced to

$$\boldsymbol{T}^{(2)}(\vec{X},t) = \hat{\boldsymbol{T}}^{(2)}\big[\boldsymbol{C}(\vec{X},t)\big]. \tag{6.147}$$

The tensor-valued function $\hat{\boldsymbol{T}}^{(2)}$ for isotropic materials may be represented according to (6.132) as

$$\boldsymbol{T}^{(2)}(\vec{X},t) = \alpha_0\boldsymbol{I} + \alpha_1\boldsymbol{C}(\vec{X},t) + \alpha_2\boldsymbol{C}^2(\vec{X},t), \tag{6.148}$$

where

$$\alpha_i = \alpha_i(I_C, II_C, III_C). \tag{6.149}$$

Equations (6.143), (6.145) and (6.148) are equivalent forms of the constitutive equation for isotropic elastic solids.

We now consider the case of small deformations when the displacement gradient $\boldsymbol{H}$ is sufficiently small such that the geometrical linearisation can be applied and it is not necessary to consider the differences between the reference and present configurations. Within the limit of geometrical linearisation, the Finger deformation tensor is equal to the Green deformation tensor, $\boldsymbol{b} \cong \boldsymbol{C}$, $\boldsymbol{R}\cdot\boldsymbol{C}\cdot\boldsymbol{R}^T \cong \boldsymbol{C}$, $\boldsymbol{R}\cdot\boldsymbol{C}^2\cdot\boldsymbol{R}^T \cong \boldsymbol{C}^2$ and the linearised Lagrangian strain tensor is equal to the linearised Eulerian strain tensor, $\widetilde{\boldsymbol{E}} \cong \tilde{\boldsymbol{e}}$. Consequently, the linearised constitutive equation (6.143) or (6.145) for $\boldsymbol{t}$ coincides with the linearised constitutive equation (6.148) for $\boldsymbol{T}^{(2)}$. Let us, for instance, linearise the constitutive equation (6.148). From (1.115) we find that

$$\begin{aligned} &\boldsymbol{C} = \boldsymbol{I} + 2\widetilde{\boldsymbol{E}}, \qquad \boldsymbol{C}^2 = \boldsymbol{I} + 4\widetilde{\boldsymbol{E}}, \qquad I_C = 3 + 2\,\mathrm{tr}\widetilde{\boldsymbol{E}}, \\ &II_C = 3 + 4\,\mathrm{tr}\widetilde{\boldsymbol{E}}, \qquad III_C = 1 + 2\,\mathrm{tr}\widetilde{\boldsymbol{E}}, \end{aligned} \tag{6.150}$$

where $\widetilde{\boldsymbol{E}}$ is the infinitesimal strain tensor. Then, the constitutive equation (6.148) may be linearised as

$$\boldsymbol{T}^{(2)} \cong \boldsymbol{t} = \lambda(\mathrm{tr}\widetilde{\boldsymbol{E}})\boldsymbol{I} + 2\mu\widetilde{\boldsymbol{E}}, \tag{6.151}$$

which is the constitutive equation for an isotropic linear elastic material, also known as *Hooke's law*, where the parameters λ and μ are called the *Lamé elastic parameters*.

6.17.2 Thermoelastic Solids

The effects of heat conduction in elastic materials is described when temperature and temperature gradients are considered (in addition to the Green deformation tensor) as the constitutive variables. From (6.108) we deduce that the simplest thermoelastic solid is defined by $N_C = N_\theta = N_g = 0$. Hence, the constitutive equations for a thermoelastic solid are

$$\boldsymbol{T}^{(2)} = \hat{\boldsymbol{T}}^{(2)}\big(\boldsymbol{C}, \theta, \vec{G}_\theta\big), \qquad \vec{Q} = \hat{\vec{Q}}\big(\boldsymbol{C}, \theta, \vec{G}_\theta\big), \qquad \varepsilon = \hat{\varepsilon}\big(\boldsymbol{C}, \theta, \vec{G}_\theta\big), \tag{6.152}$$

where the dependence on $\vec{X}$ and t has been dropped to shorten the notation. For isotropic materials, functions $\hat{\boldsymbol{T}}^{(2)}$, $\hat{\vec{Q}}$ and $\hat{\varepsilon}$ must be isotropic functions of their arguments:

$$\begin{aligned}
\boldsymbol{T}^{(2)} &= a_0\boldsymbol{I} + a_1\boldsymbol{C} + a_2\boldsymbol{C}^2 + a_3\,\vec{G}_\theta \otimes \vec{G}_\theta + a_4\,\mathrm{sym}(\boldsymbol{C}\cdot\vec{G}_\theta \otimes \vec{G}_\theta) \\
&\quad + a_5\,\mathrm{sym}(\boldsymbol{C}^2\cdot\vec{G}_\theta \otimes \vec{G}_\theta), \\
\vec{Q} &= -\kappa\,\vec{G}_\theta + c_1\,\boldsymbol{C}\cdot\vec{G}_\theta + c_2\,\boldsymbol{C}^2\cdot\vec{G}_\theta, \\
\varepsilon &= \hat{\varepsilon}\big(\theta, I_C, II_C, III_C,\ \vec{G}_\theta\cdot\vec{G}_\theta,\ \vec{G}_\theta\cdot\boldsymbol{C}\cdot\vec{G}_\theta,\ \vec{G}_\theta\cdot\boldsymbol{C}^2\cdot\vec{G}_\theta\big),
\end{aligned} \tag{6.153}$$

where the scalar functions a_i, κ and c_i depend on the same set of arguments as the function $\hat{\varepsilon}$ of the internal energy. Thus, the heat flux vector $\vec{Q}$ depends linearly on the temperature gradient $\vec{G}_\theta$, but contains two non-linear terms as the Green deformation tensor arises together with $\vec{G}_\theta$. The coefficient κ is called the *thermal conductivity*, while the remaining coefficients c_1 and c_2 do not have separate names.

6.17.3 Kelvin–Voigt Viscoelastic Solids

In general, by the term the *viscoelastic solid*, we understand a material for which stress is dependent on the strain, strain rate and stress-rate tensors only. From (6.108) we deduce that the simplest viscoelastic solid is defined by $N_C = 1$,

$$\boldsymbol{T}^{(2)} = \hat{\boldsymbol{T}}^{(2)}(\boldsymbol{C}, \dot{\boldsymbol{C}}). \tag{6.154}$$

This simple strain-rate dependent solid is called the *Kelvin–Voigt viscoelastic solid*. Since the Piola–Kirchhoff stress tensor depends on two symmetric tensors, for isotropic materials it may be represented according to (6.139) as

$$\begin{aligned} \boldsymbol{T}^{(2)} &= a_0\,\boldsymbol{I} + a_1\,\boldsymbol{C} + a_2\,\boldsymbol{C}^2 + a_3\,\dot{\boldsymbol{C}} + a_4\,\dot{\boldsymbol{C}}^2 + a_5\,\mathrm{sym}(\boldsymbol{C}\cdot\dot{\boldsymbol{C}}) + a_6\,\mathrm{sym}(\boldsymbol{C}\cdot\dot{\boldsymbol{C}}^2) \\ &\quad + a_7\,\mathrm{sym}(\boldsymbol{C}^2\cdot\dot{\boldsymbol{C}}) + a_8\,\mathrm{sym}(\boldsymbol{C}^2\cdot\dot{\boldsymbol{C}}^2), \end{aligned} \tag{6.155}$$

where

$$\begin{aligned} a_i = a_i\big(&\mathrm{tr}\,\boldsymbol{C}, \mathrm{tr}\,\boldsymbol{C}^2, \mathrm{tr}\,\boldsymbol{C}^3, \mathrm{tr}\,\dot{\boldsymbol{C}}, \mathrm{tr}\,\dot{\boldsymbol{C}}^2, \mathrm{tr}\,\dot{\boldsymbol{C}}^3, \mathrm{tr}(\boldsymbol{C}\cdot\dot{\boldsymbol{C}}), \mathrm{tr}(\boldsymbol{C}\cdot\dot{\boldsymbol{C}}^2), \\ &\mathrm{tr}(\boldsymbol{C}^2\cdot\dot{\boldsymbol{C}}), \mathrm{tr}(\boldsymbol{C}^2\cdot\dot{\boldsymbol{C}}^2)\big). \end{aligned} \tag{6.156}$$

6.17.4 Elastic Fluids

An elastic material may be a solid or a fluid. For an *elastic fluid*, the functions in the argument of the constitutive function $\hat{\boldsymbol{t}}$ in (6.112) are reduced to the dependence on the current mass density ϱ,

$$\boldsymbol{t} = \hat{\boldsymbol{t}}(\varrho), \qquad \vec{q} = \hat{\vec{q}}(\varrho), \qquad \varepsilon = \hat{\varepsilon}(\varrho), \tag{6.157}$$

where the dependence on $\vec{x}$ and t has been dropped to shorten the notation. This defines an *inviscid* (i.e., non-viscous) compressible fluid without heat conduction, also known as the *Euler* or *ideal fluid*. Under a change of observer frame, we have

$$\boldsymbol{t}^* = \hat{\boldsymbol{t}}(\varrho^*), \qquad \vec{q}^* = \hat{\vec{q}}(\varrho^*), \qquad \varepsilon^* = \hat{\varepsilon}(\varrho^*). \tag{6.158}$$

Since $\varrho^* = \varrho$, $\boldsymbol{t}^* = \boldsymbol{O}\cdot\boldsymbol{t}\cdot\boldsymbol{O}^T$, $\vec{q}^* = \boldsymbol{O}\cdot\vec{q}$ and $\varepsilon^* = \varepsilon$,

$$\boldsymbol{O}\cdot\hat{\boldsymbol{t}}(\varrho)\cdot\boldsymbol{O}^T = \hat{\boldsymbol{t}}(\varrho), \qquad \boldsymbol{O}\cdot\hat{\vec{q}}(\varrho) = \hat{\vec{q}}(\varrho) \tag{6.159}$$

for all orthogonal tensors $\boldsymbol{O}$. The only isotropic tensor and isotropic vector satisfying these identities are, respectively, an isotropic spherical tensor[7] and the zero vector,

$$\boldsymbol{t} = -p(\varrho)\boldsymbol{I}, \qquad \vec{q} = \vec{0}, \qquad \varepsilon = \hat{\varepsilon}(\varrho). \tag{6.160}$$

Therefore, the stress in an elastic fluid is always a pressure, which depends on the mass density alone. It is a matter of using the second law of thermodynamics to show that there is a relationship between the internal energy and the pressure function.

6.17.5 Thermoelastic Fluids

The effects of heat conduction in elastic fluids is described when temperature and temperature gradients are considered (in addition to the mass density) as the constitutive variables. From (6.112) we deduce that the *thermoelastic fluid* is defined by $N_C = N_\theta = N_g = 0$. Hence, the constitutive equations for a thermoelastic fluid are

$$\boldsymbol{t} = \hat{\boldsymbol{t}}\big(\theta, \vec{g}_\theta, \varrho\big), \qquad \vec{q} = \hat{\vec{q}}\big(\theta, \vec{g}_\theta, \varrho\big), \qquad \varepsilon = \hat{\varepsilon}\big(\theta, \vec{g}_\theta, \varrho\big). \tag{6.161}$$

Using the representation theorems (6.119), (6.124) and (6.132) for isotropic functions gives

$$\begin{aligned} \boldsymbol{t} &= \sigma \boldsymbol{I} + \tau\, \vec{g}_\theta \otimes \vec{g}_\theta, \\ \vec{q} &= -\kappa\, \vec{g}_\theta, \\ \varepsilon &= \hat{\varepsilon}\big(\varrho, \theta, \vec{g}_\theta \cdot \vec{g}_\theta\big), \end{aligned} \tag{6.162}$$

where

$$\sigma, \tau, \kappa = \sigma, \tau, \kappa\big(\varrho, \theta, \vec{g}_\theta \cdot \vec{g}_\theta\big). \tag{6.163}$$

[7] Every second-order tensor $\boldsymbol{A}$ can be written as the sum of a *spherical tensor* $a\boldsymbol{I}$ (i.e., a scalar multiple of the identity tensor) and a *deviator* $\boldsymbol{A}^D$:

$$\boldsymbol{A} = a\boldsymbol{I} + \boldsymbol{A}^D.$$

Under the choice of $a := \operatorname{tr} \boldsymbol{A}/3$, the trace of the deviator vanishes, i.e., $\operatorname{tr} \boldsymbol{A}^D = 0$.

6.17.6 Viscous Fluids

Considering $N_C = 1$ in (6.112), we obtain a *viscous fluid*, which is an analogue to a simple viscoelastic solid,

$$\boldsymbol{t} = \hat{\boldsymbol{t}}(\boldsymbol{d}, \varrho) \tag{6.164}$$

since $\boldsymbol{a}_1 = 2\boldsymbol{d}$. The representation theorem (6.132) for a symmetric tensor-valued isotropic function makes this constitutive equation more explicit:

$$\boldsymbol{t} = a_0\boldsymbol{I} + 2\mu_v\boldsymbol{d} + a_2\boldsymbol{d}^2, \tag{6.165}$$

where

$$a_i, \mu_v = a_i, \mu_v(\varrho, I_d, II_d, III_d) \tag{6.166}$$

and μ_v is called the *shear viscosity*.

We now wish to show that the constitutive equation (6.165) is frame-indifferent. Multiplying (6.165) from the left by $\boldsymbol{O}$ and from the right by $\boldsymbol{O}^T$, we obtain

$$\boldsymbol{O}\cdot\boldsymbol{t}\cdot\boldsymbol{O}^T = a_0\boldsymbol{O}\cdot\boldsymbol{I}\cdot\boldsymbol{O}^T + 2\mu_v\boldsymbol{O}\cdot\boldsymbol{d}\cdot\boldsymbol{O}^T + a_2\boldsymbol{O}\cdot\boldsymbol{d}\cdot\boldsymbol{O}^T\cdot\boldsymbol{O}\cdot\boldsymbol{d}\cdot\boldsymbol{O}^T. \tag{6.167}$$

Since $\boldsymbol{t}$ and $\boldsymbol{d}$ are frame-indifferent tensors, and ϱ and the invariants of $\boldsymbol{d}$ are frame-indifferent scalars, (6.167) becomes

$$\boldsymbol{t}^* = a_0\boldsymbol{I} + 2\mu_v\boldsymbol{d}^* + a_2(\boldsymbol{d}^*)^2, \tag{6.168}$$

where

$$a_i, \mu_v = a_i, \mu_v(\varrho^*, I_d^*, II_d^*, III_d^*), \tag{6.169}$$

which has the same form as in the unstarred frame. This completes the proof.

Fluids characterised by the above non-linearity are called *Stokesian* or *non-Newtonian fluids*. Linear behaviour of a viscous fluid is generally referred to as *Newtonian* behaviour. Newtonian fluids will be studied in Sect. 7.3.

6.17.7 Incompressible Viscous Fluids

For an incompressible fluid, we have the kinematic constraint that the mass density is constant and equal to a known value, $\varrho = \varrho_0$. The principle of conservation of mass then yields $I_d = \operatorname{tr}\boldsymbol{d} = \operatorname{div}\vec{v} = 0$. Combining (6.47) and (6.165), the constitutive equation for an incompressible viscous fluid is

$$\boldsymbol{t} = -\pi\boldsymbol{I} + \boldsymbol{t}^D, \tag{6.170}$$

where the spherical part of the stress tensor $\boldsymbol{t}$, that is, the tensor $a_0\boldsymbol{I}$ in (6.165), is, without loss of generality, absorbed into the constraint pressure term $-\pi\boldsymbol{I}$. The constraint pressure π is an additional unknown independent field variable, which is determined through a solution of the field equations under the constraint $\varrho = \varrho_0$. The deviatoric part $\boldsymbol{t}^D$ of the stress tensor $\boldsymbol{t}$ is given by a constitutive equation of the form

$$\boldsymbol{t}^D = 2\mu_v\boldsymbol{d} + a_2\big(\boldsymbol{d}^2 + \frac{2}{3}II_d\,\boldsymbol{I}\big), \tag{6.171}$$

where

$$\mu_v, a_2 = \mu_v, a_2(II_d, III_d) \tag{6.172}$$

and the dependencies on ϱ and I_d are dropped since ϱ is a known constant and $I_d = 0$. To show this, we take the deviatoric part of the stress tensor (6.165),

$$\begin{aligned}\boldsymbol{t}^D &= a_0\boldsymbol{I} + 2\mu_v\boldsymbol{d} + a_2\boldsymbol{d}^2 - \frac{1}{3}\mathrm{tr}\big(a_0\boldsymbol{I} + 2\mu_v\boldsymbol{d} + a_2\boldsymbol{d}^2\big)\boldsymbol{I}\\ &= a_0\boldsymbol{I} + 2\mu_v\boldsymbol{d} + a_2\boldsymbol{d}^2 - a_0\boldsymbol{I} - \frac{1}{3}a_2\mathrm{tr}(\boldsymbol{d}^2)\boldsymbol{I}\\ &= 2\mu_v\boldsymbol{d} + a_2\boldsymbol{d}^2 - \frac{1}{3}a_2\big[(\mathrm{tr}\,\boldsymbol{d})^2 - 2II_d\big]\boldsymbol{I}\,,\end{aligned}$$

where (1.91) has been used in the last step. Since $\mathrm{tr}\,\boldsymbol{d} = 0$, we obtain (6.171).[8]

6.17.8 Viscous Heat-Conducting Fluids

A *viscous heat-conducting fluid* is defined by $N_C = 1$ and $N_\theta = N_g = 0$ such that (6.112) reduces to

$$\boldsymbol{t} = \hat{\boldsymbol{t}}\big(\boldsymbol{d}, \theta, \vec{g}_\theta, \varrho\big), \qquad \vec{q} = \hat{\vec{q}}\big(\boldsymbol{d}, \theta, \vec{g}_\theta, \varrho\big), \qquad \varepsilon = \hat{\varepsilon}\big(\boldsymbol{d}, \theta, \vec{g}_\theta, \varrho\big). \tag{6.173}$$

[8]The stress tensor $\boldsymbol{t}$, as any other tensor, can be decomposed into the spherical tensor $\sigma\boldsymbol{I}$ and the deviatoric stress tensor $\boldsymbol{t}^D$:

$$\boldsymbol{t} = \sigma\boldsymbol{I} + \boldsymbol{t}^D,$$

where $\sigma = \mathrm{tr}\,\boldsymbol{t}/3$ and $\mathrm{tr}\,\boldsymbol{t}^D = 0$. The scalar σ is thus the mean of the normal-stress components and is called the *mechanical pressure*. A characteristic feature of all fluids at rest is that they cannot support shear stresses (see Sects. 7.3 and 7.7). Consequently, the deviatoric stress identically vanishes. Choosing $p := -\sigma$, we obtain $\boldsymbol{t} = -p\boldsymbol{I}$. As shown in Sect. 6.14, the stress in a fluid at rest, called *hydrostatic pressure*, depends on the density alone.

Using the representation theorems (6.119), (6.124) and (6.132) for isotropic functions yields

$$
\begin{aligned}
\boldsymbol{t} &= a_0\boldsymbol{I} + 2\mu_v\boldsymbol{d} + a_2\boldsymbol{d}^2 + a_3\,\vec{g}_\theta \otimes \vec{g}_\theta + a_4\,\mathrm{sym}(\boldsymbol{d}\cdot\vec{g}_\theta \otimes \vec{g}_\theta) \\
&\quad + a_5\,\mathrm{sym}(\boldsymbol{d}^2\cdot\vec{g}_\theta \otimes \vec{g}_\theta), \\
\vec{q} &= -\kappa\,\vec{g}_\theta + c_1\,\boldsymbol{d}\cdot\vec{g}_\theta + c_2\,\boldsymbol{d}^2\cdot\vec{g}_\theta, \\
\varepsilon &= \hat{\varepsilon}\big(\varrho, \theta, I_d, II_d, III_d, \vec{g}_\theta\cdot\vec{g}_\theta, \vec{g}_\theta\cdot\boldsymbol{d}\cdot\vec{g}_\theta, \vec{g}_\theta\cdot\boldsymbol{d}^2\cdot\vec{g}_\theta\big),
\end{aligned}
\qquad (6.174)
$$

where the scalars a_i, μ_v, κ and c_i depend on the same set of arguments as the function $\hat{\varepsilon}$ of the internal energy. The proof of frame-indifference of these constitutive equations is similar to that for the viscous fluid.

Chapter 7
Entropy Principles

The constitutive theory of materials cannot be complete without thermodynamic considerations. In thermodynamics, two concepts are essential: energy and entropy. For energy, we have already introduced the principle of conservation of energy (4.10), known as the first law of thermodynamics. Entropy, however, is a less intuitive concept, although its existence is usually inferred from a more fundamental hypothesis known as the second law of thermodynamics.

The entropy principle requires the constitutive equations to be such that the entropy inequality (4.39) is satisfied identically for any thermodynamic process. From this point of view, like the principles of material frame-indifference and material symmetry, the entropy principle also imposes restrictions on constitutive equations. Thus, the exploitation of the second law of thermodynamics is a key part of material theory and could be included in Chap. 6. Thermodynamic requirements are, however, so important that we prefer to deal with them in a separate chapter.

7.1 The Clausius–Duhem Inequality

The second law of thermodynamics states that entropy production cannot be negative. Equation (4.39) expresses this law in the Eulerian form as

$$\varrho\gamma := \varrho\dot{\eta} + \operatorname{div}\vec{s} - \varrho z \geq 0, \tag{7.1}$$

where γ is the *entropy production density* per unit mass. This law does not, however, define internal dissipation mechanisms uniquely. We must therefore add additional information to be able to draw conclusions about the material's properties. In continuum thermodynamics, there are many dissipation postulates, some of which do not even involve entropy as a field variable. In Sects. 7.1–7.4, we will confine ourselves to considerations that lead to the so-called Clausius–Duhem inequality.

Z. Martinec, *Principles of Continuum Mechanics*, Nečas Center Series,
https://doi.org/10.1007/978-3-030-05390-1_7

This inequality results from two postulates, the validity of which can only be verified by physical experiments.

- We postulate the existence of a positive valued, frame-indifferent scalar function $T = T(\theta)$, where θ is the *empirical temperature*, such that T is independent of the material for which it is defined, and changes monotonically with θ. Hence, T possesses some degree of universality as a measure of warmness and is therefore called the *absolute temperature*. The absolute temperature may differ from the empirical temperature θ, which can be measured in the Celsius scale and may, in contrast to T, take negative values. We will present one possible relationship between the absolute and empirical temperatures in Sect. 7.6.
- We shall deal with *simple thermodynamic processes* for which the entropy flux $\vec{s}$ and entropy supply z are considered in the form[1]

$$\vec{s} = \frac{\vec{q}}{T}, \qquad z = \frac{h}{T}. \tag{7.2}$$

The entropy inequality (7.1) for a simple thermodynamic process becomes

$$\varrho\dot{\eta} + \operatorname{div}\left(\frac{\vec{q}}{T}\right) - \frac{\varrho h}{T} \geq 0, \tag{7.3}$$

which is known as the *Clausius–Duhem inequality*. This inequality need not to be satisfied for all densities ϱ, motions $\vec{\chi}$ and temperatures T, but for all possible thermodynamic processes. That is, for all solutions of the principles of conservation of mass, linear and angular momenta and energy, hence for all solutions of the field equations of thermomechanics. To satisfy these additional constraints when the Clausius–Duhem inequality is applied, the body force $\vec{f}$ and heat supply h, which are free-field variables of the principles of conservation of linear momentum and energy, may be considered from two different points of view.

The first view assumes that $\vec{f}$ and h may be assigned arbitrary values. Then, arbitrary histories of motion $\vec{\chi}$ and temperature T can be chosen, and an appropriate body force $\vec{f}$ and heat supply h can still be found to satisfy identically the linear momentum and energy equations. Hence, the linear momentum and energy equations impose no constraints when the Clausius–Duhem inequality is applied. It therefore remains to satisfy the continuity equation, which may be considered as an additional constraint to the Clausius–Duhem inequality.

The second view considers $\vec{f}$ and h as prior information that cannot be altered during the solution of a given physical problem. The linear momentum and energy equations must then be considered as additional constraints on the Clausius–Duhem inequality. Although the second view is correct, the first is usually applied due

[1]This postulate is acceptable for a *one-component body*; in mixtures, a more general form of the entropy flux is necessary. For instance, an additional term is added to $(7.2)_1$ such that the entropy flux and heat flux are not collinear vectors.

to its simplicity. Here, we also start with the first view in the application of the Clausius–Duhem inequality, but we will introduce an entropy principle that employs the second view in Sect. 7.5.

The heat supply h can be eliminated from (7.3) using the energy equation (4.38), obtaining

$$\varrho(T\dot{\eta} - \dot{\varepsilon}) + \boldsymbol{t} : \boldsymbol{d} - \frac{\vec{q} \cdot \operatorname{grad} T}{T} \geq 0, \tag{7.4}$$

which is known as the *reduced Clausius–Duhem inequality* in the Eulerian form. Alternatively, this inequality can be expressed in terms of quantities referred to the reference configuration. Combining the Lagrangian form of the energy equation (4.79) and the entropy inequality (4.83) yields

$$\varrho_0(T\dot{\eta} - \dot{\varepsilon}) + \boldsymbol{T}^{(2)} : \dot{\boldsymbol{E}} - \frac{\vec{Q} \cdot \operatorname{Grad} T}{T} \geq 0. \tag{7.5}$$

7.2 Application of the Clausius–Duhem Inequality to a Thermo-Elastic Solid

We first demonstrate the application of the Clausius–Duhem inequality to a thermoelastic solid. The constitutive equations for this material are given by (6.152). Since $\boldsymbol{C} = 2\boldsymbol{E} + \boldsymbol{I}$, the dependence of (6.152) on the Green deformation tensor $\boldsymbol{C}$ is replaced by an equivalent dependence on the Lagrangian strain tensor $\boldsymbol{E}$ such that

$$\boldsymbol{T}^{(2)} = \hat{\boldsymbol{T}}^{(2)}(\boldsymbol{E}, T, \operatorname{Grad} T), \qquad \vec{Q} = \hat{\vec{Q}}(\boldsymbol{E}, T, \operatorname{Grad} T), \qquad \varepsilon = \hat{\varepsilon}(\boldsymbol{E}, T, \operatorname{Grad} T), \tag{7.6}$$

where the absolute temperature T as a measure of warmness replaces the empirical temperature θ. The Clausius–Duhem inequality introduces a new variable, the entropy density η. Since η is not determined by the field equations, a constitutive relation must be established for it. According to the principle of equipresence, we choose the same set of independent variables for η as in (7.6) such that

$$\eta = \hat{\eta}(\boldsymbol{E}, T, \operatorname{Grad} T). \tag{7.7}$$

Differentiating ε and η with respect to time according to the chain rule and substituting the result into the reduced Clausius–Duhem inequality (7.5) yields

$$\begin{aligned} &\varrho_0\Bigg[\left(T\frac{\partial\hat{\eta}}{\partial T} - \frac{\partial\hat{\varepsilon}}{\partial T}\right)\dot{T} + \left(T\frac{\partial\hat{\eta}}{\partial \boldsymbol{E}} - \frac{\partial\hat{\varepsilon}}{\partial \boldsymbol{E}} + \frac{1}{\varrho_0}\hat{\boldsymbol{T}}^{(2)}\right) : \dot{\boldsymbol{E}} \\ &+ \left(T\frac{\partial\hat{\eta}}{\partial(\operatorname{Grad} T)} - \frac{\partial\hat{\varepsilon}}{\partial(\operatorname{Grad} T)}\right) \cdot (\operatorname{Grad} T)^{\cdot}\Bigg] - \frac{\vec{Q} \cdot \operatorname{Grad} T}{T} \geq 0. \end{aligned} \tag{7.8}$$

The inequality must be satisfied for all thermodynamic processes, that is, for all solutions of the principles of conservation of mass, linear and angular momenta and energy, as well as the constitutive equations. As previously explained, the linear momentum and energy equations impose no constraints when the Clausius–Duhem inequality is applied. Also the continuity equation plays no role here since, in the Lagrangian description, the mass density in the present configuration is determined by the motion function and mass density in the reference configuration (see Sect. 4.3.1). Hence, the inequality (7.8) must hold for all fields $\vec{\chi}$ and T with no additional restrictions.

Since $\dot{T}$, $\dot{\boldsymbol{E}}$ and $(\operatorname{Grad} T)^{\cdot}$ are not considered as independent variables in the constitutive equations (7.6) and (7.7), the prefactors of $\dot{T}$, $\dot{\boldsymbol{E}}$ and $(\operatorname{Grad} T)^{\cdot}$ in (7.8) are independent of these quantities. Hence, $\dot{T}$, $\dot{\boldsymbol{E}}$ and $(\operatorname{Grad} T)^{\cdot}$ occur in (7.8) as linear functions. Since $\dot{T}$, $\dot{\boldsymbol{E}}$ and $(\operatorname{Grad} T)^{\cdot}$ may take any arbitrary values, the inequality (7.8) will not be maintained for all $\dot{T}$, $\dot{\boldsymbol{E}}$ and $(\operatorname{Grad} T)^{\cdot}$ unless their prefactors vanish, that is,

$$\frac{\partial \hat{\eta}}{\partial T} = \frac{1}{T}\frac{\partial \hat{\varepsilon}}{\partial T}, \qquad \frac{1}{\varrho_0}\hat{\boldsymbol{T}}^{(2)} = \frac{\partial \hat{\varepsilon}}{\partial \boldsymbol{E}} - T\frac{\partial \hat{\eta}}{\partial \boldsymbol{E}}, \qquad \frac{\partial \hat{\eta}}{\partial (\operatorname{Grad} T)} = \frac{1}{T}\frac{\partial \hat{\varepsilon}}{\partial (\operatorname{Grad} T)}. \tag{7.9}$$

Under these constraints, the inequality (7.8) reduces to the *residual entropy inequality*

$$-\vec{Q}\cdot \operatorname{Grad} T \geq 0, \tag{7.10}$$

which is known as *Fourier's inequality*. This result of the second law of thermodynamics states that the heat flux has an orientation opposite to the temperature gradient, meaning that heat energy flows from hot to cold.

Differentiating $(7.9)_2$ and $(7.9)_3$ with respect to $\operatorname{Grad} T$ and $\boldsymbol{E}$, respectively, and assuming the exchange of the order of differentiation of $\hat{\varepsilon}$ and $\hat{\eta}$ with respect to $\boldsymbol{E}$ and $\operatorname{Grad} T$ gives

$$\frac{\partial \hat{\boldsymbol{T}}^{(2)}}{\partial (\operatorname{Grad} T)} = \boldsymbol{0},$$

which shows that the second Piola–Kirchhoff stress tensor $\boldsymbol{T}^{(2)}$ is independent of $\operatorname{Grad} T$. In a similar manner, it can be shown that $\hat{\varepsilon}$ and $\hat{\eta}$ are functions of $\boldsymbol{E}$ and T only. The constitutive equations $(7.6)_{1,3}$ and (7.7) reduce to

$$\boldsymbol{T}^{(2)} = \hat{\boldsymbol{T}}^{(2)}(\boldsymbol{E}, T), \qquad \varepsilon = \hat{\varepsilon}(\boldsymbol{E}, T), \qquad \eta = \hat{\eta}(\boldsymbol{E}, T). \tag{7.11}$$

Taking the total differential of η,

$$d\eta = \frac{\partial \hat{\eta}}{\partial \boldsymbol{E}} : d\boldsymbol{E} + \frac{\partial \hat{\eta}}{\partial T} dT, \tag{7.12}$$

and substituting for the partial derivatives of $\hat{\eta}$ from (7.9) gives *Gibbs' relation* for a thermoelastic solid:

$$d\eta = \frac{1}{T}\Big(d\varepsilon - \frac{1}{\varrho_0}\boldsymbol{T}^{(2)} : d\boldsymbol{E}\Big). \tag{7.13}$$

If the constitutive equations for ε and $\boldsymbol{T}^{(2)}$ are known, (7.13) allows the entropy η to be determined.

Moreover, the functions $\hat{\boldsymbol{T}}^{(2)}$, $\hat{\varepsilon}$ and $\hat{\eta}$ must satisfy the constraints in (7.9). Since $(7.9)_3$ is satisfied identically, it remains to satisfy $(7.9)_{1,2}$. To accomplish this, we introduce the *Helmholtz free energy density*

$$\psi := \varepsilon - T\eta = \hat{\psi}(\boldsymbol{E}, T) \tag{7.14}$$

and compute its total differential

$$d\psi = \frac{\partial \hat{\psi}}{\partial \boldsymbol{E}} : d\boldsymbol{E} + \frac{\partial \hat{\psi}}{\partial T} dT. \tag{7.15}$$

Differentiating (7.14) with respect to $\boldsymbol{E}$ and T and substituting the result into (7.15) gives

$$d\psi = \left(\frac{\partial \hat{\varepsilon}}{\partial \boldsymbol{E}} - T\frac{\partial \hat{\eta}}{\partial \boldsymbol{E}}\right) : d\boldsymbol{E} + \left(\frac{\partial \hat{\varepsilon}}{\partial T} - T\frac{\partial \hat{\eta}}{\partial T} - \hat{\eta}\right) dT, \tag{7.16}$$

which, in view of $(7.9)_{1,2}$, becomes

$$d\psi = \frac{1}{\varrho_0}\hat{\boldsymbol{T}}^{(2)} : d\boldsymbol{E} - \hat{\eta}\, dT. \tag{7.17}$$

Comparing (7.17) with (7.15), we deduce that

$$\boldsymbol{T}^{(2)} = \varrho_0 \frac{\partial \hat{\psi}}{\partial \boldsymbol{E}}, \qquad \eta = -\frac{\partial \hat{\psi}}{\partial T}. \tag{7.18}$$

The internal energy can also be derived from the potential ψ using (7.14),

$$\varepsilon = \hat{\psi} - T\frac{\partial \hat{\psi}}{\partial T}. \tag{7.19}$$

The constraints $(7.9)_{1,2}$ are now satisfied identically, which means that ψ serves as a *thermodynamic potential* for the second Piola–Kirchhoff stress tensor, entropy and internal energy.

We now confine ourselves to a special type of thermoelastic solid for which the constitutive functions are **isotropic** and **linear** with respect to $\boldsymbol{E}$, T and Grad T. The isotropy requirement means that the thermodynamic potential ψ is a function of the temperature T and the three invariants of $\boldsymbol{E}$ given by (1.91),

$$\psi = \hat{\psi}(I_E, II_E, III_E, T). \tag{7.20}$$

To obtain the stress and entropy constitutive equations, which are linear and uncoupled in $\boldsymbol{E}$ and T, function $\hat{\psi}$ can only include polynomials up to a second degree in $\boldsymbol{E}$ and T, that is,

$$\varrho_0\hat{\psi} = \varrho_0\psi_0 + a_0 I_E + a_1(I_E)^2 + a_2 II_E + a_3 I_E T + a_4 T + a_5 T^2, \tag{7.21}$$

where ψ_0 and a_i, $i = 0, 1, \ldots, 5$, are constants independent of $\boldsymbol{E}$ and T. The third-order terms III_E, $II_E T$, $I_E^2 T$, $I_E T^2$, T^3 and higher-order terms would generate non-linear terms in the constitutive equations and are therefore omitted from $\hat{\psi}$ in the linear case. Substituting (7.21) into (7.18) and using the relationships

$$\frac{\partial I_E}{\partial \boldsymbol{E}} = \boldsymbol{I}, \qquad \frac{\partial II_E}{\partial \boldsymbol{E}} = I_E\boldsymbol{I} - \boldsymbol{E}, \tag{7.22}$$

we obtain the constitutive equations for $\boldsymbol{T}^{(2)}$ and η,

$$\boldsymbol{T}^{(2)} = (\sigma + \lambda\, \mathrm{tr}\boldsymbol{E} - \beta T)\boldsymbol{I} + 2\mu\boldsymbol{E}, \qquad \eta = \eta_0 + \frac{\beta}{\varrho_0}\mathrm{tr}\boldsymbol{E} + \frac{\gamma}{\varrho_0}T, \tag{7.23}$$

where new notations have been introduced: $\sigma = a_0$, $\lambda = 2a_1 + a_2$, $2\mu = -a_2$, $\beta = -a_3$, $\varrho_0\eta_0 = -a_4$ and $\gamma = -2a_5$. The constitutive equation for the internal energy can be derived from (7.19),

$$\varrho_0\varepsilon = \varrho_0\psi_0 + \sigma I_E + \frac{1}{2}(\lambda + 2\mu)(I_E)^2 - 2\mu II_E + \frac{1}{2}\gamma T^2. \tag{7.24}$$

To examine the residual inequality (7.10), we consider, for simplicity, a thermoelastic solid for which the constitutive function for the heat flux $\vec{Q}$ is an **isotropic** vector-valued function of $\boldsymbol{E}$, T and Grad T. In view of $(6.153)_2$, $\vec{Q}$ is expressed as

$$\vec{Q} = -\boldsymbol{\kappa} \cdot \mathrm{Grad}\, T, \tag{7.25}$$

where the *thermal conductivity tensor* $\boldsymbol{\kappa}$ has the form

$$\boldsymbol{\kappa} = \kappa\boldsymbol{I} - c_1\boldsymbol{E} - c_2\boldsymbol{E}^2. \tag{7.26}$$

Here, κ, c_1 and c_2 are functions of T, the invariants of $\boldsymbol{E}$ and Grad T, and the combined invariants of $\boldsymbol{E}$ and Grad T. Thus, $\vec{Q}$ is, in general, a non-linear function of $\boldsymbol{E}$, T and Grad T. Substituting (7.25) and (7.26) into (7.10) leads to

$$\kappa(\mathrm{Grad}\,T \cdot \mathrm{Grad}\,T) - c_1(\mathrm{Grad}\,T \cdot \boldsymbol{E} \cdot \mathrm{Grad}\,T) - c_2(\mathrm{Grad}\,T \cdot \boldsymbol{E}^2 \cdot \mathrm{Grad}\,T) \geq 0. \tag{7.27}$$

In view of the symmetry and positive definiteness of $\boldsymbol{E}$, all three terms in brackets are non-negative. To satisfy the inequality for all Grad T and $\boldsymbol{E}$, it must hold that

$$\kappa \geq 0, \qquad c_1 \leq 0, \qquad c_2 \leq 0. \tag{7.28}$$

In the special case where the heat flux $\vec{Q}$ is collinear with Grad T, $c_1 = c_2 = 0$ and (7.25) reduces to

$$\vec{Q} = -\kappa\,\mathrm{Grad}\,T, \tag{7.29}$$

where the thermal conductivity κ does not depend on $\boldsymbol{E}$, T and Grad T in the linear case. We conclude that (7.23), (7.24) and (7.29) form the constitutive equations for an isotropic linear thermoelastic solid. The inequality $\kappa \geq 0$ furthermore expresses the fact that heat flows from hot to cold in such a solid.

7.3 Application of the Clausius–Duhem Inequality to a Viscous Fluid

We next apply the Clausius–Duhem inequality to a viscous fluid. Compared to the preceding case, this is a more challenging task since the continuity equation must now be satisfied as an additional constraint on the Clausius–Duhem inequality.

The constitutive equation for the Cauchy stress tensor of a viscous fluid is given by (6.164) and the corresponding constitutive relations for the heat flux, internal energy and entropy are established according to the principle of equipresence. Hence,

$$\boldsymbol{t} = \hat{\boldsymbol{t}}(\boldsymbol{d}, \varrho), \qquad \vec{q} = \hat{\vec{q}}(\boldsymbol{d}, \varrho), \qquad \varepsilon = \hat{\varepsilon}(\boldsymbol{d}, \varrho), \qquad \eta = \hat{\eta}(\boldsymbol{d}, \varrho). \tag{7.30}$$

Differentiating ε and η with respect to time according to the chain rule and substituting the result into the reduced Clausius–Duhem inequality (7.4) yields an inequality

$$\varrho\left[\left(T\frac{\partial\hat{\eta}}{\partial\varrho} - \frac{\partial\hat{\varepsilon}}{\partial\varrho}\right)\dot{\varrho} + \left(T\frac{\partial\hat{\eta}}{\partial\boldsymbol{d}} - \frac{\partial\hat{\varepsilon}}{\partial\boldsymbol{d}}\right) : \dot{\boldsymbol{d}}\,\right] + \boldsymbol{t} : \boldsymbol{d} - \frac{\vec{q}\cdot\mathrm{grad}\,T}{T} \geq 0 \tag{7.31}$$

that must be satisfied for all thermodynamic processes. Again, only the principle of conservation of mass imposes an additional constraint on possible thermodynamic processes. Hence, $\dot{\varrho}$ cannot be chosen arbitrarily, but in accordance with the continuity equation (4.18) such that

$$\dot{\varrho} = -\varrho \operatorname{div} \vec{v} = -\varrho \operatorname{tr} \boldsymbol{d} = -\varrho \boldsymbol{I} : \boldsymbol{d}. \tag{7.32}$$

Substituting $\dot{\varrho}$ into (7.31), the first and third terms may be combined to give

$$\varrho \left(T \frac{\partial \hat{\eta}}{\partial \boldsymbol{d}} - \frac{\partial \hat{\varepsilon}}{\partial \boldsymbol{d}} \right) : \dot{\boldsymbol{d}} + \left[\boldsymbol{t} - \varrho^2 \left(T \frac{\partial \hat{\eta}}{\partial \varrho} - \frac{\partial \hat{\varepsilon}}{\partial \varrho} \right) \boldsymbol{I} \right] : \boldsymbol{d} - \frac{\vec{q} \cdot \operatorname{grad} T}{T} \geq 0, \tag{7.33}$$

which must hold for all fields ϱ, $\vec{\chi}$ and T with no additional constraints.

The quantities $\dot{\boldsymbol{d}}$ and $\operatorname{grad} T$ are not considered as independent variables in the constitutive equations (7.30), so the prefactors of $\dot{\boldsymbol{d}}$ and $\operatorname{grad} T$ are independent of these quantities. Hence, $\dot{\boldsymbol{d}}$ and $\operatorname{grad} T$ occur in (7.33) as linear functions. Since $\dot{\boldsymbol{d}}$ and $\operatorname{grad} T$ may take any arbitrary values, the inequality (7.33) will not be maintained for all $\dot{\boldsymbol{d}}$ and $\operatorname{grad} T$ unless their prefactors vanish, which gives

$$T \frac{\partial \hat{\eta}}{\partial \boldsymbol{d}} - \frac{\partial \hat{\varepsilon}}{\partial \boldsymbol{d}} = 0, \qquad \vec{q} = \vec{0}. \tag{7.34}$$

Under these constraints, (7.33) reduces to the residual inequality

$$\left[\boldsymbol{t} - \varrho^2 \left(T \frac{\partial \hat{\eta}}{\partial \varrho} - \frac{\partial \hat{\varepsilon}}{\partial \varrho} \right) \boldsymbol{I} \right] : \boldsymbol{d} \geq 0. \tag{7.35}$$

Observe that the first term in the constraint $(7.34)_1$ is linear in T; by a similar argument as for $\dot{\boldsymbol{d}}$ and $\operatorname{grad} T$, the factor multiplying T must vanish, giving

$$\frac{\partial \hat{\eta}}{\partial \boldsymbol{d}} = \frac{\partial \hat{\varepsilon}}{\partial \boldsymbol{d}} = \boldsymbol{0}. \tag{7.36}$$

In addition, the second term in (7.35) is also linear in T, so it must hold that

$$\frac{\partial \hat{\eta}}{\partial \varrho} = 0. \tag{7.37}$$

Eliminating all linear terms in T reduces the constitutive equations (7.30) to

$$\varepsilon = \hat{\varepsilon}(\varrho), \qquad \eta = \eta_0, \qquad \vec{q} = \vec{0}, \tag{7.38}$$

where η_0 is independent of ϱ and $\boldsymbol{d}$.[2] The result $\vec{q} = \vec{0}$ means the fluid specified by (7.30) does not conduct heat, which is a consequence of excluding temperature and temperature gradients from the constitutive variables. The inequality (7.35) now simplifies to

$$\gamma(\varrho, \boldsymbol{d}) \geq 0, \tag{7.39}$$

where the entropy production has the form

$$\gamma(\varrho, \boldsymbol{d}) := (\boldsymbol{t} + p\boldsymbol{I}) : \boldsymbol{d}, \tag{7.40}$$

and p is the *thermodynamic pressure*,

$$p = \hat{p}(\varrho) := \varrho^2 \frac{\partial \hat{\varepsilon}}{\partial \varrho}. \tag{7.41}$$

The last equation defines the thermodynamic pressure when the constitutive equation for the internal energy is given. The constitutive equation $p = \hat{p}(\varrho)$ is also known as the *equation of state for pressure*.

Taking the differential of $(7.38)_1$ and substituting for $\partial\hat{\varepsilon}/\partial\varrho$ from (7.41) leads to

$$d\varepsilon = \frac{p}{\varrho^2} d\varrho, \tag{7.42}$$

which can be expressed in terms of the *specific volume* $v := 1/\varrho$ and $dv = -d\varrho/\varrho^2$ to give Gibbs' relation for a viscous fluid:

$$d\varepsilon = -p\, dv. \tag{7.43}$$

To examine the residual inequality (7.39), we now define *mechanical equilibrium* as a mechanical process with a uniform (constant in space) and stationary (constant in time) velocity field. Mathematically, mechanical equilibrium is defined by

$$\vec{v}(\vec{x}, t) = \text{const}, \tag{7.44}$$

which implies that the strain-rate tensor $\boldsymbol{d}$ vanishes in mechanical equilibrium,

$$\boldsymbol{d}|_{\mathrm{E}} = \boldsymbol{0}, \tag{7.45}$$

where the subscript E denotes the state of mechanical equilibrium. Consequently,

$$\gamma(\varrho, \boldsymbol{d})|_{\mathrm{E}} = 0 \tag{7.46}$$

[2]A thermodynamic process during which the entropy remains constant is called an *isentropic process*.

in view of (7.40). Together with the residual inequality (7.39), we deduce that the entropy production γ is minimal in mechanical equilibrium and the value of this minimum is zero. According to the theorem of extrema of functions of several variables, the necessary condition for γ to reach a minimum is that the first derivatives of γ with respect to $\boldsymbol{d}$ vanish,

$$\left.\frac{\partial \gamma}{\partial \boldsymbol{d}}\right|_{\mathrm{E}} = \boldsymbol{0}. \tag{7.47}$$

Differentiating (7.40) with respect to $\boldsymbol{d}$,

$$\frac{\partial \gamma}{\partial \boldsymbol{d}} = \boldsymbol{t} + p\boldsymbol{I},$$

and evaluating the result in mechanical equilibrium yields

$$\boldsymbol{t}|_{\mathrm{E}} = -p\boldsymbol{I}, \tag{7.48}$$

which shows that, in mechanical equilibrium, the stress is spherical and determined by the thermodynamic pressure.

Comparing the constitutive equation (6.165) in mechanical equilibrium, i.e.,

$$\boldsymbol{t}|_{\mathrm{E}} = a_0|_{\mathrm{E}}\,(\varrho)\boldsymbol{I}, \tag{7.49}$$

with (7.48) shows that

$$a_0|_{\mathrm{E}}\,(\varrho) = -p(\varrho), \tag{7.50}$$

which motivates the function a_0 in (6.166) to be decomposed as

$$a_0(\varrho, I_d, II_d, III_d) = -p(\varrho) + \alpha_0(\varrho, I_d, II_d, III_d), \tag{7.51}$$

where the non-equilibrium part α_0 of stress vanishes in mechanical equilibrium,

$$\alpha_0|_{\mathrm{E}} = 0. \tag{7.52}$$

Satisfying the residual inequality (7.39) is non-trivial for the general non-linear constitutive equation (6.165), which now takes the form

$$\boldsymbol{t} = (-p + \alpha_0)\boldsymbol{I} + 2\mu_v\boldsymbol{d} + a_2\boldsymbol{d}^2, \tag{7.53}$$

where

$$\alpha_0,\ \mu_v,\ a_2 = \alpha_0,\ \mu_v,\ a_2(\varrho, I_d, II_d, III_d). \tag{7.54}$$

We will therefore consider a viscous fluid that is **linear** with respect to $\boldsymbol{d}$. In this case, (7.53) simplifies to

$$\boldsymbol{t} = (-p + \lambda_v \operatorname{tr} \boldsymbol{d})\boldsymbol{I} + 2\mu_v \boldsymbol{d}, \tag{7.55}$$

where p, λ_v and μ_v are functions of ϱ only. A fluid characterised by the constitutive equation (7.55) is known as a *compressible Newtonian viscous fluid*. The parameters λ_v and μ_v are called, respectively, the *dilatational* and *shear viscosities*.

To draw further conclusions from the residual inequality (7.39), we substitute (7.55) into (7.40) to obtain

$$\begin{aligned}\gamma(\varrho, \boldsymbol{d}) &= \lambda_v \operatorname{tr} \boldsymbol{d}(\boldsymbol{I} : \boldsymbol{d}) + 2\mu_v(\boldsymbol{d} : \boldsymbol{d}) \\ &= \lambda_v(\operatorname{tr} \boldsymbol{d})^2 + 2\mu_v(\boldsymbol{d} : \boldsymbol{d}) \\ &= \lambda_v\Big(\sum_i d_{ii}\Big)^2 + 2\mu_v \sum_i d_{ii}^2 + 2\mu_v \sum_{\substack{ij \\ i \neq j}} d_{ij}^2,\end{aligned} \tag{7.56}$$

where $\boldsymbol{d} : \boldsymbol{I} = \operatorname{tr} \boldsymbol{d}$ has been applied, and the summation indices i and j range from 1 to 3. Introducing a 9 component vector

$$\vec{z} = (d_{11}, d_{22}, d_{33}, d_{12}, d_{13}, d_{23}, d_{21}, d_{31}, d_{32}),$$

the entropy production takes the form

$$\gamma(\varrho, \boldsymbol{d}) = \vec{z} \cdot \boldsymbol{A} \cdot \vec{z}, \tag{7.57}$$

where the 9×9 matrix $\boldsymbol{A}$ is block-diagonal,

$$\boldsymbol{A} = \begin{pmatrix} \lambda_v \mathbf{1}_{3\times3} + 2\mu_v \boldsymbol{I}_{3\times3} & 0 \\ 0 & 2\mu_v \boldsymbol{I}_{6\times6} \end{pmatrix}. \tag{7.58}$$

Here, $\mathbf{1}_{3\times3}$ is the matrix composed of only ones, and $\boldsymbol{I}_{3\times3}$ and $\boldsymbol{I}_{6\times6}$ are the third- and sixth-order identity matrices. In view of (7.57), the residual inequality (7.39) is satisfied if the matrix $\boldsymbol{A}$ is positive semi-definite. According to *Sylvester's criterion*, a matrix is positive semi-definite if and only if its principal minors[3] are all non-

[3] A *principal submatrix* of a square matrix $\boldsymbol{A}$ is the matrix obtained by deleting any k rows and the corresponding k columns. The determinant of a principal submatrix is called the *principal minor* of $\boldsymbol{A}$.

negative. Applying this criterion to the matrix $\boldsymbol{A}$ gives the conditions

$$\lambda_v + 2\mu_v \geq 0,$$

$$\begin{vmatrix} \lambda_v + 2\mu_v & \lambda_v \\ \lambda_v & \lambda_v + 2\mu_v \end{vmatrix} = 4\mu_v(\lambda_v + \mu_v) \geq 0,$$

$$\begin{vmatrix} \lambda_v + 2\mu_v & \lambda_v & \lambda_v \\ \lambda_v & \lambda_v + 2\mu_v & \lambda_v \\ \lambda_v & \lambda_v & \lambda_v + 2\mu_v \end{vmatrix} = 4\mu_v^2(3\lambda_v + 2\mu_v) \geq 0,$$

$$\mu_v \geq 0.$$

The four conditions are satisfied simultaneously if

$$3\lambda_v + 2\mu_v \geq 0, \qquad \mu_v \geq 0. \tag{7.59}$$

An alternative form of (7.55) is obtained by decomposing the strain-rate tensor $\boldsymbol{d}$ into spherical and deviatoric parts, i.e., $\boldsymbol{d} = (\mathrm{tr}\,\boldsymbol{d})\,\boldsymbol{I}/3 + \boldsymbol{d}^D$, resulting in

$$\boldsymbol{t} = (-p + k_v \mathrm{tr}\,\boldsymbol{d})\boldsymbol{I} + 2\mu_v \boldsymbol{d}^D, \tag{7.60}$$

where

$$k_v := \lambda_v + \frac{2}{3}\mu_v \tag{7.61}$$

is the *bulk viscosity*. The condition (7.59) can then be expressed as

$$k_v \geq 0, \qquad \mu_v \geq 0. \tag{7.62}$$

We thus conclude that in order to satisfy the Clausius–Duhem inequality for a linear viscous fluid, the bulk viscosity k_v and shear viscosity μ_v must be non-negative functions of ϱ.

7.4 Application of the Clausius–Duhem Inequality to an Incompressible Viscous Fluid

We extend the derivation in the preceding section to the case of an incompressible viscous fluid. For such a material, the mass density ϱ remains unchanged under deformation, equal to a known value, $\varrho = \varrho_0$, and, in contrast to the preceding case, must be excluded from the list of independent constitutive variables. The stress constitutive equation for an incompressible viscous fluid is given by (6.170):

$$\boldsymbol{t} = -\pi \boldsymbol{I} + \boldsymbol{t}^D. \tag{7.63}$$

Note that, in contrast to the thermodynamic pressure $p(\varrho)$, the constraint pressure π is now an additional independent field variable. The constitutive equations for $\boldsymbol{t}^D$, $\vec{q}$, ε and η have the same form as (7.30), except that all functions are now independent of ϱ such that

$$\boldsymbol{t}^D = \hat{\boldsymbol{t}}^D(\boldsymbol{d}), \qquad \vec{q} = \hat{\vec{q}}(\boldsymbol{d}), \qquad \varepsilon = \hat{\varepsilon}(\boldsymbol{d}), \qquad \eta = \hat{\eta}(\boldsymbol{d}). \tag{7.64}$$

The inequality (7.31) then reduces to

$$\varrho\left(T\frac{\partial\hat{\eta}}{\partial \boldsymbol{d}} - \frac{\partial\hat{\varepsilon}}{\partial \boldsymbol{d}}\right) : \dot{\boldsymbol{d}} + \boldsymbol{t} : \boldsymbol{d} - \frac{\vec{q}\cdot \operatorname{grad} T}{T} \geq 0, \tag{7.65}$$

which must be satisfied for all thermodynamic processes. As before, only the principle of conservation of mass imposes an additional constraint on possible thermodynamic processes. For an incompressible material, this constraint is, however, replaced by the kinematic constraint

$$\operatorname{div}\vec{v} = \operatorname{tr}\boldsymbol{d} = 0. \tag{7.66}$$

To satisfy the inequality (7.65) subject to the constraint (7.66), we extend (7.65) such that (7.66) is explicitly introduced through a Lagrange multiplier λ,

$$\varrho\left(T\frac{\partial\hat{\eta}}{\partial \boldsymbol{d}} - \frac{\partial\hat{\varepsilon}}{\partial \boldsymbol{d}}\right) : \dot{\boldsymbol{d}} + \boldsymbol{t} : \boldsymbol{d} - \frac{\vec{q}\cdot \operatorname{grad} T}{T} + \lambda \operatorname{tr}\boldsymbol{d} \geq 0, \tag{7.67}$$

which must be satisfied for all fields $\vec{\chi}$ and T, and Lagrange multipliers λ, with no additional constraints. Since λ accounts for the additional constraint (7.66), which is independent of the field equations, λ represents an additional field variable that is independent of $\vec{\chi}$ and T. We again observe that (7.67) is linear in terms of the variables $\dot{\boldsymbol{d}}$ and $\operatorname{grad} T$, which may have arbitrarily assigned values. Thus, the prefactors of these two variables must vanish, yielding

$$\varepsilon = \varepsilon_0, \qquad \eta = \eta_0, \qquad \vec{q} = \vec{0}, \tag{7.68}$$

where ε_0 and η_0 are independent of $\boldsymbol{d}$. Under this constraint, the term $\boldsymbol{t} : \boldsymbol{d}$ occurring in (7.67) can be arranged as

$$\begin{aligned}
\boldsymbol{t} : \boldsymbol{d} &= (-\pi\boldsymbol{I} + \boldsymbol{t}^D) : \left(\frac{1}{3}(\operatorname{tr}\boldsymbol{d})\boldsymbol{I} + \boldsymbol{d}^D\right) \\
&= \frac{1}{3}\operatorname{tr}\boldsymbol{d}\,(-\pi\boldsymbol{I} + \boldsymbol{t}^D) : \boldsymbol{I} + (-\pi\boldsymbol{I} + \boldsymbol{t}^D) : \boldsymbol{d}^D \\
&= \frac{1}{3}\operatorname{tr}\boldsymbol{d}\,(-3\pi + \operatorname{tr}\boldsymbol{t}^D) - \pi\operatorname{tr}\boldsymbol{t}^D + \boldsymbol{t}^D : \boldsymbol{d}^D = -\pi\operatorname{tr}\boldsymbol{d} + \boldsymbol{t}^D : \boldsymbol{d}^D,
\end{aligned}$$

so that (7.67) becomes

$$\boldsymbol{t}^D : \boldsymbol{d}^D + (\lambda - \pi)\operatorname{tr}\boldsymbol{d} \geq 0. \tag{7.69}$$

Since $\operatorname{tr}\boldsymbol{d}$ can be assigned an arbitrary value, this inequality can only be satisfied if the Lagrange multiplier is equal to the constraint pressure,

$$\lambda = \pi. \tag{7.70}$$

Consequently, (7.69) reduces to

$$\gamma(\boldsymbol{d}^D) := \boldsymbol{t}^D : \boldsymbol{d}^D \geq 0. \tag{7.71}$$

In mechanical equilibrium, $\boldsymbol{d} = \boldsymbol{0}$, hence $\boldsymbol{d}^D = \boldsymbol{0}$, leading to

$$\gamma(\boldsymbol{0}) = 0. \tag{7.72}$$

The condition

$$\left.\frac{\partial \gamma}{\partial \boldsymbol{d}^D}\right|_{\mathrm{E}} = \boldsymbol{0} \tag{7.73}$$

for the minimum of γ in mechanical equilibrium yields $\boldsymbol{t}^D\big|_{\mathrm{E}} = 0$, or, in view of (7.63),

$$\boldsymbol{t}|_{\mathrm{E}} = -\pi \boldsymbol{I}. \tag{7.74}$$

We conclude that, for an incompressible viscous fluid, the stress is spherical in mechanical equilibrium.

For an incompressible viscous fluid that is **linear** with respect to $\boldsymbol{d}$, the constitutive equation (6.171) simplifies to

$$\boldsymbol{t}^D = 2\mu_v \boldsymbol{d}^D, \tag{7.75}$$

where the shear viscosity μ_v is independent of $\boldsymbol{d}^D$. Substituting (7.75) into (7.71) leads to

$$2\mu_v(\boldsymbol{d}^D : \boldsymbol{d}^D) \geq 0. \tag{7.76}$$

Since the last equation holds for an arbitrary (but non-zero) tensor $\boldsymbol{d}^D$, the shear viscosity must be non-negative,

$$\mu_v \geq 0. \tag{7.77}$$

Note that a fluid characterised by the constitutive equations (7.63) and (7.75) is known as an *incompressible Newtonian viscous fluid.*

7.5 The Müller–Liu Entropy Principle

Despite the widespread use of the Clausius–Duhem inequality, it contains certain limitations that might be violated in some physical situations. These limitations arise from:

- The existence of the absolute temperature T is postulated.
- The entropy flux $\vec{s}$ and the entropy supply z are chosen in the form (7.2).
- The external body force $\vec{f}$ and the external heat supply h are assumed to be assigned arbitrary values such that the conservation laws of momentum and energy are identically satisfied.

We now consider an alternative concept of the entropy inequality, formulated by Müller (1967) and Liu (1972), that imposes a different type of additional information on the entropy inequality. The Müller–Liu entropy principle is also based on the second law of thermodynamics (7.1) and, in addition, assumes that:

- The entropy η and entropy flux $\vec{s}$ are frame-indifferent quantities, for which the principle of equipresence is valid, that is, the constitutive equations for η and $\vec{s}$ depend on the same constitutive variables as the constitutive equations for $\boldsymbol{t}$, $\vec{q}$ and ε.
- The body force $\vec{f}$, heat supply h and entropy supply z do not influence the material's properties.
- There exist welded, material contact discontinuity surfaces between two materials across which the empirical temperature and the normal component of the entropy flux vector are continuous,

$$[\theta]_-^+ = 0, \qquad \left[\vec{n} \cdot \vec{s}\,\right]_-^+ = 0. \tag{7.78}$$

  Such hypothetical discontinuities are termed *ideal walls*.[4] Note that the jump condition $(7.78)_2$ on an ideal wall is a particular form of the jump entropy inequality (4.49).

The second assumption implies that the thermodynamic constraints do not depend on the presence or absence of the external supplies $\vec{f}$, h and z. Hence, it suffices to consider the thermodynamic constraints only for a *supply-free body*,

[4]This assumption is modified for a multicomponent system by the existence of semipermeable membranes.

which is a body not supplied by external sources. The entropy inequality (7.1) for such a body becomes

$$\varrho\dot{\eta} + \operatorname{div}\vec{s} \geq 0, \tag{7.79}$$

which must be satisfied for all possible thermodynamic processes that are solutions of the principles of conservation of mass, linear and angular momenta, and energy. In contrast to the Clausius–Duhem inequality, the Müller–Liu entropy principle considers all these conservation principles as additional constraints on the entropy inequality (7.79). To satisfy this inequality subject to constraints (4.18), (4.24) and (4.38) for a supply-free body, an *extended entropy inequality*

$$\varrho\dot{\eta} + \operatorname{div}\vec{s} - \lambda^{\varrho}\big(\dot{\varrho} + \varrho\operatorname{div}\vec{v}\big) - \vec{\lambda}^{v}\cdot\big(\varrho\dot{\vec{v}} - \operatorname{div}\boldsymbol{t}\big) - \lambda^{\varepsilon}\big(\varrho\dot{\varepsilon} - \boldsymbol{t} : \boldsymbol{d} + \operatorname{div}\vec{q}\,\big) \geq 0 \tag{7.80}$$

must be satisfied for all densities ϱ, motions $\vec{\chi}$ and temperatures θ, and also for all Lagrange multipliers λ^{ϱ}, $\vec{\lambda}^{v}$ and λ^{ε}, with no additional constraints.

The extended entropy inequality (7.80) can alternatively be expressed in terms of quantities referred to the reference configuration. Combining the Lagrangian form of the entropy inequality (4.83), the equation of motion (4.69) and the energy equation (4.79) for a supply-free body yields

$$\varrho_0\dot{\eta} + \operatorname{Div}\vec{S} - \vec{\Lambda}^{v}\cdot\big(\varrho_0\dot{\vec{v}} - \operatorname{Div}(\boldsymbol{T}^{(2)}\cdot\boldsymbol{F}^{T})\big) - \Lambda^{\varepsilon}\big(\varrho_0\dot{\varepsilon} - \boldsymbol{T}^{(2)} : \dot{\boldsymbol{E}} + \operatorname{Div}\vec{Q}\,\big) \geq 0. \tag{7.81}$$

The continuity equation is not included in the last constraint since, in the Lagrangian description, the mass density in the present configuration is determined by the motion and mass density in the reference configuration (see Sect. 4.3.1). The inequality (7.81) must hold for all motions $\vec{\chi}$, temperatures θ and Lagrange multipliers $\vec{\Lambda}^{v}$ and Λ^{ε}, with no additional constraints.

To express the jump conditions (7.78) in the Lagrangian form, we assume that (1) an ideal wall is a material discontinuity surface along which the velocity $\vec{W}$, defined by (4.67), vanishes, and (2) an ideal wall is a welded discontinuity surface with no tangential slip, such that the Lagrangian area element dA changes continuously across the wall. Under these assumptions, along with the entropy inequality (4.84), the jump condition $(7.78)_2$ takes the Lagrangian form $[\vec{N}\cdot\vec{S}]_-^+ = 0$. In other words, the normal component of the Lagrangian entropy flux vector is continuous across the Lagrangian representation of this ideal wall. In summary, the two jump conditions (7.78) on an ideal wall can now be expressed in terms of quantities referred to the reference configuration in the form

$$[\theta]_-^+ = 0, \qquad \big[\vec{N}\cdot\vec{S}\,\big]_-^+ = 0, \tag{7.82}$$

where $\theta = \theta(\vec{X}, t)$ is the Lagrangian representation of the empirical temperature.

Liu (1972) proved that satisfying the extended entropy inequality (7.80) for unconstrained fields is equivalent to satisfying the original entropy inequality (7.1) and the field equations for mass, linear momentum and energy simultaneously. The same statement holds for the Lagrangian form of the extended entropy inequality (7.81) and the Lagrangian form of the field equations of linear momentum and energy that satisfy the entropy inequality (4.83). It is a straightforward, but laborious, derivation to satisfy the extended entropy inequality (7.80) or (7.81) by determining the unknown Lagrange multipliers.

7.6 Application of the Müller–Liu Entropy Principle to a Thermoelastic Solid

We now demonstrate the application of the Müller–Liu entropy principle to a thermoelastic solid. The constitutive equations for this material are given by (6.152):

$$\boldsymbol{T}^{(2)} = \hat{\boldsymbol{T}}^{(2)}(\boldsymbol{E}, \theta, \operatorname{Grad}\theta), \qquad \vec{Q} = \hat{\vec{Q}}(\boldsymbol{E}, \theta, \operatorname{Grad}\theta), \qquad \varepsilon = \hat{\varepsilon}(\boldsymbol{E}, \theta, \operatorname{Grad}\theta), \tag{7.83}$$

where the dependence on the Green deformation tensor $\boldsymbol{C}$ is replaced by an equivalent dependence on the Lagrangian strain tensor $\boldsymbol{E}$. According to the principle of equipresence, we choose the same set of constitutive variables for both the entropy η and entropy flux $\vec{s}$ such that

$$\eta = \hat{\eta}(\boldsymbol{E}, \theta, \operatorname{Grad}\theta), \qquad \vec{S} = \hat{\vec{S}}(\boldsymbol{E}, \theta, \operatorname{Grad}\theta). \tag{7.84}$$

Substituting the constitutive equations (7.83) and (7.84) into the entropy inequality (7.81) and differentiating the result with respect to time according to the chain rule yields

$$\begin{aligned}
&\varrho_0\Big(\frac{\partial\hat{\eta}}{\partial E_{KL}}\dot{E}_{KL} + \frac{\partial\hat{\eta}}{\partial\theta}\dot{\theta} + \frac{\partial\hat{\eta}}{\partial\theta_{,K}}\dot{\theta}_{,K}\Big) + \frac{\partial\hat{S}_K}{\partial E_{PQ}}E_{PQ,K} + \frac{\partial\hat{S}_K}{\partial\theta}\theta_{,K} + \frac{\partial\hat{S}_K}{\partial\theta_{,P}}\theta_{,PK} \\
&-\Lambda^v_k\varrho_0\dot{v}_k + \Lambda^v_k\Big(\frac{\partial(\hat{\boldsymbol{T}}^{(2)}\cdot\boldsymbol{F}^T)_{Lk}}{\partial E_{PQ}}E_{PQ,L} + \frac{\partial(\hat{\boldsymbol{T}}^{(2)}\cdot\boldsymbol{F}^T)_{Lk}}{\partial\theta}\theta_{,L} + \frac{\partial(\hat{\boldsymbol{T}}^{(2)}\cdot\boldsymbol{F}^T)_{Lk}}{\partial\theta_{,P}}\theta_{,PL}\Big) \\
&-\Lambda^\varepsilon\varrho_0\Big(\frac{\partial\hat{\varepsilon}}{\partial E_{KL}}\dot{E}_{KL} + \frac{\partial\hat{\varepsilon}}{\partial\theta}\dot{\theta} + \frac{\partial\hat{\varepsilon}}{\partial\theta_{,K}}\dot{\theta}_{,K}\Big) + \Lambda^\varepsilon\hat{T}^{(2)}_{KL}\dot{E}_{KL} \\
&-\Lambda^\varepsilon\Big(\frac{\partial\hat{Q}_K}{\partial E_{PQ}}E_{PQ,K} + \frac{\partial\hat{Q}_K}{\partial\theta}\theta_{,K} + \frac{\partial\hat{Q}_K}{\partial\theta_{,P}}\theta_{,PK}\Big) \geq 0
\end{aligned} \tag{7.85}$$

in index notation. According to Liu's theorem (Liu 1972), the Lagrange multipliers Λ_k^v and Λ^ε are only dependent on the independent constitutive variables $\boldsymbol{E}$, θ and $\operatorname{Grad}\theta$,

$$\vec{\Lambda}^v = \hat{\vec{\Lambda}}^v(\boldsymbol{E}, \theta, \operatorname{Grad}\theta), \qquad \Lambda^\varepsilon = \hat{\Lambda}^\varepsilon(\boldsymbol{E}, \theta, \operatorname{Grad}\theta). \tag{7.86}$$

Hence, the (7.85) is linear with respect to $\dot{E}_{KL}$, $\dot{\theta}$, $\dot{\theta}_{,K}$, $\dot{v}_k$, $E_{PQ,K}$ and $\theta_{,PK}$, which are not constitutive variables, but is non-linear with respect to $\theta_{,K}$, which is a constitutive variable. To maintain this inequality for all values of the variables that enter (7.85) linearly, the prefactors of these variables must vanish. First, the linear dependence of (7.85) on $\dot{\vec{v}}$ implies that the Lagrange multiplier $\vec{\Lambda}^v$ must vanish,

$$\vec{\Lambda}^v = \vec{0}, \tag{7.87}$$

while the linear dependence on $\dot{\theta}$, $\dot{\theta}_{,K}$ and $\dot{E}_{KL}$ yields

$$\frac{\partial\hat{\eta}}{\partial\theta} = \Lambda^\varepsilon \frac{\partial\hat{\varepsilon}}{\partial\theta}, \qquad \frac{\partial\hat{\eta}}{\partial(\operatorname{Grad}\theta)} = \Lambda^\varepsilon \frac{\partial\hat{\varepsilon}}{\partial(\operatorname{Grad}\theta)},$$
$$\varrho_0 \frac{\partial\hat{\eta}}{\partial\boldsymbol{E}} - \Lambda^\varepsilon \varrho_0 \frac{\partial\hat{\varepsilon}}{\partial\boldsymbol{E}} + \Lambda^\varepsilon \hat{\boldsymbol{T}}^{(2)} = \boldsymbol{0}. \tag{7.88}$$

The linear dependence of (7.85) on $E_{PQ,K}$ and $\theta_{,PK}$, the symmetry of $\boldsymbol{E}$ and the assumption that the order of differentiation of θ with respect to X_P and X_K can be interchanged (i.e., $\theta_{,PK} = \theta_{,KP}$) result in

$$\operatorname{sym}\left(\frac{\partial\hat{\vec{S}}}{\partial\boldsymbol{E}} - \Lambda^\varepsilon \frac{\partial\hat{\vec{Q}}}{\partial\boldsymbol{E}}\right) = \boldsymbol{0}, \qquad \operatorname{sym}\left(\frac{\partial\hat{\vec{S}}}{\partial(\operatorname{Grad}\theta)} - \Lambda^\varepsilon \frac{\partial\hat{\vec{Q}}}{\partial(\operatorname{Grad}\theta)}\right) = \boldsymbol{0}. \tag{7.89}$$

Relations (7.88) and (7.89) constrain the constitutive equations for $\hat{\boldsymbol{T}}^{(2)}$, $\hat{\vec{Q}}$, $\hat{\varepsilon}$, $\hat{\eta}$ and $\hat{\vec{S}}$, but they can also be viewed as a means of determining Λ^ε. In view of (7.87)–(7.89), the inequality (7.85) reduces to the residual inequality

$$\left(\frac{\partial\hat{\vec{S}}}{\partial\theta} - \Lambda^\varepsilon \frac{\partial\hat{\vec{Q}}}{\partial\theta}\right) \cdot \operatorname{Grad}\theta \geq 0. \tag{7.90}$$

Taking the total differential of $(7.84)_1$,

$$d\eta = \frac{\partial\hat{\eta}}{\partial\boldsymbol{E}} : d\boldsymbol{E} + \frac{\partial\hat{\eta}}{\partial\theta} d\theta + \frac{\partial\hat{\eta}}{\partial(\operatorname{Grad}\theta)} d(\operatorname{Grad}\theta), \tag{7.91}$$

and substituting for the partial derivatives of $\hat{\eta}$ from (7.88) leads to

$$d\eta = \Lambda^{\varepsilon}\Big(d\varepsilon - \frac{1}{\varrho_0}\boldsymbol{T}^{(2)} : d\boldsymbol{E}\Big). \tag{7.92}$$

The relations (7.89), (7.90) and (7.92) are thermodynamic constraints on the constitutive equations for a thermoelastic solid specified by the constitutive variables $\boldsymbol{E}$, θ and $\operatorname{Grad}\theta$. These relations contain the Lagrange multiplier Λ^{ε}, which depends on the same constitutive variables. Comparing (7.92) with Gibbs' relation (7.13) suggests that $\Lambda^{\varepsilon} = 1/T$, so that the assumption $(7.2)_1$, that is, $\vec{S} = \vec{Q}/T$, might be valid in this case. A rigorous proof is, however, not available yet for a general thermoelastic solid.

We therefore confine our attention to a thermoelastic solid for which the constitutive functions for the heat flux $\vec{Q}$ and entropy flux $\vec{S}$ are **isotropic** vector-valued functions of $\boldsymbol{E}$, θ and $\operatorname{Grad}\theta$. In this case, both $\vec{Q}$ and $\vec{S}$ are expressed in the form of (7.25) as

$$\vec{Q} = -\boldsymbol{\kappa}\cdot\operatorname{Grad}\theta, \qquad \vec{S} = -\boldsymbol{\gamma}\cdot\operatorname{Grad}\theta, \tag{7.93}$$

where the symmetric tensors $\boldsymbol{\kappa}$ and $\boldsymbol{\gamma}$ are functions of $\boldsymbol{E}$, θ, the invariants of $\boldsymbol{E}$ and $\operatorname{Grad}\theta$, and their combined invariants. Moreover, we consider a thermoelastic solid for which $\vec{Q}$ and $\vec{S}$ depend **linearly** on $\operatorname{Grad}\theta$. In this case, $\boldsymbol{\kappa}$ and $\boldsymbol{\gamma}$ are independent of $\operatorname{Grad}\theta$,

$$\boldsymbol{\kappa} = \hat{\boldsymbol{\kappa}}(\boldsymbol{E},\theta), \qquad \boldsymbol{\gamma} = \hat{\boldsymbol{\gamma}}(\boldsymbol{E},\theta). \tag{7.94}$$

A solid with the material relations (7.93) and (7.94) is known as a *Fourier-type thermoelastic solid*. For such a solid, $(7.89)_2$ immediately gives

$$\boldsymbol{\gamma} = \Lambda^{\varepsilon}\boldsymbol{\kappa}, \tag{7.95}$$

which leads to

$$\vec{S} = \Lambda^{\varepsilon}\vec{Q}. \tag{7.96}$$

The entropy flux vector is thus collinear with the heat flux vector, whereby the proportionality factor is given by the Lagrange multiplier of the energy equation. On differentiating (7.95) with respect to $\operatorname{Grad}\theta$, we obtain

$$\frac{\partial\Lambda^{\varepsilon}}{\partial(\operatorname{Grad}\theta)} = \vec{0},$$

which implies that Λ^{ε} is independent of $\operatorname{Grad}\theta$. Similarly, differentiating (7.96) with respect to $\boldsymbol{E}$ and substituting the result into $(7.89)_1$ gives

$$\vec{Q}\otimes\frac{\partial\Lambda^{\varepsilon}}{\partial\boldsymbol{E}} = \boldsymbol{0},$$

which implies that Λ^ε is independent of $\boldsymbol{E}$ since $\vec{Q}$ does not, in general, vanish. In summary,

$$\Lambda^\varepsilon = \hat{\Lambda}^\varepsilon(\theta). \tag{7.97}$$

In view of (7.96) and (7.97), the residual inequality (7.90) takes the form

$$\Big(\frac{d\Lambda^\varepsilon}{d\theta}\Big)(\vec{Q}\cdot \mathrm{Grad}\,\theta) \geq 0. \tag{7.98}$$

Since the entropy production does not vanish identically in heat-conducting solids, we require that $d\Lambda^\varepsilon/d\theta \neq 0$ for any θ. Hence, $d\Lambda^\varepsilon/d\theta$ does not change sign, i.e., it is either always negative or always positive. Consequently, $\Lambda^\varepsilon(\theta)$ is a **strictly monotonic** function of θ and can be employed to define the *absolute temperature* $T(\theta)$ as

$$\Lambda^\varepsilon = \Lambda^\varepsilon(\theta) =: \frac{1}{T(\theta)}. \tag{7.99}$$

The strict monotonicity of $\Lambda^\varepsilon(\theta)$ implies that $T(\theta)$ is also a strictly monotonic function of θ. We additionally assume that $T(\theta)$ is a strictly increasing function of θ, that is, $T_A > T_B$ means that 'A is warmer than B'.[5]

This result approaches the Clausius–Duhem assumption $(7.2)_1$ very closely, however, Λ^ε is still a materially dependent function of the empirical temperature. To show that $T(\theta)$ is a meaningful absolute temperature, we must prove that $T(\theta)$ is a *universal* function of θ, not different for different materials. To show this, we consider two different thermoelastic solids that are separated by an ideal wall with the property (7.82). Let solid I be placed on the positive side and solid II on the negative side of the wall. In view of (7.96) and (7.99), the continuity condition $(7.82)_2$ of the normal component of the entropy flux vector across the wall has the form

$$\Big[\frac{1}{T(\theta)}\vec{N}\cdot\vec{Q}\Big]_-^+ = 0. \tag{7.100}$$

According to (4.54), the normal component of the Eulerian heat flux is continuous across a welded material discontinuity surface, i.e., $\vec{n}\cdot\big[\vec{q}\,\big]_-^+ = 0$. This condition, expressed in the Lagrangian form, is $\vec{N}\cdot\big[\vec{Q}\,\big]_-^+ = 0$, where the heat flux $\vec{Q}$ in the reference configuration is defined by $(4.74)_2$. Then, the jump condition (7.100) can

[5]To give an example of $T(\theta)$, let the *Celsius scale* be considered the measure of empirical temperature. Choosing

$$T(\theta) = 273.15\,\mathrm{K} + 1\,\frac{\mathrm{K}}{^{\mathrm{o}}\mathrm{C}}\,\theta,$$

we obtain the *Kelvin scale* as a measure of the absolute temperature, which can be used to replace the empirical temperature θ with the absolute temperature T.

be rewritten as

$$\left[\frac{1}{T(\theta)}\right]_{-}^{+} \vec{N} \cdot \vec{Q} = 0. \tag{7.101}$$

Since, in general, $\vec{N} \cdot \vec{Q} \neq 0$, we finally obtain

$$\left[\frac{1}{T(\theta)}\right]_{-}^{+} = 0, \qquad \text{or} \qquad T^{\mathrm{I}}(\theta) = T^{\mathrm{II}}(\theta), \tag{7.102}$$

from which we conclude that $T(\theta)$ is the same function of empirical temperature on both sides of the ideal wall. Since the solids on both sides can be arbitrary thermoelastic solids, $T(\theta)$ is materially independent within this class of materials. The same property of $T(\theta)$ can be proved for other materials. Hence, $T(\theta)$ is a material-independent, universal function of the empirical temperature θ.

To complete the Müller–Liu entropy principle for a thermoelastic solid, it remains to evaluate the residual inequality (7.98). In view of (7.99), (7.98) becomes

$$-\frac{1}{T^2}\left(\frac{dT}{d\theta}\right)(\vec{Q} \cdot \operatorname{Grad}\theta) \geq 0.$$

Since $T(\theta)$ is a strictly increasing function of θ, we then have

$$-\vec{Q} \cdot \operatorname{Grad}\theta \geq 0, \tag{7.103}$$

which, for $\vec{Q}$ given by $(7.93)_1$, takes the form

$$\operatorname{Grad}\theta \cdot \boldsymbol{\kappa} \cdot \operatorname{Grad}\theta \geq 0. \tag{7.104}$$

This states that the thermal conductivity $\boldsymbol{\kappa}$ is a positive semi-definite tensor.

One special choice of the absolute temperature T for a given empirical temperature θ is

$$T = \theta. \tag{7.105}$$

Under this choice, the results of the Müller–Liu entropy principle for a thermoelastic solid coincide with those of the Clausius–Duhem inequality. In particular, the relation $\vec{S} = \vec{Q}/T$ postulated by the Clausius–Duhem inequality follows from the Müller–Liu entropy principle when (7.96) and (7.99) are combined. However, the two principles may not provide equivalent restrictions on material properties. The Clausius–Duhem inequality imposes, in general, stronger restrictions on constitutive equations than those resulting from the application of the Müller–Liu entropy principle.

7.7 Application of the Müller–Liu Entropy Principle to a Viscous Heat-Conducting Fluid

We next apply the Müller–Liu entropy principle to a viscous heat-conducting fluid. The constitutive equations for this material are given by (6.173):

$$\boldsymbol{t} = \hat{\boldsymbol{t}}(\boldsymbol{d}, \theta, \operatorname{grad}\theta, \varrho), \qquad \vec{q} = \hat{\vec{q}}(\boldsymbol{d}, \theta, \operatorname{grad}\theta, \varrho), \qquad \varepsilon = \hat{\varepsilon}(\boldsymbol{d}, \theta, \operatorname{grad}\theta, \varrho). \tag{7.106}$$

According to the principle of equipresence, we choose the same set of constitutive variables for both the entropy η and entropy flux $\vec{s}$ such that

$$\eta = \hat{\eta}(\boldsymbol{d}, \theta, \operatorname{grad}\theta, \varrho), \qquad \vec{s} = \hat{\vec{s}}(\boldsymbol{d}, \theta, \operatorname{grad}\theta, \varrho). \tag{7.107}$$

Substituting the constitutive equations (7.106) and (7.107) into the entropy inequality (7.80) and differentiating the result with respect to time according to the chain rule yields

$$\begin{aligned}
&\varrho\left(\frac{\partial\hat{\eta}}{\partial d_{kl}}\dot{d}_{kl} + \frac{\partial\hat{\eta}}{\partial\theta}\dot{\theta} + \frac{\partial\hat{\eta}}{\partial\theta_{,k}}\dot{\theta}_{,k} + \frac{\partial\hat{\eta}}{\partial\varrho}\dot{\varrho}\right) + \frac{\partial\hat{s}_k}{\partial d_{pq}}d_{pq,k} + \frac{\partial\hat{s}_k}{\partial\theta}\theta_{,k} + \frac{\partial\hat{s}_k}{\partial\theta_{,p}}\theta_{,pk} + \frac{\partial\hat{s}_k}{\partial\varrho}\varrho_{,k}\\
&\quad - \lambda^{\varrho}(\dot{\varrho} + \varrho d_{kk}) - \lambda^{v}_{k}\varrho\,\dot{v}_k + \lambda^{v}_{k}\left(\frac{\partial\hat{t}_{lk}}{\partial d_{pq}}d_{pq,l} + \frac{\partial\hat{t}_{lk}}{\partial\theta}\theta_{,l} + \frac{\partial\hat{t}_{lk}}{\partial\theta_{,p}}\theta_{,pl} + \frac{\partial\hat{t}_{lk}}{\partial\varrho}\varrho_{,l}\right)\\
&\quad - \lambda^{\varepsilon}\varrho\left(\frac{\partial\hat{\varepsilon}}{\partial d_{kl}}\dot{d}_{kl} + \frac{\partial\hat{\varepsilon}}{\partial\theta}\dot{\theta} + \frac{\partial\hat{\varepsilon}}{\partial\theta_{,k}}\dot{\theta}_{,k} + \frac{\partial\hat{\varepsilon}}{\partial\varrho}\dot{\varrho}\right) + \lambda^{\varepsilon}\hat{t}_{lk}d_{kl}\\
&\quad - \lambda^{\varepsilon}\left(\frac{\partial\hat{q}_k}{\partial d_{pq}}d_{pq,k} + \frac{\partial\hat{q}_k}{\partial\theta}\theta_{,k} + \frac{\partial\hat{q}_k}{\partial\theta_{,p}}\theta_{,pk} + \frac{\partial\hat{q}_k}{\partial\varrho}\varrho_{,k}\right) \geq 0
\end{aligned} \tag{7.108}$$

in index notation. According to Liu's theorem (Liu 1972), the Lagrange multipliers λ^{ϱ}, λ^{v}_{k} and λ^{ε} depend only on the independent constitutive variables $\boldsymbol{d}$, θ, $\operatorname{grad}\theta$ and ϱ,

$$\lambda^{\varrho} = \hat{\lambda}^{\varrho}(\boldsymbol{d}, \theta, \operatorname{grad}\theta, \varrho), \qquad \vec{\lambda}^{v} = \hat{\vec{\lambda}}^{v}(\boldsymbol{d}, \theta, \operatorname{grad}\theta, \varrho), \qquad \lambda^{\varepsilon} = \hat{\lambda}^{\varepsilon}(\boldsymbol{d}, \theta, \operatorname{grad}\theta, \varrho). \tag{7.109}$$

The inequality (7.108) is linear with respect to $\dot{d}_{kl}, \dot{\theta}, \dot{\theta}_{,k}, \dot{\varrho}, \dot{v}_k, d_{pq,k}, \theta_{,pk}$ and $\varrho_{,k}$, which are not constitutive variables, but is non-linear with respect to d_{kl} and $\theta_{,k}$, which are the constitutive variables. To maintain this inequality for all values of the variables that enter (7.108) linearly, the prefactors of these variables must vanish. First, the linear dependence of (7.108) on $\dot{\vec{v}}$ implies that the Lagrange multiplier $\vec{\lambda}^{v}$

must vanish,

$$\vec{\lambda}^{v} = \vec{0}, \tag{7.110}$$

while the linear dependence on $\dot{\theta}$, $\dot{\theta}_{,k}$, $\dot{d}_{kl}$ and $\dot{\varrho}$ yields

$$\begin{aligned} \frac{\partial \hat{\eta}}{\partial \theta} &= \lambda^{\varepsilon} \frac{\partial \hat{\varepsilon}}{\partial \theta}, \qquad \frac{\partial \hat{\eta}}{\partial (\operatorname{grad}\theta)} = \lambda^{\varepsilon} \frac{\partial \hat{\varepsilon}}{\partial (\operatorname{grad}\theta)}, \\ \frac{\partial \hat{\eta}}{\partial \boldsymbol{d}} &= \lambda^{\varepsilon} \frac{\partial \hat{\varepsilon}}{\partial \boldsymbol{d}}, \qquad \lambda^{\varrho} = \varrho \Big(\frac{\partial \hat{\eta}}{\partial \varrho} - \lambda^{\varepsilon} \frac{\partial \hat{\varepsilon}}{\partial \varrho} \Big). \end{aligned} \tag{7.111}$$

The linear dependence of (7.108) on $d_{pq,k}$, $\theta_{,pk}$ and $\varrho_{,k}$, the symmetry of $\boldsymbol{d}$ and the assumption that the order of differentiation of θ with respect to x_p and x_k can be interchanged (i.e., $\theta_{,pk} = \theta_{,kp}$) result in

$$\operatorname{sym}\Big(\frac{\partial \hat{\vec{s}}}{\partial \boldsymbol{d}} - \lambda^{\varepsilon} \frac{\partial \hat{\vec{q}}}{\partial \boldsymbol{d}} \Big) = \boldsymbol{0}, \quad \operatorname{sym}\Big(\frac{\partial \hat{\vec{s}}}{\partial (\operatorname{grad}\theta)} - \lambda^{\varepsilon} \frac{\partial \hat{\vec{q}}}{\partial (\operatorname{grad}\theta)} \Big) = \boldsymbol{0}, \quad \frac{\partial \hat{\vec{s}}}{\partial \varrho} = \lambda^{\varepsilon} \frac{\partial \hat{\vec{q}}}{\partial \varrho}. \tag{7.112}$$

Relations (7.111) and (7.112) constrain the constitutive equations for $\hat{\vec{q}}$, $\hat{\varepsilon}$, $\hat{\eta}$ and $\hat{\vec{s}}$, but they can also be viewed as a means of determining λ^{ε}. In view of (7.110)–(7.112), the inequality (7.108) reduces to the residual inequality

$$\gamma(\boldsymbol{d}, \theta, \operatorname{grad}\theta, \varrho) \geq 0, \tag{7.113}$$

where the entropy production has the form

$$\gamma(\boldsymbol{d}, \theta, \operatorname{grad}\theta, \varrho) := (\lambda^{\varepsilon} \boldsymbol{t} - \varrho \lambda^{\varrho} \boldsymbol{I}) : \boldsymbol{d} + \Big(\frac{\partial \hat{\vec{s}}}{\partial \theta} - \lambda^{\varepsilon} \frac{\partial \hat{\vec{q}}}{\partial \theta} \Big) \cdot \operatorname{grad}\theta. \tag{7.114}$$

The constitutive functions for the heat flux $\vec{q}$ and entropy flux $\vec{s}$ are given by the isotropic vector-valued function $(6.174)_2$,

$$\vec{q} = -\boldsymbol{\kappa} \cdot \operatorname{grad}\theta, \qquad \vec{s} = -\boldsymbol{\gamma} \cdot \operatorname{grad}\theta, \tag{7.115}$$

where the symmetric tensors $\boldsymbol{\kappa}$ and $\boldsymbol{\gamma}$ are functions of $\boldsymbol{d}$, ϱ, θ, the invariants of $\boldsymbol{d}$ and $\operatorname{grad}\theta$, and their combined invariants. Moreover, we consider a heat-conducting fluid for which $\vec{q}$ and $\vec{s}$ depend **linearly** on $\operatorname{grad}\theta$. In this case, $\boldsymbol{\kappa}$ and $\boldsymbol{\gamma}$ are independent of $\operatorname{grad}\theta$, that is

$$\boldsymbol{\kappa} = \hat{\boldsymbol{\kappa}}(\boldsymbol{d}, \theta, \varrho), \qquad \boldsymbol{\gamma} = \hat{\boldsymbol{\gamma}}(\boldsymbol{d}, \theta, \varrho). \tag{7.116}$$

A fluid with the material relations (7.115) and (7.116) is known as a *Fourier-type heat-conducting fluid*. For such a fluid, $(7.112)_2$ immediately gives

$$\boldsymbol{\gamma} = \lambda^{\varepsilon}\boldsymbol{\kappa}, \tag{7.117}$$

which leads to

$$\vec{s} = \lambda^{\varepsilon}\vec{q}, \tag{7.118}$$

meaning that the entropy flux and heat flux are collinear vectors.[6] On differentiating (7.117) with respect to $\operatorname{grad}\theta$, we obtain

$$\frac{\partial \lambda^{\varepsilon}}{\partial(\operatorname{grad}\theta)} = \vec{0},$$

which implies that λ^{ε} is independent of $\operatorname{grad}\theta$. Similarly, differentiating (7.118) with respect to $\boldsymbol{d}$ and ϱ and substituting the results into $(7.112)_1$ and $(7.112)_3$, respectively, gives

$$\vec{q} \otimes \frac{\partial \lambda^{\varepsilon}}{\partial \boldsymbol{d}} = \boldsymbol{0}, \qquad \frac{\partial \lambda^{\varepsilon}}{\partial \varrho} = 0,$$

which implies that λ^{ε} is independent of $\boldsymbol{d}$ and ϱ since $\vec{q}$ does not vanish identically. In summary,

$$\lambda^{\varepsilon} = \hat{\lambda}^{\varepsilon}(\theta) \tag{7.119}$$

and the entropy production (7.114) becomes

$$\gamma(\boldsymbol{d}, \theta, \operatorname{grad}\theta, \varrho) = (\lambda^{\varepsilon}\boldsymbol{t} - \varrho\lambda^{\varrho}\boldsymbol{I}) : \boldsymbol{d} + \frac{d\lambda^{\varepsilon}}{d\theta}(\vec{q}\cdot\operatorname{grad}\theta). \tag{7.120}$$

Differentiating $(7.111)_1$ with respect to $\boldsymbol{d}$ and $(7.111)_3$ with respect to θ yields

$$\frac{\partial^2 \hat{\eta}}{\partial \boldsymbol{d}\,\partial\theta} = \lambda^{\varepsilon}\frac{\partial^2 \hat{\varepsilon}}{\partial \boldsymbol{d}\,\partial\theta}, \qquad \frac{\partial^2 \hat{\eta}}{\partial\theta\,\partial \boldsymbol{d}} = \lambda^{\varepsilon}\frac{\partial^2 \hat{\varepsilon}}{\partial\theta\,\partial \boldsymbol{d}} + \frac{d\lambda^{\varepsilon}}{d\theta}\frac{\partial \hat{\varepsilon}}{\partial \boldsymbol{d}}.$$

Assuming the exchange of the order of partial derivatives of $\hat{\eta}$ and $\hat{\varepsilon}$ with respect to $\boldsymbol{d}$ and θ, we obtain

$$\frac{d\lambda^{\varepsilon}}{d\theta}\frac{\partial \hat{\varepsilon}}{\partial \boldsymbol{d}} = \boldsymbol{0}. \tag{7.121}$$

[6] Liu (2002, Section 7.4) proved the collinearity of $\vec{q}$ and $\vec{s}$ without assuming the relation (7.116).

We now consider $d\lambda^{\varepsilon}/d\theta \neq 0$ in heat-conducting fluids. Then, (7.121) implies that $\hat{\varepsilon}$ cannot be a function of $\boldsymbol{d}$, i.e.,

$$\frac{\partial \hat{\varepsilon}}{\partial \boldsymbol{d}} = \boldsymbol{0}. \tag{7.122}$$

In view of $(7.111)_3$, the same holds for $\hat{\eta}$,

$$\frac{\partial \hat{\eta}}{\partial \boldsymbol{d}} = \boldsymbol{0}. \tag{7.123}$$

Moreover, differentiating $(7.111)_1$ and $(7.111)_2$ with respect to $\operatorname{grad}\theta$ and θ, respectively, and assuming again that $d\lambda^{\varepsilon}/d\theta \neq 0$ gives

$$\frac{\partial \hat{\varepsilon}}{\partial(\operatorname{grad}\theta)} = \vec{0}, \qquad \frac{\partial \hat{\eta}}{\partial(\operatorname{grad}\theta)} = \vec{0}, \tag{7.124}$$

which state that the constitutive functions $\hat{\varepsilon}$ and $\hat{\eta}$ cannot depend on $\boldsymbol{d}$ and $\operatorname{grad}\theta$. In view of $(7.111)_4$, the same holds for the Lagrange multiplier λ^{ϱ}. Hence, the constitutive equations $(7.106)_3$, $(7.107)_1$ and $(7.109)_1$ reduce to

$$\varepsilon = \hat{\varepsilon}(\theta, \varrho), \qquad \eta = \hat{\eta}(\theta, \varrho), \qquad \lambda^{\varrho} = \hat{\lambda}^{\varrho}(\theta, \varrho), \tag{7.125}$$

where functions $\hat{\varepsilon}$, $\hat{\eta}$ and $\hat{\lambda}^{\varrho}$ are still subject to constraints $(7.111)_{1,4}$.

To derive the thermodynamic restrictions imposed on the constitutive equations (7.106) from the entropy inequality (7.113), we first define *thermodynamic equilibrium* as a time-independent thermodynamic process with uniform (constant in space) and stationary (constant in time) velocity and temperature fields. Mathematically, the thermodynamic equilibrium is defined by

$$\vec{v}(\vec{x}, t) = \text{const}, \qquad \theta(\vec{x}, t) = \text{const}, \tag{7.126}$$

which implies that the strain-rate tensor $\boldsymbol{d}$ and the temperature gradient $\operatorname{grad}\theta$ vanish in thermodynamic equilibrium,

$$\boldsymbol{d}|_{\mathrm{E}} = \boldsymbol{0}, \qquad \operatorname{grad}\theta|_{\mathrm{E}} = \vec{0}, \tag{7.127}$$

where the subscript E denotes the state of thermodynamic equilibrium. Consequently,

$$\gamma(\boldsymbol{d}, \theta, \operatorname{grad}\theta, \varrho)|_{\mathrm{E}} = 0 \tag{7.128}$$

in view of (7.120). Considering (7.128) and the residual inequality (7.113), we deduce that the entropy production γ is minimal in thermodynamic equilibrium and the value of this minimum is zero. According to the theorem of extrema of functions of several variables, the necessary condition for γ to reach a minimum is that the first derivatives of γ with respect to $\boldsymbol{d}$ and grad θ vanish,

$$\frac{\partial \gamma}{\partial \boldsymbol{d}}\bigg|_{\mathrm{E}} = \boldsymbol{0}, \qquad \frac{\partial \gamma}{\partial (\operatorname{grad}\theta)}\bigg|_{\mathrm{E}} = \vec{0}, \tag{7.129}$$

Differentiating (7.120) with respect to $\boldsymbol{d}$ and grad θ, respectively, and evaluating the result in thermodynamic equilibrium yields

$$\frac{\partial \gamma}{\partial \boldsymbol{d}}\bigg|_{\mathrm{E}} = \lambda^{\varepsilon}\boldsymbol{t}\big|_{\mathrm{E}} - \varrho\lambda^{\varrho}\boldsymbol{I}, \qquad \frac{\partial \gamma}{\partial (\operatorname{grad}\theta)}\bigg|_{\mathrm{E}} = \frac{d\lambda^{\varepsilon}}{d\theta}\,\vec{q}\,|_{\mathrm{E}}\,.$$

Using this in (7.129) gives

$$\boldsymbol{t}|_{\mathrm{E}} = -p\boldsymbol{I}, \qquad \vec{q}\,|_{\mathrm{E}} = \vec{0}, \tag{7.130}$$

where p is the *thermodynamic pressure*,

$$p = p(\theta, \varrho) := -\frac{\varrho\lambda^{\varrho}}{\lambda^{\varepsilon}}, \tag{7.131}$$

which, upon substituting for λ^{ϱ} from $(7.111)_4$, can be expressed as

$$p = -\varrho^2\Big(\frac{1}{\lambda^{\varepsilon}}\frac{\partial \hat{\eta}}{\partial \varrho} - \frac{\partial \hat{\varepsilon}}{\partial \varrho}\Big). \tag{7.132}$$

For a viscous heat-conducting fluid, the stress is spherical in thermodynamic equilibrium, and determined by the entropy and internal energy. The constitutive equation (7.115) already accounts for the result $\vec{q}\,|_{\mathrm{E}} = \vec{0}$. The entropy production (7.120) now becomes

$$\gamma(\boldsymbol{d}, \theta, \operatorname{grad}\theta, \varrho) = \lambda^{\varepsilon}(\boldsymbol{t} + p\boldsymbol{I}) : \boldsymbol{d} + \frac{d\lambda^{\varepsilon}}{d\theta}(\vec{q}\cdot \operatorname{grad}\theta). \tag{7.133}$$

Taking the total differential of $(7.125)_2$,

$$d\eta = \frac{\partial \hat{\eta}}{\partial \theta}d\theta + \frac{\partial \hat{\eta}}{\partial \varrho}d\varrho, \tag{7.134}$$

substituting for the partial derivatives of $\hat{\eta}$ from $(7.111)_{1,4}$,

$$d\eta = \lambda^{\varepsilon}d\varepsilon + \frac{\lambda^{\varrho}}{\varrho}d\varrho, \tag{7.135}$$

and for λ^{ϱ} from (7.131) results in

$$d\eta = \lambda^{\varepsilon}\Big(d\varepsilon - \frac{p}{\varrho^2}d\varrho\Big). \tag{7.136}$$

Comparing this with Gibbs' relation for the same material (e.g. Hutter and Jöhnk 2004, Eq. (5.7.27)) suggests that $\lambda^{\varepsilon}(\theta)$ can be used to define the *absolute temperature* $T(\theta)$ as

$$\lambda^{\varepsilon} = \lambda^{\varepsilon}(\theta) =: \frac{1}{T(\theta)}, \tag{7.137}$$

hence,

$$\lambda^{\varepsilon}(\theta) > 0. \tag{7.138}$$

Moreover, the requirement $d\lambda^{\varepsilon}/d\theta \neq 0$ for all θ, used to derive (7.122), implies that $d\lambda^{\varepsilon}/d\theta$ maintains either the positive or negative sign for all θ. Thus, $\lambda^{\varepsilon}(\theta)$ and $T(\theta)$ are **strictly monotonic** functions of θ. As for a thermoelastic solid (see the preceding section), we additionally assume that $T(\theta)$ is a strictly increasing function of θ.

To draw further conclusions from the residual inequality (7.113), we confine ourselves to a special type of viscous heat-conducting fluid for which the constitutive equations are **linear** with respect to $\boldsymbol{d}$ and $\operatorname{grad}\theta$. In this case, the constitutive equations $(6.174)_1$ and $(7.115)_1$ reduce to

$$\boldsymbol{t} = (-p + \lambda_v \operatorname{tr}\boldsymbol{d})\boldsymbol{I} + 2\mu_v\boldsymbol{d}, \qquad \vec{q} = -\kappa \operatorname{grad}\theta, \tag{7.139}$$

where

$$p,\ \lambda_v,\ \mu_v,\ \kappa = p,\ \lambda_v,\ \mu_v,\ \kappa(\theta, \varrho). \tag{7.140}$$

The fluid characterised by the constitutive equation $(7.139)_1$ is known as a *Newtonian viscous fluid*, and has already been introduced by (7.55), while equation $(7.139)_2$ is known as *Fourier's law of heat conduction*. Recall that the parameters λ_v and μ_v are, respectively, the *dilatational* and *shear viscosities*, while the parameter κ is the *thermal conductivity*.

Substituting (7.139) into (7.133) gives

$$\gamma(\boldsymbol{d}, \theta, \operatorname{grad}\theta, \varrho) = \lambda^{\varepsilon}\Big[\lambda_v(\operatorname{tr}\boldsymbol{d})^2 + 2\mu_v(\boldsymbol{d}:\boldsymbol{d}) - \frac{1}{\lambda^{\varepsilon}}\frac{d\lambda^{\varepsilon}}{d\theta}\kappa(\operatorname{grad}\theta)^2\Big]. \tag{7.141}$$

In view of (7.138), λ^{ε} can be dropped from γ and the residual inequality (7.113) becomes

$$\gamma_1(\boldsymbol{d}, \theta, \operatorname{grad}\theta, \varrho) := \lambda_v(\operatorname{tr}\boldsymbol{d})^2 + 2\mu_v(\boldsymbol{d}:\boldsymbol{d}) - \frac{1}{\lambda^{\varepsilon}}\frac{d\lambda^{\varepsilon}}{d\theta}\kappa(\operatorname{grad}\theta)^2 \geq 0. \tag{7.142}$$

Introducing a 12 component vector

$$\vec{z} = \big(d_{11}, d_{22}, d_{33}, d_{12}, d_{13}, \cdots, d_{32}, (\operatorname{grad}\theta)_1, (\operatorname{grad}\theta)_2, (\operatorname{grad}\theta)_3\big),$$

the entropy production γ_1 can be expressed as

$$\gamma_1(\boldsymbol{d}, \theta, \operatorname{grad}\theta, \varrho) = \vec{z} \cdot \boldsymbol{A} \cdot \vec{z}, \tag{7.143}$$

where the 12×12 matrix $\boldsymbol{A}$ is block-diagonal,

$$\boldsymbol{A} = \begin{pmatrix} \lambda_v \mathbf{1}_{3\times3} + 2\mu_v \boldsymbol{I}_{3\times3} & 0 & 0 \\ 0 & 2\mu_v \boldsymbol{I}_{6\times6} & 0 \\ 0 & 0 & -\dfrac{1}{\lambda^\varepsilon}\dfrac{d\lambda^\varepsilon}{d\theta}\kappa \boldsymbol{I}_{3\times3} \end{pmatrix}. \tag{7.144}$$

Here, $\mathbf{1}_{3\times3}$ is the matrix composed of only ones, and $\boldsymbol{I}_{3\times3}$ and $\boldsymbol{I}_{6\times6}$ are the third- and sixth-order identity matrices. In view of (7.143), the inequality (7.142) is satisfied if the matrix $\boldsymbol{A}$ is positive semi-definite. As in Sect. 7.3, we use Sylvester's criterion to determine the conditions for $\boldsymbol{A}$ to be positive semi-definite. The non-negativeness of the principal minors of the 9×9 upper-left submatrix of $\boldsymbol{A}$ gives the condition (7.59), i.e.,

$$3\lambda_v + 2\mu_v \geq 0, \qquad \mu_v \geq 0, \tag{7.145}$$

or, equivalently,

$$k_v \geq 0, \qquad \mu_v \geq 0, \tag{7.146}$$

while the non-negativeness of the principal minors of the 3×3 bottom-right submatrix of $\boldsymbol{A}$ gives

$$-\frac{1}{\lambda^\varepsilon}\frac{d\lambda^\varepsilon}{d\theta}\kappa \geq 0. \tag{7.147}$$

Moreover, in view of (7.137),

$$-\frac{1}{\lambda^\varepsilon}\frac{d\lambda^\varepsilon}{d\theta} = \frac{1}{T}\frac{dT}{d\theta} > 0 \tag{7.148}$$

since $T(\theta)$ is a strictly increasing function of θ. We finally have

$$\kappa \geq 0. \tag{7.149}$$

We conclude that in order to satisfy the Müller–Liu entropy principle for a linear viscous heat-conducting fluid, the bulk viscosity k_v, the shear viscosity μ_v and the thermal conductivity κ must be non-negative functions of ϱ and θ.

To show that λ^{ε} is independent of the material properties within the class of viscous heat-conducting fluids, we use the existence of an ideal wall with the property (7.78). The proof that the absolute temperature has no jump on an ideal wall between two different viscous heat-conducting fluids is similar to that in Sect. 7.6.

Chapter 8
Classical Linear Elasticity

The classical linearised theory of elasticity provides a model that is useful for studying infinitesimal deformations of an elastic material. In this chapter, we briefly remark on the relationship between this linearised theory for infinitesimal deformations and the exact theory of elastic simple materials.

8.1 Linear Elastic Solids

In Sect. 7.2, we stated the most general constitutive equations for a classical thermoelastic solid in (7.18), (7.19) and (7.25). If thermal effects are neglected, the Helmholtz free energy density ψ and the second Piola–Kirchhoff stress tensor $\boldsymbol{T}^{(2)}$ are functions of strain alone and the solid is termed *elastic* (Sect. 6.17.1). For an elastic solid, the constitutive equation $(7.18)_1$ reduces to

$$\boldsymbol{T}^{(2)} = \frac{\partial W}{\partial \boldsymbol{E}}, \tag{8.1}$$

where W is the *elastic strain energy density* (strain energy per unit undeformed volume), which relates to ψ as

$$W := \varrho_0 \psi = \hat{W}(\boldsymbol{E}). \tag{8.2}$$

We should note that an elastic material is sometimes defined by the existence of a strain energy density such that the stress is determined by (8.1). A material defined

Z. Martinec, *Principles of Continuum Mechanics*, Nečas Center Series,
https://doi.org/10.1007/978-3-030-05390-1_8

in this way is called a *hyperelastic* material.[1] The stress is still a unique function of strain, so this 'energy approach' is compatible with our earlier definition of elastic behaviour in Sect. 6.17.1. A *hypoelastic material* is distinct from a hyperelastic material in that, except under special circumstances, its constitutive equation cannot be derived from a strain energy density function. The following text considers hyperelastic materials only.

We can derive various approximation theories for the constitutive equation (8.1) for non-linear elastic solids. Expanding W about the configuration κ_0 to which the strain $\boldsymbol{E}$ is referred, we obtain

$$W(\boldsymbol{E}) = W(\mathbf{0}) + \frac{\partial W(\mathbf{0})}{\partial \boldsymbol{E}} : \boldsymbol{E} + \frac{1}{2}\boldsymbol{E} : \frac{\partial^2 W(\mathbf{0})}{\partial \boldsymbol{E}\,\partial \boldsymbol{E}} : \boldsymbol{E} + O(|\boldsymbol{E}|^3), \tag{8.3}$$

which gives

$$\boldsymbol{T}^{(2)} = \frac{\partial W(\mathbf{0})}{\partial \boldsymbol{E}} + \frac{\partial^2 W(\mathbf{0})}{\partial \boldsymbol{E}\,\partial \boldsymbol{E}} : \boldsymbol{E} + O(|\boldsymbol{E}|^2) \tag{8.4}$$

or, equivalently,

$$\boldsymbol{T}^{(2)} = \boldsymbol{T}_0 + \boldsymbol{\mathcal{C}} : \boldsymbol{E} + O(|\boldsymbol{E}|^2), \tag{8.5}$$

upon substitution into (8.1), where $\boldsymbol{T}_0$ is the stress in the configuration κ_0 to which the strain $\boldsymbol{E}$ is referred,

$$\boldsymbol{T}_0 := \frac{\partial W(\mathbf{0})}{\partial \boldsymbol{E}}. \tag{8.6}$$

In classical linear theory, the configuration κ_0 is used as the reference and the stress $\boldsymbol{T}_0$ is a tensor function of the referential coordinates, $\boldsymbol{T}_0 = \boldsymbol{T}_0(\vec{X})$. The fourth-order tensor $\boldsymbol{\mathcal{C}}$ introduced in (8.5) is defined by

$$\boldsymbol{\mathcal{C}} := \frac{\partial^2 W(\mathbf{0})}{\partial \boldsymbol{E}\,\partial \boldsymbol{E}}, \qquad \mathcal{C}_{KLMN} := \frac{\partial^2 W(\mathbf{0})}{\partial E_{KL}\,\partial E_{MN}}. \tag{8.7}$$

The symmetry of both the stress and strain tensors implies that

$$\mathcal{C}_{KLMN} = \mathcal{C}_{LKMN} = \mathcal{C}_{KLNM}, \tag{8.8}$$

[1] For *isothermal elastic processes*, the strain energy density relates to the Helmholtz free energy density by (8.2). For *isentropic elastic processes*, the strain energy density relates to the internal energy density ε as

$$W := \varrho_0 \varepsilon.$$

which reduces the $3^4 = 81$ components of $\mathcal{C}$ to at most 36 distinct coefficients $\mathcal{C}_{KLMN}$. Moreover, as a consequence of the equality of the mixed partial derivatives of W in $(8.7)_2$, the coefficients $\mathcal{C}_{KLMN}$ satisfy the additional symmetry relation

$$\mathcal{C}_{KLMN} = \mathcal{C}_{MNKL}, \tag{8.9}$$

which is known as the *Maxwell relation*. Thus, the existence of strain energy density reduces the number of distinct coefficients $\mathcal{C}_{KLMN}$ from 36 to 21. Further reductions for special types of elastic behaviour are obtained from material symmetry properties. Note that the same material may, in general, have different coefficients $\mathcal{C}_{KLMN}$ for different configurations κ_0.

Using classical linear theory, we consider only infinitesimal deformations from the reference configuration κ_0; hence, the last term on the right-hand side of the constitutive equation (8.5) can be dropped. Moreover, the strain tensor $\boldsymbol{E}$ is replaced by the infinitesimal strain tensor $\widetilde{\boldsymbol{E}}$, so that (8.5) simplifies to

$$\boldsymbol{T}^{(2)} = \boldsymbol{T}_0 + \mathcal{C} : \widetilde{\boldsymbol{E}}. \tag{8.10}$$

Within the framework of the classical infinitesimal model, we may study the effects of an infinitesimal deformation superimposed upon a reference configuration κ_0 with a finite pre-stress $\boldsymbol{T}_0$. Then, (8.10) is taken as the **exact** constitutive equation defining the classical linear (infinitesimal) model of elasticity. However, within the framework of the general theory of elastic simple materials, (8.10) is an approximation to the exact stress relation (8.1) for small deformations.

Dahlen and Tromp (1998) noticed that we have considerable flexibility in the choice of $\mathcal{C}$. The only requirement is that we satisfy the symmetries (8.8) and (8.9). These relations are satisfied by a fourth-order tensor $\mathcal{Z}$ of the form

$$\begin{aligned}\mathcal{Z}_{KLMN} = \mathcal{C}_{KLMN} &+ a\left[T_{0,KL}\delta_{MN} + T_{0,MN}\delta_{KL}\right] \\ &+ b\left[T_{0,KM}\delta_{LN} + T_{0,LN}\delta_{KM}\right] \\ &+ c\left[T_{0,KN}\delta_{ML} + T_{0,ML}\delta_{KN}\right],\end{aligned} \tag{8.11}$$

or, in tensor notation,[2]

$$\begin{aligned}\mathcal{Z} = \mathcal{C} &+ a\left[(\boldsymbol{T}_0 \otimes \boldsymbol{I})^{1234} + (\boldsymbol{I} \otimes \boldsymbol{T}_0)^{1234}\right] \\ &+ b\left[(\boldsymbol{T}_0 \otimes \boldsymbol{I})^{1324} + (\boldsymbol{I} \otimes \boldsymbol{T}_0)^{1324}\right] \\ &+ c\left[(\boldsymbol{T}_0 \otimes \boldsymbol{I})^{1432} + (\boldsymbol{I} \otimes \boldsymbol{T}_0)^{1432}\right],\end{aligned} \tag{8.12}$$

[2]The symbols $(\ \ )^{1324}$, $(\ \ )^{1432}$, etc. denote the transpose of a quadric, e.g. $(\vec{a} \otimes \vec{b} \otimes \vec{c} \otimes \vec{d})^{1324} = \vec{a} \otimes \vec{c} \otimes \vec{b} \otimes \vec{d}$.

where a, b and c are arbitrary scalars. The expressions in square brackets are the only three linear combinations of permutations of $\boldsymbol{T}_0 \otimes \boldsymbol{I}$ satisfying the symmetries (8.8) and (8.9). Any choice of a, b and c defines the behaviour of a linear elastic solid. We may thus substitute (8.12) into (8.10) to obtain

$$\begin{aligned}\boldsymbol{T}^{(2)} &= \boldsymbol{T}_0 + \mathcal{Z} : \widetilde{\boldsymbol{E}} \\ &= \boldsymbol{T}_0 + \mathcal{C} : \widetilde{\boldsymbol{E}} + a\big[(\operatorname{tr}\widetilde{\boldsymbol{E}})\boldsymbol{T}_0 + (\boldsymbol{T}_0 : \widetilde{\boldsymbol{E}})\boldsymbol{I}\big] + (b+c)\big(\boldsymbol{T}_0 \cdot \widetilde{\boldsymbol{E}} + \widetilde{\boldsymbol{E}} \cdot \boldsymbol{T}_0\big).\end{aligned} \tag{8.13}$$

The linearised constitutive equations for the first Piola–Kirchhoff stress tensor $\boldsymbol{T}^{(1)}$ and the Cauchy stress tensor $\boldsymbol{t}$ can be obtained by substituting (8.13) into $(3.26)_2$ and $(3.30)_2$, respectively,

$$\begin{aligned}\boldsymbol{T}^{(1)} &= \boldsymbol{T}_0 + \boldsymbol{T}_0 \cdot \boldsymbol{H} + \mathcal{C} : \widetilde{\boldsymbol{E}} + a\big[(\operatorname{tr}\widetilde{\boldsymbol{E}})\boldsymbol{T}_0 + (\boldsymbol{T}_0 : \widetilde{\boldsymbol{E}})\boldsymbol{I}\big] \\ &\quad + (b+c)\big(\boldsymbol{T}_0 \cdot \widetilde{\boldsymbol{E}} + \widetilde{\boldsymbol{E}} \cdot \boldsymbol{T}_0\big),\end{aligned} \tag{8.14}$$

$$\begin{aligned}\boldsymbol{t} &= \boldsymbol{T}_0 - (\operatorname{tr}\boldsymbol{H})\,\boldsymbol{T}_0 + \boldsymbol{H}^T \cdot \boldsymbol{T}_0 + \boldsymbol{T}_0 \cdot \boldsymbol{H} + \mathcal{C} : \widetilde{\boldsymbol{E}} \\ &\quad + a\big[(\operatorname{tr}\widetilde{\boldsymbol{E}})\boldsymbol{T}_0 + (\boldsymbol{T}_0 : \widetilde{\boldsymbol{E}})\boldsymbol{I}\big] + (b+c)\big(\boldsymbol{T}_0 \cdot \widetilde{\boldsymbol{E}} + \widetilde{\boldsymbol{E}} \cdot \boldsymbol{T}_0\big).\end{aligned} \tag{8.15}$$

Using the decomposition (1.111), we rewrite the last two equations as

$$\begin{aligned}\boldsymbol{T}^{(1)} &= \boldsymbol{T}_0 - \boldsymbol{T}_0 \cdot \widetilde{\boldsymbol{R}} + \mathcal{C} : \widetilde{\boldsymbol{E}} + a\big[(\operatorname{tr}\widetilde{\boldsymbol{E}})\boldsymbol{T}_0 + (\boldsymbol{T}_0 : \widetilde{\boldsymbol{E}})\boldsymbol{I}\big] + (b+c)\big(\widetilde{\boldsymbol{E}} \cdot \boldsymbol{T}_0\big) \\ &\quad + (b+c+1)\big(\boldsymbol{T}_0 \cdot \widetilde{\boldsymbol{E}}\big),\end{aligned} \tag{8.16}$$

$$\begin{aligned}\boldsymbol{t} &= \boldsymbol{T}_0 + \widetilde{\boldsymbol{R}} \cdot \boldsymbol{T}_0 - \boldsymbol{T}_0 \cdot \widetilde{\boldsymbol{R}} + \mathcal{C} : \widetilde{\boldsymbol{E}} + (a-1)(\operatorname{tr}\widetilde{\boldsymbol{E}})\boldsymbol{T}_0 + a(\boldsymbol{T}_0 : \widetilde{\boldsymbol{E}})\boldsymbol{I} \\ &\quad + (b+c+1)\big(\boldsymbol{T}_0 \cdot \widetilde{\boldsymbol{E}} + \widetilde{\boldsymbol{E}} \cdot \boldsymbol{T}_0\big).\end{aligned} \tag{8.17}$$

The first terms in (8.16) and (8.17) can be identified as the rotated initial stress, whereas the remaining terms represent the stress perturbation due to the infinitesimal strain $\widetilde{\boldsymbol{E}}$.

Substituting (8.6) and (8.7) into (8.3), and the tensor $\mathcal{Z}$ for the tensor $\mathcal{C}$, the strain energy density can be rewritten, correct to second order in $\widetilde{\boldsymbol{E}}$, in the form

$$\begin{aligned}W(\widetilde{\boldsymbol{E}}) &= W(\boldsymbol{0}) + \boldsymbol{T}_0 : \widetilde{\boldsymbol{E}} + \frac{1}{2}\,\widetilde{\boldsymbol{E}} : \mathcal{Z} : \widetilde{\boldsymbol{E}} \\ &= W(\boldsymbol{0}) + \boldsymbol{T}_0 : \widetilde{\boldsymbol{E}} + \frac{1}{2}\,\widetilde{\boldsymbol{E}} : \mathcal{C} : \widetilde{\boldsymbol{E}} + a(\operatorname{tr}\widetilde{\boldsymbol{E}})(\boldsymbol{T}_0 : \widetilde{\boldsymbol{E}}) \\ &\quad + (b+c)\operatorname{tr}(\widetilde{\boldsymbol{E}} \cdot \boldsymbol{T}_0 \cdot \widetilde{\boldsymbol{E}}).\end{aligned} \tag{8.18}$$

The terms proportional to $\boldsymbol{T}_0$ represent the work done against the initial stress, whereas the term $\frac{1}{2}\widetilde{\boldsymbol{E}} : \mathcal{C} : \widetilde{\boldsymbol{E}}$ is the classical elastic energy density in the absence of an initial stress.

8.2 The Elastic Tensor

A stress-free configuration of a body is called a *natural configuration*.[3] If such a configuration is used as the reference, then $\boldsymbol{T}_0 = \boldsymbol{0}$, and (8.13)–(8.15) simplify to

$$\boldsymbol{\tau} = \boldsymbol{\mathcal{C}} : \boldsymbol{\varepsilon}, \tag{8.19}$$

where the infinitesimal stress tensor $\boldsymbol{\tau}$ can be interpreted as $\boldsymbol{t}$, $\boldsymbol{T}^{(1)}$ or $\boldsymbol{T}^{(2)}$, and the infinitesimal strain tensor $\boldsymbol{\varepsilon}$ as either the infinitesimal Lagrangian or Eulerian strain tensors, $\widetilde{\boldsymbol{E}}$ or $\tilde{\boldsymbol{e}}$. This classical linear elastic stress-strain relation is known as the *generalised Hooke's law* and the fourth-order tensor $\boldsymbol{\mathcal{C}}$ is called the *elastic tensor*. The linear stress-strain constitutive relation for infinitesimal deformations of an anisotropic solid in its natural, stress-free configuration are specified by 21 independent components of $\boldsymbol{\mathcal{C}}$. In the more general case of an incremental stress superimposed upon the zero-order initial stress $\boldsymbol{T}_0$, the stress-strain relation is specified by 27 coefficients: the 21 components of $\boldsymbol{\mathcal{C}}$ and 6 components of the initial stress tensor $\boldsymbol{T}_0$. Equations (8.16) and (8.17) constitute the generalisation of Hooke's law to the case of a pre-stressed elastic material.

In the special case of $\boldsymbol{T}_0 = \boldsymbol{0}$, the strain energy density simplifies to

$$W(\boldsymbol{\varepsilon}) = W(\boldsymbol{0}) + \frac{1}{2}\boldsymbol{\tau} : \boldsymbol{\varepsilon}. \tag{8.20}$$

8.3 Isotropic Linear Elastic Solids

The highest symmetry of a solid is reached if the solid possesses no preferred direction with respect to its elastic properties. This also means that the linear elasticity $\boldsymbol{\mathcal{C}}$ is invariant under orthogonal transformations of the coordinate system. We refer to such a solid as *isotropic*. We showed in Sect. 6.17.1 that the number of independent elastic constants reduces to two and the constitutive equation for an isotropic linear elastic solid has the form

$$\boldsymbol{\tau} = \lambda\vartheta\boldsymbol{I} + 2\mu\boldsymbol{\varepsilon}, \tag{8.21}$$

where λ and μ are the Lamé elastic parameters introduced in (6.151),

$$\vartheta := \operatorname{tr}\boldsymbol{\varepsilon} = \operatorname{div}\vec{u}, \tag{8.22}$$

[3] We should note that there are certain kinds of bodies, for example, solids in different phases or polycrystalline metals, that may remain stress-free in different configurations that are not related by a rigid-body motion. These bodies may, in addition, possess different material symmetries in these different natural (or stress-free) configurations (e.g. Rajagopal and Srinivasa 2004).

and $\vec{u}$ is the displacement vector. We will now show, independently of the considerations in Sect. 6.17.1, that the generalised Hooke's law (8.19) reduces to the form (8.21) for an isotropic linear elastic solid.

In linear theory, it is not necessary to distinguish between the referential and spatial coordinate systems, therefore we use Cartesian coordinates x_k for these coinciding coordinate systems. An orthogonal transformation of x_k coordinates onto $x'_{k'}$ coordinates is expressed by the transformation equation (5.4),

$$x'_{k'} = Q_{k'k}x_k + c'_{k'}, \tag{8.23}$$

which is subject to the orthogonality conditions,

$$Q_{k'k}Q_{l'k} = \delta_{k'l'}, \qquad Q_{k'k}Q_{k'l} = \delta_{kl}. \tag{8.24}$$

The inverse transformation to (8.23) is

$$x_k = Q_{k'k}x'_{k'} + c_k, \tag{8.25}$$

where $c_k = c'_{k'}Q_{k'k}$. Differentiating (8.23) and (8.25) with respect to x_k and $x'_{k'}$, respectively, yields

$$\frac{\partial x'_{k'}}{\partial x_k} = Q_{k'k}, \qquad \frac{\partial x_k}{\partial x'_{k'}} = Q_{k'k}, \tag{8.26}$$

such that the transformation rules (3.19) and (1.67) for the infinitesimal stress tensor $\boldsymbol{\tau}$ and infinitesimal strain tensor $\boldsymbol{\varepsilon}$ become

$$\tau'_{k'l'} = Q_{k'k}Q_{l'l}\tau_{kl}, \qquad \varepsilon'_{k'l'} = Q_{k'k}Q_{l'l}\varepsilon_{kl}. \tag{8.27}$$

The proof of Hooke's law (8.21) for an isotropic linear elastic solid is accomplished in four steps. First, we show that for an isotropic linear elastic solid, the principal axes of the infinitesimal stress and infinitesimal strain tensors coincide. To prove this, we take, without loss of generality, the coordinate axes x_k in the principal directions of $\boldsymbol{\varepsilon}$. Then, $\varepsilon_{12} = \varepsilon_{13} = \varepsilon_{23} = 0$. We now show that $\tau_{23} = 0$. First,

$$\tau_{23} = A\varepsilon_{11} + B\varepsilon_{22} + C\varepsilon_{33}$$

with $A := C_{2311}$, $B := C_{2322}$, and $C := C_{2333}$. We now rotate the coordinate system about the x_3 axis by 180° such that $x'_1 = -x_1$, $x'_2 = -x_2$ and $x'_3 = x_3$, and

$$\boldsymbol{Q} = \begin{pmatrix} -1 & 0 & 0 \\ 0 & -1 & 0 \\ 0 & 0 & 1 \end{pmatrix},$$

which, in view of (8.27), yields

$$\begin{aligned}\tau'_{23} &= Q_{2k}Q_{3l}\tau_{kl} = -\tau_{23},\\ \varepsilon'_{11} &= Q_{1k}Q_{1l}\varepsilon_{kl} = \varepsilon_{11},\\ \varepsilon'_{22} &= Q_{2k}Q_{2l}\varepsilon_{kl} = \varepsilon_{22},\\ \varepsilon'_{33} &= \varepsilon_{33}.\end{aligned}$$

The relationship

$$\tau'_{23} = A\varepsilon'_{11} + B\varepsilon'_{22} + C\varepsilon'_{33} = A\varepsilon_{11} + B\varepsilon_{22} + C\varepsilon_{33} = \tau_{23}$$

is now a consequence of isotropy since the constants A, B and C do not depend on the choice of the reference configuration. Thus,

$$-\tau_{23} = \tau'_{23} = \tau_{23},$$

which implies that $\tau_{23} = 0$. Similarly, it can be shown that $\tau_{12} = \tau_{13} = 0$. Hence, we have proved that the principal axes of infinitesimal stress and infinitesimal strain coincide for an isotropic solid.

Second, consider the component τ_{11}. Taking the coordinate axes in the principal directions of strain, we obtain

$$\tau_{11} = a_1\varepsilon_{11} + b_1\varepsilon_{22} + c_1\varepsilon_{33},$$

with $a_1 := C_{1111}$, $b_1 := C_{1122}$ and $c_1 := C_{1133}$. We now rotate the coordinate system about the x_1 axis by 90° such that $x'_1 = x_1$, $x'_2 = x_3$ and $x'_3 = -x_2$, and

$$\boldsymbol{Q} = \begin{pmatrix} 1 & 0 & 0\\ 0 & 0 & 1\\ 0 & -1 & 0\end{pmatrix},$$

which gives

$$\begin{aligned}\tau'_{11} &= Q_{1k}Q_{1l}\tau_{kl} = \tau_{11},\\ \varepsilon'_{11} &= \varepsilon_{11},\\ \varepsilon'_{22} &= Q_{2k}Q_{2l}\varepsilon_{kl} = \varepsilon_{33},\\ \varepsilon'_{33} &= \varepsilon_{22}.\end{aligned}$$

In view of isotropy, the constants a_1, b_1 and c_1 do not depend on the reference configuration, hence

$$\tau'_{11} = a_1\varepsilon'_{11} + b_1\varepsilon'_{22} + c_1\varepsilon'_{33}.$$

Substituting for ε'_{kk}, we obtain

$$\tau'_{11} = a_1\varepsilon_{11} + b_1\varepsilon_{33} + c_1\varepsilon_{22},$$

which implies that $b_1 = c_1$ because $\tau_{11} = a_1\varepsilon_{11} + b_1\varepsilon_{22} + c_1\varepsilon_{33}$. We can thus write τ_{11} as

$$\tau_{11} = a_1\varepsilon_{11} + b_1\left(\varepsilon_{22} + \varepsilon_{33}\right) = \lambda_1\vartheta + 2\mu_1\varepsilon_{11},$$

where

$$\lambda_1 := b_1 = C_{1122}, \qquad 2\mu_1 := a_1 - b_1 = C_{1111} - C_{1122}.$$

The same relationship can be obtained for subscripts 2 and 3:

$$\begin{aligned}\tau_{22} &= \lambda_2\vartheta + 2\mu_2\varepsilon_{22},\\ \tau_{33} &= \lambda_3\vartheta + 2\mu_3\varepsilon_{33},\end{aligned}$$

where $\lambda_2 := C_{2222}$, $2\mu_2 := C_{2211} - C_{2222}$, $\lambda_3 := C_{3322}$ and $2\mu_3 := C_{3311} - C_{3322}$.

Third, we now rotate the coordinate system about the x_3 axis by 90° such that $x'_1 = x_2$, $x'_2 = -x_1$, and $x'_3 = x_3$, and

$$\boldsymbol{Q} = \begin{pmatrix} 0 & 1 & 0\\ -1 & 0 & 0\\ 0 & 0 & 1\end{pmatrix},$$

which gives

$$\begin{aligned}\tau'_{11} &= \tau_{22}, & \varepsilon'_{11} &= \varepsilon_{22},\\ \tau'_{22} &= \tau_{11}, & \varepsilon'_{22} &= \varepsilon_{11},\\ \tau'_{33} &= \tau_{33}, & \varepsilon'_{33} &= \varepsilon_{33}.\end{aligned}$$

Combining the results of the previous steps, we can write

$$\begin{aligned}\tau'_{22} &= \lambda_2\vartheta' + 2\mu_2\varepsilon'_{22}\\ &= \lambda_2\vartheta + 2\mu_2\varepsilon'_{22}\\ &= \lambda_2\vartheta + 2\mu_2\varepsilon_{11}\end{aligned}$$

$$\stackrel{!}{=} \tau_{11}$$
$$= \lambda_1 \vartheta + 2\mu_1 \varepsilon_{11},$$

which implies that $\lambda_2 = \lambda_1$ and $\mu_2 = \mu_1$. Similarly, it can be shown that $\lambda_3 = \lambda_1$ and $\mu_3 = \mu_1$. In summary,

$$\tau_{11} = \lambda\vartheta + 2\mu\varepsilon_{11},$$
$$\tau_{22} = \lambda\vartheta + 2\mu\varepsilon_{22},$$
$$\tau_{33} = \lambda\vartheta + 2\mu\varepsilon_{33},$$
$$\tau_{kl} = 0 \quad \text{for} \quad k \neq j,$$

which is the generalised Hooke's law for an isotropic solid in principal directions. This may also be written as

$$\tau_{kl} = \lambda\vartheta\delta_{kl} + 2\mu\varepsilon_{kl}. \tag{8.28}$$

Fourth, we now rotate the coordinate system about axes x_k coinciding with the principal directions of strain to an arbitrary coordinate system with axes $x'_{k'}$, and show that Hooke's law (8.28) also holds in the rotated coordinate system $x'_{k'}$. Denoting the transformation matrix of this rotation by Q_{kl}, the stress and strain tensors transform according to (8.27). Multiplying (8.28) by $Q_{k'k}Q_{l'l}$ and using the transformation rule (8.27) and the orthogonality property (8.24), we have

$$\tau'_{k'l'} = \lambda\vartheta\delta_{k'l'} + 2\mu\varepsilon'_{k'l'}.$$

Moreover,

$$\vartheta' = \varepsilon'_{k'k'} = Q_{k'k}Q_{k'l}\varepsilon_{kl} = \delta_{kl}\varepsilon_{kl} = \varepsilon_{kk} = \vartheta,$$

which finally gives

$$\tau'_{k'l'} = \lambda\vartheta'\delta_{k'l'} + 2\mu\varepsilon'_{k'l'}, \tag{8.29}$$

demonstrating that Hooke's law (8.21) also holds in the rotated coordinates $x'_{k'}$.

We note, without proof, that the most general form of a fourth-order isotropic tensor $\mathcal{A}$ is

$$\mathcal{A}_{klmn} = \lambda\delta_{kl}\delta_{mn} + \mu\left(\delta_{km}\delta_{ln} + \delta_{kn}\delta_{lm}\right) + \nu\left(\delta_{km}\delta_{ln} - \delta_{kn}\delta_{lm}\right), \tag{8.30}$$

where λ, μ and ν are scalars. If $\mathcal{A}$ is replaced by the linear elasticity $\mathcal{C}$, the symmetry relations $\mathcal{C}_{klmn} = \mathcal{C}_{lkmn} = \mathcal{C}_{klnm}$ imply that ν must be zero since by interchanging

k and l in

$$\nu\,(\delta_{km}\delta_{ln} - \delta_{kn}\delta_{lm}) = \nu\,(\delta_{lm}\delta_{kn} - \delta_{ln}\delta_{km})$$

gives $\nu = -\nu$, and therefore $\nu = 0$. Thus, the linear elasticity for an isotropic solid has the form

$$\mathcal{C}_{klmn} = \lambda\delta_{kl}\delta_{mn} + \mu\,(\delta_{km}\delta_{ln} + \delta_{kn}\delta_{lm})\,. \tag{8.31}$$

Hooke's law for an isotropic solid takes a particularly simple form in the spherical and deviatoric parts of $\boldsymbol{\varepsilon}$ and $\boldsymbol{\tau}$. Let us decompose the infinitesimal strain tensor $\boldsymbol{\varepsilon}$ into its spherical and deviatoric parts,

$$\boldsymbol{\varepsilon} = \frac{1}{3}\vartheta\boldsymbol{I} + \boldsymbol{\varepsilon}^D, \tag{8.32}$$

where the *mean normal strain* ϑ is defined by (8.22) and the *deviatoric strain tensor* $\boldsymbol{\varepsilon}^D$ is a trace-free, symmetric tensor,

$$\boldsymbol{\varepsilon}^D := \boldsymbol{\varepsilon} - \frac{1}{3}\vartheta\boldsymbol{I}. \tag{8.33}$$

The same decomposition of the infinitesimal stress tensor $\boldsymbol{\tau}$ is

$$\boldsymbol{\tau} = \sigma\boldsymbol{I} + \boldsymbol{\tau}^D, \tag{8.34}$$

where the scalar σ and tensor $\boldsymbol{\tau}^D$ are the mechanical pressure and deviatoric stress, respectively,

$$\sigma := \frac{1}{3}\mathrm{tr}\,\boldsymbol{\tau}, \qquad \boldsymbol{\tau}^D := \boldsymbol{\tau} - \sigma\boldsymbol{I}. \tag{8.35}$$

Substituting (8.32) and (8.34) into (8.21), Hooke's law for an isotropic linear elastic solid is written separately for the spherical and deviatoric parts of the stress and strain as

$$\sigma = k\vartheta, \qquad \boldsymbol{\tau}^D = 2\mu\boldsymbol{\varepsilon}^D, \tag{8.36}$$

where

$$k := \lambda + \frac{2}{3}\mu \tag{8.37}$$

is referred to as the *elastic bulk modulus*. Finally, substituting (8.21) into (8.20) gives the strain energy density for an isotropic linear elastic solid,

$$W(\boldsymbol{\varepsilon}) = W(\boldsymbol{0}) + \frac{1}{2}\lambda\vartheta^2 + \mu(\boldsymbol{\varepsilon} : \boldsymbol{\varepsilon}), \tag{8.38}$$

which, following on from (8.33) and (8.37), can be written as

$$W(\boldsymbol{\varepsilon}) = W(\mathbf{0}) + \frac{1}{2}k\vartheta^2 + \mu(\boldsymbol{\varepsilon}^D : \boldsymbol{\varepsilon}^D). \tag{8.39}$$

8.4 Constraints on Elastic Coefficients

In this section, we state several experimental constraints that are placed upon elastic moduli so that they adequately represent real materials.

We first assume that Hooke's law (8.21) for an isotropic linear elastic solid is invertible for $\boldsymbol{\varepsilon}$. Applying the trace operator to this equation gives

$$\operatorname{tr}\boldsymbol{\tau} = (3\lambda + 2\mu)\operatorname{tr}\boldsymbol{\varepsilon}. \tag{8.40}$$

Solving (8.21) for $\boldsymbol{\varepsilon}$ and substituting for $\operatorname{tr}\boldsymbol{\varepsilon}$ from (8.40) gives the inverse form of Hooke's law,

$$\boldsymbol{\varepsilon} = \frac{\boldsymbol{\tau}}{2\mu} - \frac{\lambda}{2\mu(3\lambda + 2\mu)}\operatorname{tr}\boldsymbol{\tau}\,\boldsymbol{I}, \tag{8.41}$$

which requires that

$$\mu \neq 0, \qquad 3\lambda + 2\mu \neq 0, \tag{8.42}$$

in order to $\boldsymbol{\varepsilon}$ to be uniquely determined by $\boldsymbol{\tau}$.

In addition, two inequalities for the Lamé elastic parameters are deduced from the following experiments:

(i) Hydrostatic Pressure Experimental observations indicate that the volume of an elastic solid diminishes under hydrostatic pressure. For hydrostatic compression, the stress tensor has the form

$$\boldsymbol{\tau} = -p\boldsymbol{I}, \qquad p > 0. \tag{8.43}$$

From $(8.36)_1$ we see that

$$p = -k\vartheta, \tag{8.44}$$

where $\vartheta := \operatorname{tr}\boldsymbol{\varepsilon} = \operatorname{div}\vec{u} = J - 1 = (dv - dV)/dV$ is the *cubical dilatation*. In the case of hydrostatic pressure, $dv < dV$, hence $\vartheta < 0$, thus we must have $k > 0$, which implies that

$$3\lambda + 2\mu > 0. \tag{8.45}$$

(ii) Simple Shear Consider a simple constant shear in which $\tau_{12} \neq 0$ and otherwise $\tau_{kl} = 0$. In this case, (8.41) gives

$$\tau_{12} = 2\mu\varepsilon_{12}. \tag{8.46}$$

Experimental observations of small deformations of elastic solids subjected to simple shear indicate that τ_{12} and ε_{12} have the same direction. Consequently,

$$\mu > 0. \tag{8.47}$$

Under conditions (8.45) and (8.47), and provided that the strain energy density vanishes at the unstrained natural configuration (i.e., $W(\mathbf{0}) = 0$), the strain energy density $W(\boldsymbol{\varepsilon})$ for an isotropic linear elastic solid is a positive-definite, quadratic function of the infinitesimal strain $\boldsymbol{\varepsilon}$. This fact is fundamental to providing the uniqueness of a solution to boundary-value problems for elastic bodies in *elastic equilibrium*.[4]

The conditions (8.45) and (8.47) may be sufficient, but not necessary, to obtain plausible results from infinitesimal field theory for isotropic linear elastic solids. For example, it can be shown that

$$\lambda + 2\mu > 0 \tag{8.48}$$

together with (8.47) give positive wave speeds in an isotropic linear elastic solid. A number of special a priori inequalities have been proposed based on certain 'reasonable' physical expectations when an isotropic linear elastic solid is subjected to pressure, tensions and shears. For a discussion of these, we refer the reader to Truesdell and Noll (1965).

In the classical infinitesimal theory of elasticity, other material parameters are often used in place of the Lamé elastic parameters λ and μ. The relationships between some of these parameters are

$$\begin{aligned}
&E := \mu(3\lambda + 2\mu)/(\lambda + \mu), && \nu := \lambda/2(\lambda + \mu),\\
&\lambda = E\nu/(1 + \nu)(1 - 2\nu) = 2G\nu/(1 - 2\nu), && \mu \equiv G = E/2(1 + \nu),\\
&k = \lambda + \tfrac{2}{3}\mu = E/3(1 - 2\nu),
\end{aligned} \tag{8.49}$$

[4]The state of an elastic body in which each volume element of the body is in equilibrium under the combined effects of elastic stresses and externally applied body forces.

where E is called *Young's modulus*, ν is *Poisson's ratio* and G is the *elastic shear modulus* or *modulus of rigidity*, which, as noted, is identical to the Lamé parameter μ. The constraints

$$G > 0, \qquad -1 < \nu < \frac{1}{2}, \tag{8.50}$$

are equivalent to (8.45) and (8.47).

8.5 Field Equations

The field equations in continuum mechanics consist of the conservation principles, valid for all continua, and a constitutive equation, valid for a particular material. Here, we will consider isotropic linear elastic or linear thermoelastic solids. The field equations may be expressed in either the Lagrangian or the Eulerian form, though the two forms are identical in the linearised theory of elasticity.

8.5.1 Isotropic Linear Elastic Solids

Within the framework of infinitesimal strain theory, the Lagrangian form of the continuity equation (4.63) can be arranged as

$$\varrho = \frac{\varrho_0}{J} = \frac{\varrho_0}{1 + \operatorname{tr} \widetilde{\boldsymbol{E}}} + O(|\widetilde{\boldsymbol{E}}|^2) = \varrho_0 + O(|\widetilde{\boldsymbol{E}}|). \tag{8.51}$$

Using this approximation and provided, in addition, that the body force $\vec{f}$ is also infinitesimally small, the equation of motion (4.24) in the present configuration coincides with the equation of motion (4.69) in the reference configuration such that

$$\operatorname{div} \boldsymbol{\tau} + \varrho_0 \vec{f} = \varrho_0 \frac{\partial^2 \vec{u}}{\partial t^2}, \tag{8.52}$$

where we have substituted the Lagrangian representation of velocity, $\vec{v} = \partial \vec{u}/\partial t$. The continuity equation plays no role here because the principle of conservation of mass results in the statement $\varrho_0 = \varrho_0(\vec{X})$.

We substitute Hooke's law (8.21) for an isotropic linear elastic solid into the equation of motion (8.52). Using vector differential identities (see Appendix A), the divergence of the stress tensor $\boldsymbol{\tau}$ can be arranged as

$$\begin{aligned}
\operatorname{div} \boldsymbol{\tau} &= \operatorname{div} (\lambda \vartheta \boldsymbol{I} + 2\mu \boldsymbol{\varepsilon}) \\
&= \operatorname{grad} (\lambda \vartheta) + 2\mu \operatorname{div} \boldsymbol{\varepsilon} + 2 \operatorname{grad} \mu \cdot \boldsymbol{\varepsilon}
\end{aligned}$$

$$
\begin{aligned}
&= \lambda \,\mathrm{grad}\,\mathrm{div}\,\vec{u} + \mathrm{div}\,\vec{u}\,\mathrm{grad}\,\lambda + \mu\,\mathrm{div}\big(\mathrm{grad}\,\vec{u} + \mathrm{grad}^T\vec{u}\big)\\
&\quad + \mathrm{grad}\,\mu \cdot \big(\mathrm{grad}\,\vec{u} + \mathrm{grad}^T\vec{u}\big)\\
&= (\lambda + \mu)\,\mathrm{grad}\,\mathrm{div}\,\vec{u} + \mu\,\nabla^2\vec{u} + \mathrm{div}\,\vec{u}\,\mathrm{grad}\,\lambda + \mathrm{grad}\,\mu \cdot \big(\mathrm{grad}\,\vec{u} + \mathrm{grad}^T\vec{u}\big)\\
&= (\lambda + 2\mu)\,\mathrm{grad}\,\mathrm{div}\,\vec{u} - \mu\,\mathrm{rot}\,\mathrm{rot}\,\vec{u} + \mathrm{div}\,\vec{u}\,\mathrm{grad}\,\lambda\\
&\quad + \mathrm{grad}\,\mu \cdot \big(\mathrm{grad}\,\vec{u} + \mathrm{grad}^T\vec{u}\big).
\end{aligned}
$$

The linearised equation of motion (8.52) can now be expressed in the form

$$
\begin{aligned}
&(\lambda + 2\mu)\,\mathrm{grad}\,\mathrm{div}\,\vec{u} - \mu\,\mathrm{rot}\,\mathrm{rot}\,\vec{u} + \mathrm{div}\,\vec{u}\,\mathrm{grad}\,\lambda\\
&\qquad + \mathrm{grad}\,\mu \cdot \big(\mathrm{grad}\,\vec{u} + \mathrm{grad}^T\vec{u}\big) + \varrho_0\vec{f} = \varrho_0\frac{\partial^2\vec{u}}{\partial t^2},
\end{aligned}
\tag{8.53}
$$

which is known as the *Navier–Cauchy equation* of an isotropic linear elastic solid. It consists of three second-order differential equations for the three displacement components. Note that the vector form (8.53) is independent of the coordinate system and may be expressed in orthogonal curvilinear coordinates by the method presented in Appendix C.

For a homogeneous solid, λ and μ are constants (independent of $\vec{X}$) and the Navier–Cauchy equation simplifies to

$$
(\lambda + 2\mu)\,\mathrm{grad}\,\mathrm{div}\,\vec{u} - \mu\,\mathrm{rot}\,\mathrm{rot}\,\vec{u} + \varrho_0\vec{f} = \varrho_0\frac{\partial^2\vec{u}}{\partial t^2},
\tag{8.54}
$$

or, alternatively,

$$
(\lambda + \mu)\,\mathrm{grad}\,\mathrm{div}\,\vec{u} + \mu\nabla^2\vec{u} + \varrho_0\vec{f} = \varrho_0\frac{\partial^2\vec{u}}{\partial t^2}.
\tag{8.55}
$$

Note that the term *elastostatics* refers to the case where the inertial force $\varrho_0\partial^2\vec{u}/\partial t^2$ is neglected. Then, (8.52) does not involve displacements and an alternative formulation of elastostatic problems in terms of stresses alone is possible (Nečas and Hlaváček 1981).

8.5.2 *Incompressible, Isotropic Linear Elastic Solids*

For incompressible solids,

$$
\mathrm{div}\,\vec{u} = 0,
\tag{8.56}
$$

and the mass density remains unchanged, equal to a known value, $\varrho = \varrho_0$. Hooke's law (8.21) modifies to

$$\boldsymbol{\tau} = -\pi \boldsymbol{I} + 2\mu\boldsymbol{\varepsilon}, \tag{8.57}$$

where the constraint pressure π is a new unknown field variable. The Navier–Cauchy equation of motion (8.53) is

$$-\operatorname{grad}\pi + \mu\,\nabla^2\vec{u} + \operatorname{grad}\mu \cdot \left(\operatorname{grad}\vec{u} + \operatorname{grad}^T\vec{u}\right) + \varrho_0\vec{f} = \varrho_0\frac{\partial^2\vec{u}}{\partial t^2}. \tag{8.58}$$

Hence, (8.56) and (8.58) represent four differential equations for four unknown field variables: constraint pressure π and the three components of the displacement vector $\vec{u}$.

8.5.3 *Isotropic Linear Thermoelastic Solids*

Equations (7.23), (7.24) and (7.29) form the constitutive equations for an isotropic linear thermoelastic solid. Substituting $(7.23)_1$ and (7.24) into the Lagrangian form (4.79) of the energy equation leads to

$$2\mu\left[I_E(I_E)^{\cdot} - (II_E)^{\cdot}\right] + \gamma T\,\dot{T} = -\beta T(I_E)^{\cdot} + 2\mu\boldsymbol{E} : \dot{\boldsymbol{E}} + \operatorname{Div}\vec{Q} + \varrho_0 h, \tag{8.59}$$

where the heat flux $\vec{Q}$ is given by (7.29). The material time derivatives of the invariants of $\boldsymbol{E}$ are

$$\begin{aligned}(I_E)^{\cdot} &= \operatorname{Div}\dot{\vec{u}},\\ (II_E)^{\cdot} &= I_E(I_E)^{\cdot} - \boldsymbol{E} : \dot{\boldsymbol{E}},\end{aligned} \tag{8.60}$$

which simplifies (8.59) to

$$\gamma T\,\dot{T} = -\beta T \operatorname{Div}\dot{\vec{u}} + \operatorname{Div}\vec{Q} + \varrho_0 h. \tag{8.61}$$

This is a non-linear equation with respect to temperature, which can be linearised by assuming that the instantaneous absolute temperature T differs from a reference temperature T_0 by a small quantity, that is,

$$T = T_0 + T_1, \qquad |T_1| \ll T_0. \tag{8.62}$$

The linearisation yields

$$\gamma T_0\dot{T}_1 = -\beta T_0 \operatorname{Div}\dot{\vec{u}} + \operatorname{Div}\vec{Q} + \varrho_0 h. \tag{8.63}$$

Considering the case where $\operatorname{Div}\dot{\vec{u}} = 0$ and $h = 0$, the heat conduction equation is not coupled with the equation of motion, and it has the form

$$\operatorname{Div}\vec{Q} = \varrho_0 c_v \dot{T}_1, \tag{8.64}$$

where the *specific heat at constant volume* c_v is introduced. Comparing (8.63) and (8.64), $\gamma T_0 = \varrho_0 c_v$, and the energy equation (8.63) becomes

$$\varrho_0 c_v \dot{T}_1 = -\beta T_0 \operatorname{Div}\dot{\vec{u}} + \operatorname{Div}\vec{Q} + \varrho_0 h. \tag{8.65}$$

This equation, known as the *coupled heat conduction equation*, relates the rate of change of temperature, strain and heat conduction.

Under the decomposition (8.62), the linear constitutive equation $(7.23)_1$ for the second Piola–Kirchhoff stress tensor becomes

$$\boldsymbol{T}^{(2)} = (\sigma - \beta\, T_0)\boldsymbol{I} + \lambda(\operatorname{tr}\boldsymbol{E})\boldsymbol{I} + 2\mu\boldsymbol{E} - \beta\, T_1\boldsymbol{I}. \tag{8.66}$$

where $\sigma - \beta\, T_0 = 0$ if the reference configuration is stress free. Within the framework of linear theory, we consider only infinitesimal deformations from the reference configuration. Then, $\boldsymbol{T}^{(2)}$ and $\boldsymbol{E}$ can be replaced by the infinitesimal stress tensor $\boldsymbol{\tau}$ and infinitesimal strain tensor $\boldsymbol{\varepsilon}$, respectively. The constitutive equation (8.66), known as the *Duhamel–Neumann* extension of Hooke's law, may be inverted to give

$$\boldsymbol{E} = -\frac{1}{3\lambda + 2\mu}(\sigma - \beta\, T_0)\boldsymbol{I} + \frac{1}{2\mu}\boldsymbol{T}^{(2)} - \frac{\lambda}{2\mu(3\lambda + 2\mu)}\operatorname{tr}\boldsymbol{T}^{(2)}\,\boldsymbol{I} + \frac{\beta}{3\lambda + 2\mu}T_1\boldsymbol{I}, \tag{8.67}$$

where

$$\alpha := \frac{\beta}{3\lambda + 2\mu} \tag{8.68}$$

is the *thermal expansion coefficient*.

The coupled heat conduction equation (8.65), along with the equation of motion (8.52) and thermoelastic stress-strain equation (8.66), constitutes the basic set of field equations for coupled thermoelastic problems. There are, however, problems in which the heat conduction equation can be further simplified. For instance, if the reference configuration of the thermoelastic solid coincides with the *thermostatic equilibrium*, then $\operatorname{Grad} T_0 = \vec{0}$ and $\vec{Q} = \kappa \operatorname{Grad} T_1$ in the second term on the right-hand side of (8.65). Another simplification arises when the isotropic linear thermoelastic solid is also *incompressible*. Then, both $\operatorname{Div}\vec{u} = 0$ and $\operatorname{Div}\dot{\vec{u}} = 0$ hold such that the first term on the right-hand side of (8.65) vanishes. In this case, the heat conduction equation is decoupled from the equation of motion and the thermoelastic problem is decomposed into two separate problems that are solved consecutively, but independently.

Chapter 9
Infinitesimal Deformation of a Body with a Finite Pre-stress

We devote this chapter to deriving the linearised equations and jump conditions that govern infinitesimal elastic deformations of a body initially in static equilibrium. The property of elasticity refers only to incremental deformations, while the state of initial stress may be due to any physical cause. Here, we consider that the initial static stress is caused by self-gravitation, that is, the gravitational forces acting between the material particles of a body. The principal assumption is that the initial static stress caused by self-gravitation is accounted for in a self-consistent treatment of a gravitating body with a three-dimensional density structure.

9.1 Equations for the Initial Configuration of a Body

We consider a material body that occupies an initial configuration κ_0 at time $t = 0$. We assume that the body is composed of solid and fluid regions, and denote the volume of solid regions by V_S, and the volume of fluid regions by V_F. The total volume of the body in the configuration κ_0 is denoted by $V = V_S \cup V_F$. The solid and fluid regions in V are separated by non-intersecting internal discontinuity surfaces (or, simply, discontinuities). We denote the internal welded solid–solid discontinuities by Σ_S, and the internal fluid–solid discontinuities by Σ_F. The union of all internal discontinuities in the configuration κ_0 is denoted by $\Sigma = \Sigma_S \cup \Sigma_F$.

The initial configuration κ_0 will be used as the reference. We suppose that this configuration does *not* correspond to the natural, stress-free configuration of the body, but that the body is pre-stressed by a finite stress such that the body is in *static equilibrium*. In this *pre-stressed configuration*, the Cauchy stress and the two Piola–Kirchhoff stresses coincide; we denote this initial static stress by $\boldsymbol{t}_0(\vec{X})$, where $\vec{X}$ is the position of a material particle in the initial configuration. Static equilibrium is

Z. Martinec, *Principles of Continuum Mechanics*, Nečas Center Series,
https://doi.org/10.1007/978-3-030-05390-1_9

guaranteed by the linear momentum equation (4.69) with no inertial force,

$$\operatorname{Div} \boldsymbol{t}_0 + \varrho_0 \vec{f}_0 = 0 \qquad \text{in } V - \Sigma, \tag{9.1}$$

where $\varrho_0(\vec{X})$ and $\vec{f}_0(\vec{X})$ are, respectively, the mass density and the body force per unit mass in κ_0. Since the body contains fluid regions, it is convenient to decompose the initial static stress $\boldsymbol{t}_0$ into the spherical and deviatoric parts,

$$\boldsymbol{t}_0 = \sigma_0 \boldsymbol{I} + \boldsymbol{t}_0^D, \tag{9.2}$$

where $\sigma_0 := \frac{1}{3}\operatorname{tr} \boldsymbol{t}_0$ is the mechanical pressure and $\operatorname{tr} \boldsymbol{t}_0^D = 0$. The equation of static equilibrium then takes the form

$$\operatorname{Grad} \sigma_0 + \operatorname{Div} \boldsymbol{t}_0^D + \varrho_0 \vec{f}_0 = 0 \qquad \text{in } V - \Sigma. \tag{9.3}$$

Since a fluid is unable to support shear stresses when it is in static equilibrium, the static deviatoric stress $\boldsymbol{t}_0^D$ vanishes in fluid regions and (9.3) simplifies to

$$\operatorname{Grad} \sigma_0 + \varrho_0 \vec{f}_0 = 0 \qquad \text{in } V_{\mathrm{F}} - \Sigma_{\mathrm{F}}. \tag{9.4}$$

The internal discontinuities Σ within the body are assumed to be material surfaces. The jump condition at welded discontinuities between two solids, and at slipping discontinuities between a solid and fluid, is given by (4.50),

$$[\vec{N} \cdot \boldsymbol{t}_0]_-^+ = \vec{0} \qquad \text{on } \Sigma, \tag{9.5}$$

where $\vec{N}$ is the outward unit normal vector to the discontinuity Σ. We introduce the normal stress τ_0 as the normal component of the traction vector $\vec{N} \cdot \boldsymbol{t}_0$ on the discontinuity Σ,

$$\tau_0 := \vec{N} \cdot \boldsymbol{t}_0 \cdot \vec{N}. \tag{9.6}$$

In view of (9.5), τ_0 passes through Σ continuously such that

$$[\tau_0]_-^+ = 0 \qquad \text{on } \Sigma. \tag{9.7}$$

The relationship between τ_0 and σ_0 on Σ is

$$\tau_0 = \sigma_0 + \vec{N} \cdot \boldsymbol{t}_0^D \cdot \vec{N} \qquad \text{on } \Sigma, \tag{9.8}$$

which follows from (9.2) and (9.6).

In addition, there are no shear stresses in the static equilibrium on the fluid side on a fluid–solid discontinuity, and the stress vector $\vec{N} \cdot \boldsymbol{t}_0$ points in the direction of the normal vector $\vec{N}$. The stress vector can then be expressed on both sides of the discontinuity in the form (4.57),

$$\vec{N} \cdot \boldsymbol{t}_0 = \tau_0 \vec{N} \qquad \text{on } \Sigma_{\mathrm{F}}, \tag{9.9}$$

where $\tau_0 = \sigma_0$ on the fluid side.

9.2 Superimposed Infinitesimal Deformations

We now consider a time- and space-dependent infinitesimal deformation field, characterised by the displacement field $\vec{U}(\vec{X}, t)$, that is superimposed upon the initial configuration κ_0. The superimposed displacement field brings the body into another time-dependent configuration κ_t. Since the displacement field is small, the present configuration κ_t is close to the initial configuration κ_0, which justifies using the same coordinates to describe the position of particles in both κ_0 and κ_t. We use the Lagrangian description of motion and denote by $\vec{x}$ the instantaneous position of a particle in the configuration κ_t initially located at the position $\vec{X}$ in the configuration κ_0. In view of $(1.100)_1$,

$$\vec{x} = \vec{X} + \vec{U}(\vec{X}, t). \tag{9.10}$$

In the next sections, we will use the principle of geometrical linearisation (see Sect. 1.12) and develop field equations and jump conditions correct to first order in $|\boldsymbol{H}|$, where $\boldsymbol{H} = \operatorname{Grad} \vec{U}$. In addition, we will assume that the ratio of an incremental stress caused by the infinitesimal deformation (9.10) to the initial static stress is also of the order of $|\boldsymbol{H}|$ (see Sect. 9.5).

9.3 Lagrangian and Eulerian Increments

Time changes of any physical quantity $\mathcal{Q}$ that has a non-zero initial static value can be described by either the *Eulerian* or *Lagrangian* increments, q^E or Q^L,

$$\begin{aligned} q^E(\vec{x}, t) &:= q(\vec{x}, t) - q(\vec{x}, 0), \\ Q^L(\vec{X}, t) &:= Q(\vec{X}, t) - q(\vec{X}, 0), \end{aligned} \tag{9.11}$$

where $q(\vec{x}, t)$ and $Q(\vec{X}, t)$ are the Eulerian and Lagrangian representations of $\mathcal{Q}$, respectively, and $q(\vec{x}, 0)$ and $q(\vec{X}, 0)$ are the initial static values of $\mathcal{Q}$ in the configurations κ_t and κ_0, respectively. The former are related by (1.24) and the latter by

$$\begin{aligned} q(\vec{x}, 0) &= q(\vec{X} + \vec{U}, 0) \\ &= q(\vec{X}, 0) + \vec{U} \cdot \operatorname{Grad} q(\vec{X}, 0) + O\big(|\vec{U}|^2\big). \end{aligned} \tag{9.12}$$

Note that the initial static values of $\mathcal{Q}$ at the configuration κ_0 can be denoted by either $q(\vec{X}, 0)$ or $Q(\vec{X}, 0)$, that is, $q(\vec{X}, 0) \equiv Q(\vec{X}, 0)$. Substituting (9.10)–(9.12) into $(1.24)_1$ yields

$$\begin{aligned} Q(\vec{X}, t) &= q(\vec{x}(\vec{X}, t), t) \\ &= q(\vec{X} + \vec{U}, 0) + q^E(\vec{X} + \vec{U}, t) \\ &= q(\vec{X}, 0) + \vec{U} \cdot \operatorname{Grad} q(\vec{X}, 0) + q^E(\vec{X} + \vec{U}, t) + O\big(|\vec{U}|^2\big) \\ &= q(\vec{X}, 0) + \vec{U} \cdot \operatorname{Grad} q(\vec{X}, 0) + q^E(\vec{X}, t) + O\big(|\vec{U}|^2\big) \\ &\stackrel{!}{=} q(\vec{X}, 0) + Q^L(\vec{X}, t). \end{aligned}$$

Hence, correct to first order in $|\vec{U}|$, the Lagrangian and Eulerian increments are related by

$$Q^L = q^E + \vec{U} \cdot \operatorname{Grad} q_0(\vec{X}) \,, \tag{9.13}$$

where $q_0(\vec{X}) \equiv q(\vec{X}, 0)$. In first-order theory, the increments Q^L and q^E can be considered to be functions of either $\vec{X}$ or $\vec{x}$. For this reason, we have dropped the dependence of the increments on $\vec{X}$ and $\vec{x}$ from the notation.

9.4 Linearised Continuity Equations

In view of (9.11), the Eulerian and Lagrangian increments in mass density are

$$\begin{aligned} \varrho^E &:= \varrho(\vec{x}, t) - \varrho_0(\vec{x}), \\ \mathcal{Q}^L &:= \mathcal{Q}(\vec{X}, t) - \varrho_0(\vec{X}), \end{aligned} \tag{9.14}$$

where $\varrho(\vec{x}, t)$ and $\mathcal{Q}(\vec{X}, t)$ are, respectively, the Eulerian and Lagrangian representations of mass density, which are related by (1.24).

The two increments can be expressed in terms of $\vec{U}$ by linearising the Eulerian and Lagrangian forms of the principle of conservation of mass. Integrating the Eulerian continuity equation (4.20) with respect to time gives

$$\int_0^t \frac{\partial \varrho}{\partial t} dt + \int_0^t \operatorname{div}(\varrho \vec{v}) dt = 0,$$

where the first integral results in the difference $\varrho(\vec{x}, t) - \varrho(\vec{x}, 0)$, which is equal to ϱ^E in view of $(9.14)_1$, so that

$$\varrho^E = -\int_0^t \operatorname{div}(\varrho\vec{v})dt. \tag{9.15}$$

Using $(1.116)_2$, the derivative of the Eulerian variables with respect to the spatial coordinates is now expressed in terms of the derivative of the Lagrangian variables with respect to the referential coordinates as

$$\operatorname{div}(\varrho\vec{v}) = \operatorname{Div}(\mathcal{Q}\vec{V}) - \boldsymbol{H} : \operatorname{Grad}(\mathcal{Q}\vec{V}) + O(|\boldsymbol{H}|^2),$$

where the Lagrangian velocity $\vec{V} = \vec{V}(\vec{X}, t)$ is defined by (2.3). Substituting for $\mathcal{Q}$ from $(9.14)_2$ and including the term $\boldsymbol{H} :$ Grad $(\mathcal{Q}\vec{V})$ into $O(|\boldsymbol{H}|^2)$ leads to

$$\operatorname{div}(\varrho\vec{v}) = \operatorname{Div}[(\varrho_0(\vec{X}) + \mathcal{Q}^L)\vec{V}] + O(|\boldsymbol{H}|^2).$$

Including, in addition, the term $\mathcal{Q}^L\vec{V}$ into $O(|\boldsymbol{H}|^2)$, expressing $\vec{V}(\vec{X}, t)$ in terms of $\vec{U}(\vec{X}, t)$ using $(2.5)_1$ and substituting the result into (9.15), we obtain

$$\varrho^E = -\int_0^t \operatorname{Div}\left[\varrho_0(\vec{X})\frac{\partial\vec{U}}{\partial t}\right]dt + O(|\boldsymbol{H}|^2). \tag{9.16}$$

The integration over time can now be readily performed; choosing $\vec{U}(\vec{X}, 0) = 0$ results in

$$\varrho^E = -\operatorname{Div}\left[\varrho_0(\vec{X})\vec{U}\right], \tag{9.17}$$

which is correct to first order in $|\boldsymbol{H}|$.

The linearised form of $\mathcal{Q}^L$ is obtained by substituting $(1.115)_2$ and $(9.14)_2$ into the Lagrangian form of the principle of conservation of mass (4.63):

$$\mathcal{Q}^L = -\varrho_0(\vec{X}) \operatorname{Div}\vec{U}, \tag{9.18}$$

which is correct to first order in $|\boldsymbol{H}|$. Using the differential identity (A.12), it is easy to verify that (9.17) and (9.18) are consistent with the general expression (9.13), that is, the increments ϱ^E and $\mathcal{Q}^L$ satisfy the first-order relation

$$\mathcal{Q}^L = \varrho^E + \vec{U} \cdot \operatorname{Grad}\varrho_0(\vec{X}). \tag{9.19}$$

9.5 Increments in Stress

We now consider the state of stress in the present configuration κ_t. This stress, referred to the area element in the initial configuration κ_0, is characterised by the first Piola–Kirchhoff stress tensor $\boldsymbol{T}^{(1)}$ (see Sect. 3.5). Since the initial static stress for $\boldsymbol{T}^{(1)}$ (and also for the Cauchy stress tensor $\boldsymbol{t}$) is $\boldsymbol{t}_0(\vec{X})$, we express $\boldsymbol{T}^{(1)}$ in a form analogous to $(9.11)_2$

$$\boldsymbol{T}^{(1)}(\vec{X}, t) = \boldsymbol{t}_0(\vec{X}) + \boldsymbol{T}^{(1),L}, \tag{9.20}$$

where $\boldsymbol{T}^{(1),L}$ is the Lagrangian increment in the first Piola–Kirchhoff stress tensor (or, simply, the *incremental first Piola–Kirchhoff stress tensor*) caused by the infinitesimal displacement $\vec{U}(\vec{X}, t)$. We assume that the increment in stresses caused by the infinitesimal deformation is also small, of the order of $|\boldsymbol{H}|$:

$$\frac{|\boldsymbol{T}^{(1),L}|}{|\boldsymbol{t}_0|} \approx O(|\boldsymbol{H}|). \tag{9.21}$$

The Lagrangian Cauchy stress tensor $\boldsymbol{t}$ is related to $\boldsymbol{T}^{(1)}$ by $(3.22)_2$,

$$\boldsymbol{t}(\vec{X}, t) = J^{-1}\boldsymbol{F} \cdot \boldsymbol{T}^{(1)}(\vec{X}, t), \tag{9.22}$$

where J and $\boldsymbol{F}$ are the Jacobian and the deformation gradient of the infinitesimal deformation (9.10), respectively. Correct to first order in $|\boldsymbol{H}|$, (9.22) is expressed using $(3.30)_1$ as

$$\boldsymbol{t} = (1 - \operatorname{tr} \boldsymbol{H})\, \boldsymbol{T}^{(1)} + \boldsymbol{H}^T \cdot \boldsymbol{T}^{(1)} + O(|\boldsymbol{H}|^2). \tag{9.23}$$

Substituting (9.20) into (9.23) yields the decomposition

$$\boldsymbol{t}(\vec{X}, t) = \boldsymbol{t}_0(\vec{X}) + \boldsymbol{t}^L, \tag{9.24}$$

where $\boldsymbol{t}^L$ is the Lagrangian increment in the Lagrangian Cauchy stress tensor (or, simply, the *incremental Lagrangian Cauchy stress tensor*),

$$\boldsymbol{t}^L = \boldsymbol{T}^{(1),L} - (\operatorname{tr} \boldsymbol{H})\boldsymbol{t}_0(\vec{X}) + \boldsymbol{H}^T \cdot \boldsymbol{t}_0(\vec{X}) + O(|\boldsymbol{H}|^2). \tag{9.25}$$

Finally, substituting $\boldsymbol{T}^{(1),L}$ from (9.25) into (9.20) leads to

$$\boldsymbol{T}^{(1)}(\vec{X}, t) = \boldsymbol{t}_0(\vec{X}) + (\operatorname{tr} \boldsymbol{H})\boldsymbol{t}_0(\vec{X}) - \boldsymbol{H}^T \cdot \boldsymbol{t}_0(\vec{X}) + \boldsymbol{t}^L + O(|\boldsymbol{H}|^2). \tag{9.26}$$

Therefore, knowledge of the displacement gradient, initial stress distribution and incremental Lagrangian Cauchy stress tensor determines the first Piola–Kirchhoff stress tensor.

In addition to $\boldsymbol{t}^L$, we can introduce the Eulerian increment $\boldsymbol{t}^E$ in the Eulerian Cauchy stress tensor (or, simply, the *incremental Eulerian Cauchy stress tensor*) as

$$\boldsymbol{t}^E := \boldsymbol{t}(\vec{x}, t) - \boldsymbol{t}_0(\vec{x}), \tag{9.27}$$

where $\boldsymbol{t}(\vec{x}, t)$ is the Eulerian Cauchy stress tensor. The two increments in the Cauchy stress tensor are then related by the first-order relation (9.13),

$$\boldsymbol{t}^L = \boldsymbol{t}^E + \vec{U} \cdot \operatorname{Grad} \boldsymbol{t}_0(\vec{X}), \tag{9.28}$$

which shows that the symmetry of $\boldsymbol{t}^L$ has not been violated by the linearisation since the tensor $\vec{U} \cdot \operatorname{Grad} \boldsymbol{t}_0$ is symmetric.

9.6 Linearised Equations of Motion

The exact form of the equation of motion in the Lagrangian description is given by (4.69),

$$\operatorname{Div} \boldsymbol{T}^{(1)} + \varrho_0 \vec{F} = \varrho_0 \frac{\partial^2 \vec{U}}{\partial t^2} \qquad \text{in } V - \Sigma\,, \tag{9.29}$$

where $\vec{F}(\vec{X}, t)$ is the Lagrangian representation of the body force. A change of the body force can be described by either the Eulerian increment $\vec{f}^E$ or the Lagrangian increment $\vec{F}^L$,

$$\begin{aligned} \vec{f}^E &:= \vec{f}(\vec{x}, t) - \vec{f}_0(\vec{x}), \\ \vec{F}^L &:= \vec{F}(\vec{X}, t) - \vec{f}_0(\vec{X}), \end{aligned} \tag{9.30}$$

where $\vec{f}(\vec{x}, t)$ is the Eulerian representation of the body force, $\vec{f}(\vec{x}, t) = \vec{F}(\vec{X}(\vec{x}, t), t)$. The increments in body force satisfy the first-order relation (9.13),

$$\vec{F}^L = \vec{f}^E + \vec{U} \cdot \operatorname{Grad} \vec{f}_0(\vec{X}). \tag{9.31}$$

To obtain the equation of motion in the Lagrangian increments, we substitute (9.20) and $(9.30)_2$ into (9.29) and subtract the static equilibrium equation (9.1):

$$\operatorname{Div} \boldsymbol{T}^{(1),L} + \varrho_0 \vec{F}^L = \varrho_0 \frac{\partial^2 \vec{U}}{\partial t^2} \qquad \text{in } V - \Sigma. \tag{9.32}$$

which is exact in the initial configuration κ_0 of the body, that is, everywhere in the volume $V - \Sigma$.

We will now use this form to express the equation of motion in terms of the incremental Lagrangian Cauchy stress tensor $\boldsymbol{t}^L$. We substitute $\boldsymbol{T}^{(1),L}$ from (9.25) into (9.32) and neglect the second-order terms in $|\boldsymbol{H}|$, obtaining

$$\mathrm{Div}\, \boldsymbol{t}^L + \mathrm{Div}\left[(\mathrm{tr}\, \boldsymbol{H})\boldsymbol{t}_0\right] - \mathrm{Div}\left(\boldsymbol{H}^T \cdot \boldsymbol{t}_0\right) + \varrho_0 \vec{F}^L = \varrho_0 \frac{\partial^2 \vec{U}}{\partial t^2} \qquad \text{in } V - \Sigma. \tag{9.33}$$

Using the differential identities (A.15), (A.25) and (A.27), and recalling that $\mathrm{tr}\, \boldsymbol{H} = \mathrm{Div}\, \vec{U}$, the sum of the second and the third terms on the left simplifies to

$$\mathrm{Div}\left[(\mathrm{tr}\, \boldsymbol{H})\boldsymbol{t}_0\right] - \mathrm{Div}\left(\boldsymbol{H}^T \cdot \boldsymbol{t}_0\right) = (\mathrm{tr}\, \boldsymbol{H})\mathrm{Div}\, \boldsymbol{t}_0 - \boldsymbol{H} : \mathrm{Grad}\, \boldsymbol{t}_0.$$

In view of this result and the static linear momentum equation (9.1), (9.33) becomes

$$\mathrm{Div}\, \boldsymbol{t}^L + \varrho_0 \vec{F}^L + \varrho^L \vec{f}_0 - \boldsymbol{H} : \mathrm{Grad}\, \boldsymbol{t}_0 = \varrho_0 \frac{\partial^2 \vec{U}}{\partial t^2} \qquad \text{in } V - \Sigma. \tag{9.34}$$

Decomposing the initial static stress $\boldsymbol{t}_0$ into the spherical and deviatoric parts according to (9.2), the fourth term on the left is

$$\boldsymbol{H} : \mathrm{Grad}\, \boldsymbol{t}_0 = \boldsymbol{H} \cdot \mathrm{Grad}\, \sigma_0 + \boldsymbol{H} : \mathrm{Grad}\, \boldsymbol{t}_0^D.$$

Finally, eliminating $\mathrm{Grad}\, \sigma_0$ by means of (9.3), the linearised form of the Lagrangian equation of motion in $V - \Sigma$, is

$$\mathrm{Div}\, \boldsymbol{t}^L + \varrho_0(\vec{F}^L + \boldsymbol{H} \cdot \vec{f}_0) + \varrho^L \vec{f}_0 - \boldsymbol{H} : \mathrm{Grad}\, \boldsymbol{t}_0^D + \boldsymbol{H} \cdot \mathrm{Div}\, \boldsymbol{t}_0^D = \varrho_0 \frac{\partial^2 \vec{U}}{\partial t^2}, \tag{9.35}$$

which simplifies to

$$\mathrm{Div}\, \boldsymbol{t}^L + \varrho_0(\vec{F}^L + \boldsymbol{H} \cdot \vec{f}_0) + \varrho^L \vec{f}_0 = \varrho_0 \frac{\partial^2 \vec{U}}{\partial t^2} \qquad \text{in } V_\mathrm{F} - \Sigma_\mathrm{F} \tag{9.36}$$

in the fluid regions because $\boldsymbol{t}_0^D = \boldsymbol{0}$.

The equation of motion (9.34) can be rewritten in terms of the incremental Eulerian Cauchy stress tensor $\boldsymbol{t}^E$. Applying the divergence operator to (9.28), using the differential identity (A.26) and substituting for $\mathrm{Div}\, \boldsymbol{t}_0$ from the static linear momentum equation (9.1), we obtain

$$\mathrm{Div}\, \boldsymbol{t}^L = \mathrm{Div}\, \boldsymbol{t}^E + \boldsymbol{H} : \mathrm{Grad}\, \boldsymbol{t}_0 - \vec{U} \cdot \mathrm{Grad}\, (\varrho_0 \vec{f}_0). \tag{9.37}$$

In view of this result, the differential identity (A.2) and the relation (9.31) between $\vec{F}^L$ and $\vec{f}^E$, (9.34) becomes

$$\text{Div}\, \boldsymbol{t}^E + \varrho_0 \vec{f}^E + \varrho^E \vec{f}_0 = \varrho_0 \frac{\partial^2 \vec{U}}{\partial t^2}, \tag{9.38}$$

where ϱ^E is the Eulerian increment in mass density. Strictly speaking, this equation is valid at a point $\vec{x}$ within the volume $v(t) - \sigma(t)$ of the present configuration κ_t of the body rather than at a point $\vec{X}$ within the volume $V - \Sigma$ of the initial configuration κ_0. However, correct to first order in $|\boldsymbol{H}|$, this distinction is insignificant.

The final form of the equation of motion is

$$\begin{aligned} \text{Div}\, \boldsymbol{t}^L + \text{Grad}\,(\varrho_0 \vec{U} \cdot \vec{f}_0) + \varrho_0 \vec{f}^E + \varrho^E \vec{f}_0 + \text{Grad}\,(\vec{U} \cdot \text{Div}\, \boldsymbol{t}_0^D) \\ - \text{Div}\,(\vec{U} \cdot \text{Grad}\, \boldsymbol{t}_0^D) = \varrho_0 \frac{\partial^2 \vec{U}}{\partial t^2}, \end{aligned} \tag{9.39}$$

which is derived from (9.33) using the identity

$$\begin{aligned} -\boldsymbol{H} : \text{Grad}\, \boldsymbol{t}_0^D + \boldsymbol{H} \cdot \text{Div}\, \boldsymbol{t}_0^D = \text{Grad}\,(\vec{U} \cdot \text{Div}\, \boldsymbol{t}_0^D) - \text{Div}\,(\vec{U} \cdot \text{Grad}\, \boldsymbol{t}_0^D) \\ + \text{Grad}\,(\varrho_0 \vec{f}_0) \cdot \vec{U} - \vec{U} \cdot \text{Grad}\,(\varrho_0 \vec{f}_0). \end{aligned} \tag{9.40}$$

The various forms of the equation of motion valid in the volume $V - \Sigma$ are summarised in Table 9.1.

Table 9.1 Summary of the various forms of the linearised equation of motion valid in the volume $V - \Sigma$

Field variables	Linearised equation of motion
$\boldsymbol{T}^{(1),L}, \vec{F}^L, \vec{U}$	$\text{Div}\, \boldsymbol{T}^{(1),L} + \varrho_0 \vec{F}^L = \varrho_0 \frac{\partial^2 \vec{U}}{\partial t^2}$ (exact)
$\boldsymbol{t}^L, \vec{F}^L, \vec{U}$	$\text{Div}\, \boldsymbol{t}^L + \varrho_0 \vec{F}^L + \mathcal{Q}^L \vec{f}_0 - \text{Grad}\, \vec{U} : \text{Grad}\, \boldsymbol{t}_0 = \varrho_0 \frac{\partial^2 \vec{U}}{\partial t^2}$
	$\text{Div}\, \boldsymbol{t}^L + \varrho_0 (\vec{F}^L + \text{Grad}\, \vec{U} \cdot \vec{f}_0) + \mathcal{Q}^L \vec{f}_0$
	$-\text{Grad}\, \vec{U} : \text{Grad}\, \boldsymbol{t}_0^D + \text{Grad}\, \vec{U} \cdot \text{Div}\, \boldsymbol{t}_0^D = \varrho_0 \frac{\partial^2 \vec{U}}{\partial t^2}$
$\boldsymbol{t}^L, \vec{f}^E, \vec{U}$	$\text{Div}\, \boldsymbol{t}^L + \varrho_0 (\vec{f}^E + \vec{U} \cdot \text{Grad}\, \vec{f}_0) + \mathcal{Q}^L \vec{f}_0 - \text{Grad}\, \vec{U} : \text{Grad}\, \boldsymbol{t}_0 = \varrho_0 \frac{\partial^2 \vec{U}}{\partial t^2}$
	$\text{Div}\, \boldsymbol{t}^L + \text{Grad}\,(\varrho_0 \vec{U} \cdot \vec{f}_0) + \varrho_0 \vec{f}^E + \varrho^E \vec{f}_0$
	$+\text{Grad}\,(\vec{U} \cdot \text{Div}\, \boldsymbol{t}_0^D) - \text{Div}\,(\vec{U} \cdot \text{Grad}\, \boldsymbol{t}_0^D) = \varrho_0 \frac{\partial^2 \vec{U}}{\partial t^2}$
$\boldsymbol{t}^E, \vec{f}^E, \vec{U}$	$\text{Div}\, \boldsymbol{t}^E + \varrho_0 \vec{f}^E + \varrho^E \vec{f}_0 = \varrho_0 \frac{\partial^2 \vec{U}}{\partial t^2}$

9.7 Linearised Jump Conditions

The linearised equation of motion must be supplemented by kinematic and dynamic jump conditions at the internal discontinuities.

9.7.1 Kinematic Jump Conditions

The kinematic jump condition on a welded solid–solid discontinuity is given by (4.53). Transforming this Eulerian form to the Lagrangian form, we obtain

$$[\vec{V}]_{-}^{+} = \vec{0} \qquad \text{on } \Sigma_{\mathrm{S}}, \tag{9.41}$$

where $\vec{V}(\vec{X},t)$ the Lagrangian representation (2.3) of the velocity. Substituting for $\vec{V}(\vec{X},t)$ from $(2.5)_1$, integrating the result with respect to time and choosing $\vec{U}(\vec{X},0) = \vec{0}$, the jump condition (9.41) becomes

$$\left[\vec{U}\right]_{-}^{+} = \vec{0} \qquad \text{on } \Sigma_{\mathrm{S}}, \tag{9.42}$$

which is exact.

Fluid–solid discontinuity may experience tangential slip. Hence, the jump condition (4.53) must be replaced by (4.55). Considering the transformation rule (1.131) between the unit normal vector $\vec{n}$ to the deformed discontinuity σ and the unit normal vector $\vec{N}$ to the undeformed discontinuity Σ, and integrating (4.55) with respect to time, results in

$$[\vec{N}\cdot\vec{U}]_{-}^{+} = 0 \qquad \text{on } \Sigma_{\mathrm{F}}, \tag{9.43}$$

which is correct to first order in $|\boldsymbol{H}|$. This first-order condition guarantees that there is no separation or interpenetration of the material on either side of the Σ_{F}.

9.7.2 Dynamic Jump Conditions

The dynamic jump condition on the welded discontinuities Σ_{S} can be deduced from the jump condition (4.71). Realising that Σ_{S} is a material non-slipping discontinuity across which the area element dA in the initial configuration changes continuously, the scalar factor dA can be dropped from (4.71), leading to

$$[\vec{N}\cdot\boldsymbol{T}^{(1)}]_{-}^{+} = \vec{0} \qquad \text{on } \Sigma_{\mathrm{S}}. \tag{9.44}$$

Substituting $\boldsymbol{T}^{(1)}$ from (9.20) and subtracting the static jump condition (9.5) gives

$$[\vec{N} \cdot \boldsymbol{T}^{(1),L}]_-^+ = \vec{0} \qquad \text{on } \Sigma_S, \tag{9.45}$$

which is exact.

The jump condition (9.45) for $\vec{N} \cdot \boldsymbol{T}^{(1),L}$ can be expressed in terms of the incremental Lagrangian Cauchy stress vector $\vec{N} \cdot \boldsymbol{t}^L$. Multiplying (9.25) by $\vec{N}$ from the left gives

$$\vec{N} \cdot \boldsymbol{t}^L = \vec{N} \cdot \boldsymbol{T}^{(1),L} - (\text{tr}\,\boldsymbol{H})(\vec{N} \cdot \boldsymbol{t}_0) + \vec{N} \cdot (\boldsymbol{H}^T \cdot \boldsymbol{t}_0), \tag{9.46}$$

where the last term on the right can be rewritten as

$$\vec{N} \cdot (\boldsymbol{H}^T \cdot \boldsymbol{t}_0) = (\boldsymbol{H}^T \cdot \boldsymbol{t}_0)^T \cdot \vec{N} = (\boldsymbol{t}_0^T \cdot \boldsymbol{H}) \cdot \vec{N} = (\boldsymbol{t}_0 \cdot \boldsymbol{H}) \cdot \vec{N}$$

due to the symmetry of the tensor $\boldsymbol{t}_0$. Expressing tr $\boldsymbol{H}$ and $\boldsymbol{H}$ by means of (B.27) and (B.28) gives

$$(\text{tr}\,\boldsymbol{H})(\vec{N} \cdot \boldsymbol{t}_0) - (\boldsymbol{t}_0 \cdot \boldsymbol{H}) \cdot \vec{N} = \text{Div}_\Sigma \vec{U}(\vec{N} \cdot \boldsymbol{t}_0) - (\boldsymbol{t}_0 \cdot \text{Grad}_\Sigma \vec{U}) \cdot \vec{N},$$

so that (9.46) becomes

$$\vec{N} \cdot \boldsymbol{t}^L = \vec{N} \cdot \boldsymbol{T}^{(1),L} - \text{Div}_\Sigma \vec{U}(\vec{N} \cdot \boldsymbol{t}_0) + (\boldsymbol{t}_0 \cdot \text{Grad}_\Sigma \vec{U}) \cdot \vec{N}, \tag{9.47}$$

which shows that the continuity of $\vec{N} \cdot \boldsymbol{T}^{(1),L}$, $\vec{N} \cdot \boldsymbol{t}_0$ and $\vec{U}$ on Σ_S, expressed by the respective jump conditions (9.45), (9.5) and (9.42), does not guarantee the continuity of the last term on the right on Σ_S, thus the continuity of $\vec{N} \cdot \boldsymbol{t}^L$.

Finding the dynamic jump condition on slipping discontinuities, such as the fluid–solid discontinuities, is more complicated. Let a slipping discontinuity referred to the initial and present configurations be denoted by Σ and σ, respectively. Imagine two area elements $\vec{n}^+ da^+$ and $\vec{n}^- da^-$ that lie on the positive and negative sides of σ at point $\vec{x}$, respectively, as illustrated in Fig. 9.1. According to the principle of continuity, an area element changes continuously across a discontinuity in the present configuration. Hence, $\vec{n}^+ da^+ = \vec{n}^- da^-$. Let these area elements be images of the area elements $\vec{N}^+ dA^+$ and $\vec{N}^- dA^-$ placed at points $\vec{X}^+$ and $\vec{X}^-$ on the positive and negative sides of the discontinuity Σ, respectively. Due to tangential slip along the discontinuity, $\vec{X}^+$ and $\vec{X}^-$ differ by the offset $[\vec{U}]_-^+ = \vec{X}^- - \vec{X}^+$.

The dynamic jump condition (4.71) simplifies to

$$\left[\vec{N} dA \cdot \boldsymbol{T}^{(1)}\right]_-^+ = \vec{0} \qquad \text{on } \Sigma \tag{9.48}$$

because the slipping discontinuity is assumed to be material. The continuity of the Eulerian area element on σ, i.e., $[da]_-^+ = 0$ on σ, and (1.130) give

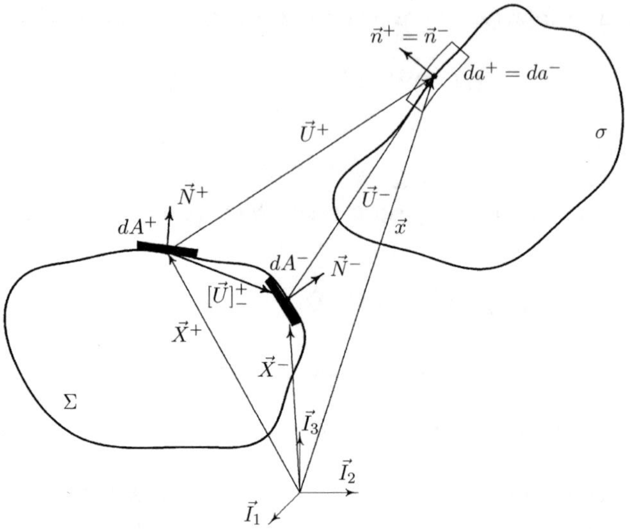

Fig. 9.1 The area elements of a slipping discontinuity at the initial (left) and present (right) configurations

$\left[(1 + \mathrm{Div}_\Sigma \vec{U})dA\right]_-^+ = 0$ on Σ, or

$$dA^+ - dA^- = -\mathrm{Div}_\Sigma [\vec{U}]_-^+ dA, \tag{9.49}$$

where the superscripts $\pm$ denote evaluation at $\vec{X}^\pm$. As we are only concerned with first-order accuracy in $|\boldsymbol{H}|$, dA in the term on the right can be taken at either $\vec{X}^+$ or $\vec{X}^-$.

Moreover, the continuity (9.5) of the initial static stress vector $\vec{N} \cdot \boldsymbol{t}_0$ on Σ in the initial configuration implies that the difference between $\vec{N}^+ \cdot \boldsymbol{t}_0^+$ and $\vec{N}^- \cdot \boldsymbol{t}_0^-$ caused by the infinitesimal deformation (9.10) is small and can be expressed by a first-order Taylor approximation

$$\vec{N}^+ \cdot \boldsymbol{t}_0^+ = \vec{N}^- \cdot \boldsymbol{t}_0^- - [\vec{U}]_-^+ \cdot \mathrm{Grad}_\Sigma (\vec{N} \cdot \boldsymbol{t}_0) \qquad \text{on } \Sigma, \tag{9.50}$$

where $\mathrm{Grad}_\Sigma (\vec{N} \cdot \boldsymbol{t}_0)$ can be taken at either $\vec{X}^+$ or $\vec{X}^-$. Note that $\vec{U}$ in (9.50) can be replaced by its tangential part $\vec{U}_\Sigma$. Substituting $\boldsymbol{T}^{(1)}$ from (9.20) and using (9.49) and (9.50), the jump condition (9.48), expressed correct to first order in $|\boldsymbol{H}|$, is

$$\left[\vec{N} \cdot \boldsymbol{T}^{(1),L} dA\right]_-^+ - \mathrm{Div}_\Sigma [\vec{U}]_-^+ (\vec{N} \cdot \boldsymbol{t}_0)^- dA - [\vec{U}]_-^+ \cdot \mathrm{Grad}_\Sigma (\vec{N} \cdot \boldsymbol{t}_0) dA^- = \vec{0} \quad \text{on } \Sigma. \tag{9.51}$$

To the same degree of approximation, dA in the first and third terms on the left can be taken at either $\vec{X}^+$ or $\vec{X}^-$ and, consequently, the scalar factor dA can be dropped from (9.51). Although the stress vector $\vec{N}\cdot\boldsymbol{t}_0$ is not continuous across Σ, see (9.50), it is valid to first order in $|\vec{U}|$ to consider $\vec{N}\cdot\boldsymbol{t}_0$ continuous in the second term of (9.51). Then, (9.51) becomes

$$\left[\vec{N}\cdot\boldsymbol{T}^{(1),L}\right]_-^+ - \mathrm{Div}_\Sigma[\vec{U}]_-^+(\vec{N}\cdot\boldsymbol{t}_0) - [\vec{U}]_-^+\cdot\mathrm{Grad}_\Sigma(\vec{N}\cdot\boldsymbol{t}_0) = \vec{0} \qquad \text{on } \Sigma. \tag{9.52}$$

Using the differential identity (B.14)$_5$ finally gives

$$\left[\vec{N}\cdot\boldsymbol{T}^{(1),L} - \mathrm{Div}_\Sigma[\vec{U}\otimes(\vec{N}\cdot\boldsymbol{t}_0)]\right]_-^+ = \vec{0} \qquad \text{on } \Sigma, \tag{9.53}$$

which guarantees the continuity of the stress vector across a slipping discontinuity. At a welded discontinuity with no tangential slip, the displacement $\vec{U}$ is continuous and condition (9.53) reduces to (9.45).

On a fluid–solid discontinuity, the initial stress vector has, in addition, the form (9.9), therefore, (9.53) simplifies to

$$\left[\vec{N}\cdot\boldsymbol{T}^{(1),L} - \mathrm{Div}_\Sigma(\tau_0\vec{U}\otimes\vec{N})\right]_-^+ = \vec{0} \qquad \text{on } \Sigma_\mathrm{F}, \tag{9.54}$$

which, using the differential identity (B.14)$_5$, gives

$$\left[\vec{N}\cdot\boldsymbol{T}^{(1),L} - \vec{N}\,\mathrm{Div}_\Sigma(\tau_0\vec{U}) - \tau_0(\vec{U}\cdot\mathrm{Grad}_\Sigma\vec{N})\right]_-^+ = \vec{0} \qquad \text{on } \Sigma_\mathrm{F}. \tag{9.55}$$

The kinematic condition (9.43) requires the normal component of $\vec{U}$ to be continuous across Σ_F, i.e., changes in $\vec{N}\cdot\vec{U}$ must be equally large on both sides of the discontinuity Σ_F:

$$\left[\mathrm{Grad}_\Sigma(\vec{N}\cdot\vec{U})\right]_-^+ = \vec{0} \qquad \text{on } \Sigma_\mathrm{F}. \tag{9.56}$$

Employing the identity (B.14)$_2$ and the symmetry (B.16) of the surface curvature tensor, $(\mathrm{Grad}_\Sigma\vec{N})^T = \mathrm{Grad}_\Sigma\vec{N}$, (9.56) becomes

$$\left[\mathrm{Grad}_\Sigma\vec{U}\cdot\vec{N} + \vec{U}\cdot\mathrm{Grad}_\Sigma\vec{N}\right]_-^+ = \vec{0} \qquad \text{on } \Sigma_\mathrm{F}. \tag{9.57}$$

This relation enables us to change the order of the multiplicand and the multiplier of the scalar product in the last term of (9.55). Using, in addition, the continuity of the normal stress τ_0, (9.55) takes the form

$$\left[\vec{N}\cdot\boldsymbol{T}^{(1),L} - \vec{N}\,\mathrm{Div}_\Sigma(\tau_0\vec{U}) + \tau_0(\mathrm{Grad}_\Sigma\vec{U}\cdot\vec{N})\right]_-^+ = \vec{0} \qquad \text{on } \Sigma_F. \tag{9.58}$$

Moreover, since the vector $\mathrm{Grad}_\Sigma\vec{U}\cdot\vec{N}$ is tangent to Σ_F,

$$\mathrm{Grad}_\Sigma\vec{U}\cdot\vec{N} = (\mathrm{Grad}_\Sigma\vec{U}\cdot\vec{N})\cdot\left(\boldsymbol{I} - \vec{N}\otimes\vec{N}\right), \tag{9.59}$$

see (B.24), the tangential component of the condition (9.58) is

$$\left[\vec{N}\cdot\boldsymbol{T}^{(1),L}\cdot\left(\boldsymbol{I} - \vec{N}\otimes\vec{N}\right) + \tau_0(\mathrm{Grad}_\Sigma\vec{U}\cdot\vec{N})\right]_-^+ = \vec{0} \qquad \text{on } \Sigma_F. \tag{9.60}$$

If the fluid–solid discontinuity is also frictionless, the Eulerian Cauchy stress vector $\vec{n}\cdot\boldsymbol{t}$ must point in the direction of the unit normal vector $\vec{n}$ to the deformed σ_F from the fluid side (denoted by σ_F^-), i.e., its projection onto σ_F must vanish from the fluid side,

$$\vec{n}\cdot\boldsymbol{t}\cdot\left(\boldsymbol{I} - \vec{n}\otimes\vec{n}\right) = \vec{0} \qquad \text{on } \sigma_F^-. \tag{9.61}$$

Using (1.126), (9.24) and (9.25), the Eulerian Cauchy stress vector $\vec{n}\cdot\boldsymbol{t}$ at the deformed discontinuity σ_F can be expressed in terms of the first Piola–Kirchhoff stress tensor on the undeformed discontinuity Σ_F. Correct to first order in $|\boldsymbol{H}|$, it holds

$$\begin{aligned}
\vec{n}\cdot\boldsymbol{t} &= \left[\vec{N} + (\vec{N}\cdot\boldsymbol{H}\cdot\vec{N})\vec{N} - \boldsymbol{H}\cdot\vec{N}\right]\cdot\left[\boldsymbol{t}_0 + \boldsymbol{T}^{(1),L} - (\mathrm{tr}\,\boldsymbol{H})\boldsymbol{t}_0 + \boldsymbol{H}^T\cdot\boldsymbol{t}_0\right]\\
&= \vec{N}\cdot\boldsymbol{t}_0 + \vec{N}\cdot\boldsymbol{T}^{(1),L} - (\mathrm{tr}\,\boldsymbol{H})(\vec{N}\cdot\boldsymbol{t}_0) + \vec{N}\cdot\boldsymbol{H}^T\cdot\boldsymbol{t}_0\\
&\quad + (\vec{N}\cdot\boldsymbol{H}\cdot\vec{N})(\vec{N}\cdot\boldsymbol{t}_0) - (\boldsymbol{H}\cdot\vec{N})\cdot\boldsymbol{t}_0\\
&= \vec{N}\cdot\boldsymbol{t}_0 + \vec{N}\cdot\boldsymbol{T}^{(1),L} - \left(\mathrm{tr}\,\boldsymbol{H} - \vec{N}\cdot\boldsymbol{H}\cdot\vec{N}\right)(\vec{N}\cdot\boldsymbol{t}_0),
\end{aligned}$$

where we have abbreviated $\boldsymbol{t}_0 \equiv \boldsymbol{t}_0(\vec{X})$ and used the identity $\vec{N}\cdot\boldsymbol{H}^T\cdot\boldsymbol{t}_0 = (\boldsymbol{H}\cdot\vec{N})\cdot\boldsymbol{t}_0$. Equation (1.128) shows that the last term can be further expressed in terms of the surface divergence of the displacement $\vec{U}$ as

$$\vec{n}\cdot\boldsymbol{t} = \vec{N}\cdot\boldsymbol{t}_0 + \vec{N}\cdot\boldsymbol{T}^{(1),L} - (\mathrm{Div}_\Sigma\vec{U})(\vec{N}\cdot\boldsymbol{t}_0) \qquad \text{on } \Sigma_F. \tag{9.62}$$

In view of (1.131), the dyadic product $\vec{n}\otimes\vec{n}$ occurring in the jump condition (9.61) is written, correct to first order in $|\boldsymbol{H}|$, in the form

$$\vec{n}\otimes\vec{n} = \vec{N}\otimes\vec{N} - (\mathrm{Grad}_\Sigma\vec{U}\cdot\vec{N})\otimes\vec{N} - \vec{N}\otimes(\mathrm{Grad}_\Sigma\vec{U}\cdot\vec{N}). \tag{9.63}$$

Substituting (9.62) and (9.63) into (9.61) yields

$$\begin{aligned}&\big[\vec{N}\cdot \boldsymbol{t}_0+\vec{N}\cdot \boldsymbol{T}^{(1),L}-(\mathrm{Div}_\Sigma \vec{U})(\vec{N}\cdot \boldsymbol{t}_0)\big]\cdot\big(\boldsymbol{I}-\vec{N}\otimes\vec{N}\big)\\&+(\vec{N}\cdot \boldsymbol{t}_0)\cdot\big[(\mathrm{Grad}_\Sigma \vec{U}\cdot\vec{N})\otimes\vec{N}+\vec{N}\otimes(\mathrm{Grad}_\Sigma \vec{U}\cdot\vec{N})\big]=\vec{0},\end{aligned}\tag{9.64}$$

which is valid on Σ_F^-. At this discontinuity, the initial stress vector $\vec{N}\cdot \boldsymbol{t}_0$ has the form (9.9), which, together with (9.59), simplifies (9.64) to

$$\big(\vec{N}\cdot \boldsymbol{T}^{(1),L}\big)\cdot\big(\boldsymbol{I}-\vec{N}\otimes\vec{N}\big)+\tau_0(\mathrm{Grad}_\Sigma \vec{U}\cdot\vec{N})=\vec{0} \qquad \text{on } \Sigma_F^-. \tag{9.65}$$

In view of (9.60), the last condition must also hold on the solid side of Σ_F, i.e., on Σ_F^+.

Defining the auxiliary vector

$$\vec{\tau}^L := \vec{N}\cdot \boldsymbol{T}^{(1),L}-\vec{N}\,\mathrm{Div}_\Sigma(\tau_0\vec{U})+\tau_0(\mathrm{Grad}_\Sigma \vec{U}\cdot\vec{N}), \tag{9.66}$$

the jump conditions (9.58) and (9.65) can be written in a compact form

$$\big[\vec{\tau}^L\big]_-^+=\vec{0}, \tag{9.67}$$

$$\vec{\tau}^L\cdot\big(\boldsymbol{I}-\vec{N}\otimes\vec{N}\big)=\vec{0}. \tag{9.68}$$

Hence, $\vec{\tau}^L$ is a continuous, normal vector on Σ_F. At a solid–solid discontinuity Σ_S, the last two terms on the right-hand side of (9.66) are continuous and (9.67) coincides with (9.44). Therefore, a frictionless fluid–solid discontinuity differs from a welded solid–solid discontinuity by the normality condition (9.68). The complete set of the jump conditions is summarised in Table 9.2.

Table 9.2 Summary of exact (E) and linearised (L) jump conditions

Discontinuity type	Jump conditions	
Σ_S: solid–solid	$[\vec{U}]_-^+=\vec{0}$	(E)
	$[\vec{N}\cdot \boldsymbol{T}^{(1),L}]_-^+=\vec{0}$	(E)
Σ_F: fluid–solid	$[\vec{N}\cdot\vec{U}]_-^+=0$	(L)
	$[\vec{\tau}^L]_-^+=0$	(L)
	$\vec{\tau}^L\cdot(\boldsymbol{I}-\vec{N}\otimes\vec{N})=\vec{0}$	(L)
$\vec{\tau}^L=\vec{N}\cdot \boldsymbol{T}^{(1),L}-\vec{N}\,\mathrm{Div}_\Sigma(\tau_0\vec{U})+\tau_0(\mathrm{Grad}_\Sigma \vec{U}\cdot\vec{N})$		

9.8 Linearised Elastic Constitutive Equations

The linearised principles of conservation derived in the preceding sections are valid regardless of the constitution of the material. To complete these equations, we specify the constitutive relations between the incremental stress and displacement gradient. Here, we will assume that the material behaviour for the infinitesimal superimposed deformation is linear and elastic, as given in Sect. 8.1.

The linearised constitutive equations for the incremental first Piola–Kirchhoff stress tensor $\boldsymbol{T}^{(1),L}$ and the incremental Lagrangian Cauchy stress tensor $\boldsymbol{t}^L$ can be obtained from (8.16) and (8.17) using the decompositions (9.20) and (9.24):

$$\begin{aligned}\boldsymbol{T}^{(1),L} &= \boldsymbol{C} : \widetilde{\boldsymbol{E}} - \boldsymbol{t}_0 \cdot \widetilde{\boldsymbol{R}} + a\big[(\operatorname{tr}\widetilde{\boldsymbol{E}})\boldsymbol{t}_0 + (\boldsymbol{t}_0 : \widetilde{\boldsymbol{E}})\boldsymbol{I}\big] + (b+c)\big(\widetilde{\boldsymbol{E}} \cdot \boldsymbol{t}_0\big) \\ &\quad + (b+c+1)\big(\boldsymbol{t}_0 \cdot \widetilde{\boldsymbol{E}}\big), \qquad (9.69)\\ \boldsymbol{t}^L &= \boldsymbol{C} : \widetilde{\boldsymbol{E}} + \widetilde{\boldsymbol{R}} \cdot \boldsymbol{t}_0 - \boldsymbol{t}_0 \cdot \widetilde{\boldsymbol{R}} + (a-1)(\operatorname{tr}\widetilde{\boldsymbol{E}})\boldsymbol{t}_0 + a(\boldsymbol{t}_0 : \widetilde{\boldsymbol{E}})\boldsymbol{I} \\ &\quad + (b+c+1)\big(\boldsymbol{t}_0 \cdot \widetilde{\boldsymbol{E}} + \widetilde{\boldsymbol{E}} \cdot \boldsymbol{t}_0\big), \qquad (9.70)\end{aligned}$$

where $\widetilde{\boldsymbol{E}}$ and $\widetilde{\boldsymbol{R}}$ are the linearised Lagrangian strain and rotation tensors, respectively. Any choice of the scalars a, b and c defines the behaviour of a linear elastic solid. Dahlen and Tromp (1998) chose

$$a = -b = -c = \frac{1}{2}, \qquad (9.71)$$

which simplifies (9.69) and (9.70) to

$$\boldsymbol{T}^{(1),L} = \boldsymbol{C} : \widetilde{\boldsymbol{E}} - \boldsymbol{t}_0 \cdot \widetilde{\boldsymbol{R}} - \widetilde{\boldsymbol{E}} \cdot \boldsymbol{t}_0 + \frac{1}{2}(\operatorname{tr}\widetilde{\boldsymbol{E}})\boldsymbol{t}_0 + \frac{1}{2}(\boldsymbol{t}_0 : \widetilde{\boldsymbol{E}})\boldsymbol{I}, \qquad (9.72)$$

$$\boldsymbol{t}^L = \boldsymbol{C} : \widetilde{\boldsymbol{E}} + \widetilde{\boldsymbol{R}} \cdot \boldsymbol{t}_0 - \boldsymbol{t}_0 \cdot \widetilde{\boldsymbol{R}} - \frac{1}{2}(\operatorname{tr}\widetilde{\boldsymbol{E}})\boldsymbol{t}_0 + \frac{1}{2}(\boldsymbol{t}_0 : \widetilde{\boldsymbol{E}})\boldsymbol{I}. \qquad (9.73)$$

Decomposing the initial static stress into the spherical and deviatoric parts according to (9.2) gives

$$\begin{aligned}\boldsymbol{T}^{(1),L} &= \boldsymbol{C} : \widetilde{\boldsymbol{E}} - \sigma_0\big[\widetilde{\boldsymbol{E}} + \widetilde{\boldsymbol{R}} - (\operatorname{tr}\widetilde{\boldsymbol{E}})\boldsymbol{I}\big] - \boldsymbol{t}_0^D \cdot \widetilde{\boldsymbol{R}} - \widetilde{\boldsymbol{E}} \cdot \boldsymbol{t}_0^D \\ &\quad + \frac{1}{2}(\operatorname{tr}\widetilde{\boldsymbol{E}})\boldsymbol{t}_0^D + \frac{1}{2}(\boldsymbol{t}_0^D : \widetilde{\boldsymbol{E}})\boldsymbol{I}, \qquad (9.74)\\ \boldsymbol{t}^L &= \boldsymbol{C} : \widetilde{\boldsymbol{E}} + \widetilde{\boldsymbol{R}} \cdot \boldsymbol{t}_0^D - \boldsymbol{t}_0^D \cdot \widetilde{\boldsymbol{R}} - \frac{1}{2}(\operatorname{tr}\widetilde{\boldsymbol{E}})\boldsymbol{t}_0^D + \frac{1}{2}(\boldsymbol{t}_0^D : \widetilde{\boldsymbol{E}})\boldsymbol{I}. \qquad (9.75)\end{aligned}$$

Note that $\boldsymbol{t}^L$ does not depend explicitly on the initial spherical pressure σ_0, but only on the elastic tensor $\boldsymbol{C}$ and the initial deviatoric stress $\boldsymbol{t}_0^D$. In fact, being the only choice that eliminates the explicit dependence of $\boldsymbol{t}^L$ on σ_0, this was Dahlen and Tromp (1998)'s motivation for setting (9.71) in (9.69) and (9.70).

9.9 The Gravitational Force

We have not yet specified the body force acting in the body $\mathcal{B}$; we will now assume that this is the *gravitational attraction*. Since gravitation is a *conservative force*, the Eulerian gravitational force per unit mass $\vec{g}(\vec{x}, t)$ is derivable from an *Eulerian gravitational potential* $\phi(\vec{x}, t)$:

$$\vec{g} = -\operatorname{grad}\phi. \tag{9.76}$$

The potential ϕ satisfies Poisson's equation inside the volume $v(t)$ that is filled by a material body at time t, i.e., over all points $\vec{x}$ with an Eulerian mass density distribution $\varrho(\vec{x}, t)$,

$$\operatorname{div}\operatorname{grad}\phi = 4\pi G\varrho \qquad \text{in } v(t) - \sigma(t), \tag{9.77}$$

where G is Newton's gravitational constant. At an internal discontinuity across which the mass density ϱ has a jump, Poisson's equation (9.77) must be supplemented by jump conditions of the form

$$[\phi]_-^+ = 0 \qquad \text{on } \sigma(t), \tag{9.78}$$

$$\left[\vec{n}\cdot\operatorname{grad}\phi\right]_-^+ = 0 \qquad \text{on } \sigma(t). \tag{9.79}$$

Note that the continuity of the scalar fields ϕ and $\vec{n}\cdot\operatorname{grad}\phi$ implies the continuity of the gravitational attraction vector $\vec{g} = -\operatorname{grad}\phi$.

As any other variable, the gravitational potential and gravitational attraction have both an Eulerian and a Lagrangian representation. The Lagrangian variables Φ and $\vec{G}$ associated with the Eulerian variables ϕ and $\vec{g}$ are defined by

$$\Phi(\vec{X}, t) := \phi\big(\vec{x}(\vec{X}, t), t\big), \qquad \vec{G}(\vec{X}, t) := \vec{g}\big(\vec{x}(\vec{X}, t), t\big). \tag{9.80}$$

Using $(1.48)_3$, it is straightforward to rewrite Poisson's equation (9.77) in the Lagrangian form,

$$\operatorname{Div}(J\boldsymbol{B}\cdot\operatorname{Grad}\Phi) = 4\pi G\varrho_0 \qquad \text{in } V - \Sigma, \tag{9.81}$$

where V is the volume of the material body in the initial configuration κ_0 and $\boldsymbol{B}$ is the Piola deformation tensor defined by $(1.62)_2$. We use $(1.48)_1$ and (1.83) to transform the jump conditions (9.78) and (9.79) to the undeformed discontinuity Σ, obtaining

$$[\Phi]_-^+ = 0 \qquad \text{on } \Sigma, \tag{9.82}$$

$$\left[\frac{\vec{N}\cdot\boldsymbol{B}}{\sqrt{\vec{N}\cdot\boldsymbol{B}\cdot\vec{N}}}\cdot \operatorname{Grad}\Phi\right]_-^+ = 0 \qquad \text{on } \Sigma. \tag{9.83}$$

9.9.1 Equations for the Initial Configuration

Let the mass density $\varrho_0(\vec{X})$ of the body in the initial configuration κ_0 generate the initial gravitational potential $\phi_0(\vec{X})$, and let

$$\vec{g}_0 = -\operatorname{Grad}\phi_0 \tag{9.84}$$

be the initial gravitational attraction. The potential ϕ_0 satisfies Poisson's equation[1] inside the body

$$\nabla^2\phi_0 = 4\pi G\varrho_0 \qquad \text{in } V-\Sigma, \tag{9.85}$$

along with the jump conditions at internal discontinuities across which the mass density ϱ_0 is discontinuous,

$$[\phi_0]_-^+ = 0 \qquad \text{on } \Sigma, \tag{9.86}$$

$$\left[\vec{N}\cdot\operatorname{Grad}\phi_0\right]_-^+ = 0 \qquad \text{on } \Sigma. \tag{9.87}$$

Outside the body, the mass density ϱ_0 is assumed to vanish and the potential is harmonic,

$$\nabla^2\phi_0 = 0 \qquad \text{outside } V. \tag{9.88}$$

9.9.2 Increments in Gravitation

In accordance with (9.11), the Eulerian and Lagrangian increments in gravitational potential are defined by

$$\begin{aligned}\phi^E &:= \phi(\vec{x},t) - \phi_0(\vec{x}),\\ \Phi^L &:= \Phi(\vec{X},t) - \phi_0(\vec{X}),\end{aligned} \tag{9.89}$$

[1] Note that $\nabla^2 \equiv \operatorname{Div}\operatorname{Grad}$.

where the Eulerian and Lagrangian representations of the gravitational potential, $\phi(\vec{x}, t)$ and $\Phi(\vec{X}, t)$, are related to each other by $(9.80)_1$. The two increments in gravitational potential satisfy the first-order relation (9.13):

$$\Phi^L = \phi^E + \vec{U} \cdot \operatorname{Grad} \phi_0(\vec{X}). \tag{9.90}$$

Applying the gradient operator to $(9.89)_1$, the Eulerian gravitation attraction is

$$\begin{aligned}\vec{g}(\vec{x}, t) &= -\operatorname{grad} \phi(\vec{x}, t) = -\operatorname{grad} \phi_0(\vec{x}) - \operatorname{grad} \phi^E = \vec{g}_0(\vec{x}) - \operatorname{grad} \phi^E \\ &= \vec{g}_0(\vec{x}) - \operatorname{Grad} \phi^E + \boldsymbol{H} \cdot \operatorname{Grad} \phi^E,\end{aligned}$$

where $(1.116)_1$ has been used in the last step. Defining the Eulerian increment in gravitation in accordance with $(9.11)_1$ such that

$$\vec{g}^E := \vec{g}(\vec{x}, t) - \vec{g}_0(\vec{x}), \tag{9.91}$$

and neglecting the term $\boldsymbol{H} \cdot \operatorname{Grad} \phi^E$ because it is of higher order than $O(|\boldsymbol{H}|)$ shows that $\vec{g}^E$ is expressible as the gradient of the corresponding potential increment,

$$\vec{g}^E = -\operatorname{Grad} \phi^E. \tag{9.92}$$

To express the Lagrangian increment in gravitation $\vec{G}^L$ in terms of the Lagrangian increment in potential Φ^L, write the Lagrangian representation of the gravitational attraction as

$$\begin{aligned}\vec{G}(\vec{X}, t) &= \vec{g}(\vec{x}(\vec{X}, t), t) = \vec{g}(\vec{X} + \vec{U}, t) = \vec{g}_0(\vec{X} + \vec{U}, t) + \vec{g}^E \\ &= \vec{g}_0(\vec{X}) + \vec{U} \cdot \operatorname{Grad} \vec{g}_0(\vec{X}) + \vec{g}^E \\ &\stackrel{!}{=} \vec{g}_0(\vec{X}) + \vec{G}^L\end{aligned}$$

demonstrating that the two increments in gravitation are related by

$$\vec{G}^L = \vec{g}^E + \vec{U} \cdot \operatorname{Grad} \vec{g}_0(\vec{X}), \tag{9.93}$$

which is consistent with the general relation (9.13). Inserting $\vec{g}^E$ from (9.92) and $\vec{g}_0(\vec{X})$ from (9.84) gives

$$\vec{G}^L = -\operatorname{Grad} \phi^E - \vec{U} \cdot \operatorname{Grad} \operatorname{Grad} \phi_0(\vec{X}). \tag{9.94}$$

Alternatively, applying the gradient operator to (9.90) and using the identity (A.3) gives

$$\operatorname{Grad} \Phi^L = \operatorname{Grad} \phi^E + \operatorname{Grad} \vec{U} \cdot \operatorname{Grad} \phi_0(\vec{X}) + \operatorname{Grad} \operatorname{Grad} \phi_0(\vec{X}) \cdot \vec{U}. \tag{9.95}$$

Substituting $\text{Grad}\,\phi^E$ from (9.94) and using the symmetry of the tensor $\text{Grad}\,\text{Grad}\,\phi_0$, the Lagrangian increment in gravitation can be expressed as

$$\vec{G}^L = -\text{Grad}\,\Phi^L + \text{Grad}\,\vec{U} \cdot \text{Grad}\,\phi_0(\vec{X}). \tag{9.96}$$

The form-difference between (9.92) and (9.96) shows that the Eulerian description of the gravitational field is more convenient than the Lagrangian description.

9.9.3 Linearised Poisson's Equations

To obtain the linearised forms of Poisson's equation, we substitute the linearised form of the Laplace operator $(1.116)_3$ into the exact relation (9.77),

$$\nabla^2\Phi - 2\boldsymbol{H} : \text{Grad}\;\text{Grad}\,\Phi - \text{Div}\,\boldsymbol{H} \cdot \text{Grad}\,\Phi = 4\pi G \mathcal{Q} \qquad \text{in } V - \Sigma, \tag{9.97}$$

which is correct to first order in $|\boldsymbol{H}|$. Substituting Φ from $(9.89)_2$ and $\mathcal{Q}$ from $(9.14)_2$, and subtracting Poisson's equation (9.85) from the result, we obtain

$$\nabla^2\Phi^L - 2\boldsymbol{H} : \text{Grad}\;\text{Grad}\,\phi_0 - \text{Div}\,\boldsymbol{H} \cdot \text{Grad}\,\phi_0 = 4\pi G \mathcal{Q}^L \qquad \text{in } V - \Sigma, \tag{9.98}$$

where $\mathcal{Q}^L$ is the Lagrangian increment in mass density. Using the differential identities (A.40) and (A.41), and the symmetry of the tensor $\text{Grad}\,\text{Grad}\,\phi_0$, we derive the identity

$$\nabla^2(\vec{U} \cdot \text{Grad}\,\phi_0) = 2\boldsymbol{H} : \text{Grad}\,\text{Grad}\,\phi_0 + \text{Div}\,\boldsymbol{H} \cdot \text{Grad}\,\phi_0 + \vec{U} \cdot \text{Grad}\,(\nabla^2\phi_0). \tag{9.99}$$

Substituting $2\boldsymbol{H} : \text{Grad}\;\text{Grad}\,\phi_0 + \text{Div}\,\boldsymbol{H} \cdot \text{Grad}\,\phi_0$ from (9.99) into (9.98), using Poisson's equation (9.85) for the initial potential and substituting $\mathcal{Q}^L$ from (9.18), Poisson's equation for Φ^L is

$$\nabla^2\Phi^L + \nabla^2(\vec{U} \cdot \vec{g}_0) = -4\pi G\,\text{Div}\,(\varrho_0\vec{U}) \qquad \text{in } V - \Sigma. \tag{9.100}$$

The linearised Poisson's equation can also be expressed in terms of the Eulerian increments in gravitational potential and mass density. In view of (9.17), the right-hand side of (9.100) is a multiple of the Eulerian increment in mass density.

Moreover, taking into account (9.90) and $\Phi^L = \phi^E - \vec{U} \cdot \vec{g}_0$, we obtain

$$\nabla^2 \phi^E = 4\pi G \varrho^E \qquad \text{in } V - \Sigma. \tag{9.101}$$

9.9.4 Linearised Jump Conditions for Potential Increments

The exact continuity condition (9.82) for the potential at the undeformed discontinuity Σ can be expressed in terms of the Lagrangian increment in potential as

$$\left[\phi_0(\vec{X}) + \Phi^L\right]_-^+ = 0 \qquad \text{on } \Sigma. \tag{9.102}$$

At a welded discontinuity Σ_S, which does not experience tangential slip, the initial potential ϕ_0 is continuous and (9.102) reduces to

$$\left[\Phi^L\right]_-^+ = 0 \qquad \text{on } \Sigma_S. \tag{9.103}$$

The corresponding result for ϕ^E is obtained from (9.90):

$$\left[\phi^E + \vec{U} \cdot \text{Grad}\phi_0\right]_-^+ = 0 \qquad \text{on } \Sigma_S, \tag{9.104}$$

which simplifies to

$$\left[\phi^E\right]_-^+ = 0 \qquad \text{on } \Sigma_S \tag{9.105}$$

because the displacement and initial gravitation are continuous on Σ_S.

At a fluid–solid discontinuity Σ_F, which may exhibit tangential slip, ϕ_0^+ and ϕ_0^- are related by an expansion analogous to (9.50),

$$\phi_0^+ = \phi_0^- - \left[\vec{U}\right]_-^+ \cdot \text{Grad}_\Sigma \phi_0 \qquad \text{on } \Sigma_F, \tag{9.106}$$

which is correct to first order in $|\vec{U}|$. To the same degree of approximation, $\text{Grad}_\Sigma \phi_0$ is taken at either $\vec{X}^+$ or $\vec{X}^-$. Substituting (9.106) into (9.102) yields

$$\left[\Phi^L - \vec{U}_\Sigma \cdot \text{Grad}_\Sigma \phi_0\right]_-^+ = 0 \qquad \text{on } \Sigma_F, \tag{9.107}$$

where $\vec{U}$ has been replaced by $\vec{U}_\Sigma$ since $\text{Grad}_\Sigma \phi_0$ is tangent to Σ_F. Although the normal component of the initial gravitation is not continuous across Σ_F, see (9.119) in the following section, it is valid to first order in $|\vec{U}|$ to consider $\vec{N} \cdot \text{Grad}\phi_0$

continuous in the product $(\vec{N} \cdot \vec{U})(\vec{N} \cdot \text{Grad}\phi_0)$. Moreover, using the continuity of the normal component of displacement along Σ_F, we subtract the term $(\vec{N} \cdot \vec{U})$ $(\vec{N} \cdot \text{Grad}\phi_0)$ from the left-hand side of (9.107) for convenience to obtain

$$\left[\Phi^L - \vec{U} \cdot \text{Grad}\, \phi_0\right]_-^+ = 0 \qquad \text{on } \Sigma_F. \tag{9.108}$$

The corresponding result for ϕ^E is obtained from the relation (9.90) between the increments ϕ^E and Φ^L; the jump condition (9.108) transforms to a condition of the same form as (9.105). Hence, for both types of discontinuity, the Eulerian increment in potential satisfies the jump condition

$$\left[\phi^E\right]_-^+ = 0 \qquad \text{on } \Sigma, \tag{9.109}$$

where $\Sigma = \Sigma_S \cup \Sigma_F$.

9.9.5 Linearised Jump Conditions for Increments in Gravitation

To derive the linearised form of the jump condition for increments in gravitation, we use the geometrical linearisation $(1.115)_1$ for $\boldsymbol{F}^{-1}$ and (1.124) for $(\vec{N} \cdot \boldsymbol{B} \cdot \vec{N})^{-1/2}$. The exact jump condition (9.83) then becomes

$$\Big[\vec{N} \cdot \text{Grad}\, \Phi - \vec{N} \cdot \boldsymbol{H} \cdot \text{Grad}\, \Phi - \text{Grad}\, \Phi \cdot \boldsymbol{H} \cdot \vec{N} \\ + (\vec{N} \cdot \boldsymbol{H} \cdot \vec{N})(\vec{N} \cdot \text{Grad}\, \Phi)\Big]_-^+ = 0 \qquad \text{on } \Sigma \tag{9.110}$$

correct to first order in $|\boldsymbol{H}|$. To the same degree of approximation, substituting Φ from $(9.89)_2$ gives

$$\Big[-\vec{N} \cdot \vec{g}_0 + \vec{N} \cdot \text{Grad}\, \Phi^L + \vec{N} \cdot \boldsymbol{H} \cdot \vec{g}_0 + \vec{g}_0 \cdot \boldsymbol{H} \cdot \vec{N} \\ - (\vec{N} \cdot \boldsymbol{H} \cdot \vec{N})(\vec{N} \cdot \vec{g}_0)\Big]_-^+ = 0, \tag{9.111}$$

where the initial gravitation $\vec{g}_0$ is introduced for convenience. Using (B.27), the last two terms on the left sum to

$$\vec{g}_0 \cdot \boldsymbol{H} \cdot \vec{N} - (\vec{N} \cdot \boldsymbol{H} \cdot \vec{N})(\vec{N} \cdot \vec{g}_0) = \vec{g}_0 \cdot \text{Grad}_\Sigma \vec{U} \cdot \vec{N}.$$

For convenience, let us add and subtract the expression $\vec{N} \cdot \mathrm{Grad}\, \vec{g}_0 \cdot \vec{U}$ to the left-hand side of (9.111) and use the differential identity (A.3). Then,

$$\Big[-\vec{N}\cdot\vec{g}_0+\vec{N}\cdot\mathrm{Grad}\,(\Phi^L+\vec{U}\cdot\vec{g}_0)-\vec{N}\cdot\mathrm{Grad}\,\vec{g}_0\cdot\vec{U}+\vec{g}_0\cdot\mathrm{Grad}_\Sigma\,\vec{U}\cdot\vec{N}\Big]_-^+ = 0. \qquad (9.112)$$

The symmetry of the tensor $\mathrm{Grad}\,\vec{g}_0$ implies that $\vec{N} \cdot \mathrm{Grad}\,\vec{g}_0 \cdot \vec{U} = \vec{U} \cdot \mathrm{Grad}\,\vec{g}_0 \cdot \vec{N}$. Moreover, expressing $\mathrm{Grad}\,\vec{g}_0 \cdot \vec{N}$ according to (B.29),

$$\mathrm{Grad}\,\vec{g}_0 \cdot \vec{N} = \mathrm{Grad}_\Sigma\,\vec{g}_0 \cdot \vec{N} + \vec{N}(\vec{N} \cdot \mathrm{Grad}\,\vec{g}_0 \cdot \vec{N}), \qquad (9.113)$$

the jump condition (9.112) becomes

$$\Big[-\vec{N}\cdot\vec{g}_0+\vec{N}\cdot\mathrm{Grad}\,(\Phi^L+\vec{U}\cdot\vec{g}_0)-(\vec{N}\cdot\mathrm{Grad}\,\vec{g}_0\cdot\vec{N})(\vec{N}\cdot\vec{U})$$
$$-\vec{U}\cdot\mathrm{Grad}_\Sigma\,\vec{g}_0\cdot\vec{N}+\vec{g}_0\cdot\mathrm{Grad}_\Sigma\,\vec{U}\cdot\vec{N}\Big]_-^+ = 0. \qquad (9.114)$$

In view of (9.85) and (B.32) applied to $\vec{g}_0$, we have

$$\vec{N} \cdot \mathrm{Grad}\,\vec{g}_0 \cdot \vec{N} = -4\pi G\varrho_0 - \mathrm{Div}_\Sigma\,\vec{g}_0. \qquad (9.115)$$

Continuity of the initial gravitation $\vec{g}_0$ along Σ_S implies continuity of $\mathrm{Div}_\Sigma\,\vec{g}_0$. However, $\vec{g}_0$ is not continuous along Σ_F due to tangential slip. Since the slip is of order of $|\vec{U}|$, the jump of $\mathrm{Div}_\Sigma\,\vec{g}_0$ across Σ_F is also of order of $|\vec{U}|$. Since the normal component of displacement is continuous along Σ_F (and Σ_S), correct to first order in $|\vec{U}|$, the product $\mathrm{Div}_\Sigma\,\vec{g}_0\,(\vec{N}\cdot\vec{U})$ can also be considered continuous along Σ_F (and Σ_S). In view of this approximation and (9.115), the jump condition (9.114) becomes

$$\Big[-\vec{N}\cdot\vec{g}_0+\vec{N}\cdot\mathrm{Grad}\,(\Phi^L+\vec{U}\cdot\vec{g}_0)+4\pi G\varrho_0(\vec{N}\cdot\vec{U})$$
$$-\vec{U}\cdot\mathrm{Grad}_\Sigma\,\vec{g}_0\cdot\vec{N}+\vec{g}_0\cdot\mathrm{Grad}_\Sigma\,\vec{U}\cdot\vec{N}\Big]_-^+ = 0, \qquad (9.116)$$

valid at all discontinuities Σ.

At a solid–solid discontinuity Σ_S, the displacement and initial gravitation are continuous and (9.116) simplifies to

$$\Big[\vec{N}\cdot\mathrm{Grad}\,(\Phi^L+\vec{U}\cdot\vec{g}_0)+4\pi G\varrho_0(\vec{N}\cdot\vec{U})\Big]_-^+ = 0 \qquad \text{on } \Sigma_\mathrm{S}. \qquad (9.117)$$

The corresponding results for ϕ^E is obtained from the relation (9.90) between the increments ϕ^E and Φ^L,

$$\Big[\vec{N}\cdot\mathrm{Grad}\,\phi^E+4\pi G\varrho_0(\vec{N}\cdot\vec{U})\Big]_-^+ = 0 \qquad \text{on } \Sigma_\mathrm{S}. \qquad (9.118)$$

At a fluid–solid discontinuity Σ_F, which admits tangential slip, the quantities $\vec{N}^{\pm} \cdot \vec{g}_0^{\pm}$ are related by an expansion analogous to (9.106):

$$\vec{N}^{+} \cdot \vec{g}_0^{+} = \vec{N}^{-} \cdot \vec{g}_0^{-} - \left[\vec{U}\right]_{-}^{+} \cdot \text{Grad}_{\Sigma}\,(\vec{N} \cdot \vec{g}_0) \qquad \text{on } \Sigma_F, \tag{9.119}$$

which is correct to first order in $|\vec{U}|$. Substituting (9.119) into (9.116) yields

$$\Big[\vec{N} \cdot \text{Grad}\,(\Phi^L + \vec{U} \cdot \vec{g}_0) + 4\pi G\varrho_0(\vec{N} \cdot \vec{U}) + \vec{U} \cdot \text{Grad}_{\Sigma}\,(\vec{N} \cdot \vec{g}_0)$$
$$- \vec{U} \cdot \text{Grad}_{\Sigma}\,\vec{g}_0 \cdot \vec{N} + \vec{g}_0 \cdot \text{Grad}_{\Sigma}\,\vec{U} \cdot \vec{N}\Big]_{-}^{+} = 0. \tag{9.120}$$

Using the identity (A.3), the third term is

$$\Big[\vec{N} \cdot \text{Grad}\,(\Phi^L + \vec{U} \cdot \vec{g}_0) + 4\pi G\varrho_0(\vec{N} \cdot \vec{U}) + \vec{U} \cdot \text{Grad}_{\Sigma}\,\vec{N} \cdot \vec{g}_0$$
$$+ \vec{g}_0 \cdot \text{Grad}_{\Sigma}\,\vec{U} \cdot \vec{N}\Big]_{-}^{+} = 0, \tag{9.121}$$

where the sum of the last two terms on the left is

$$\begin{aligned}\vec{U} \cdot \text{Grad}_{\Sigma}\,\vec{N} \cdot \vec{g}_0 + \vec{g}_0 \cdot \text{Grad}_{\Sigma}\,\vec{U} \cdot \vec{N} &= \vec{g}_0 \cdot \text{Grad}_{\Sigma}\,\vec{N} \cdot \vec{U} + \vec{g}_0 \cdot \text{Grad}_{\Sigma}\,\vec{U} \cdot \vec{N} \\ &= \vec{g}_0 \cdot \text{Grad}_{\Sigma}\,(\vec{N} \cdot \vec{U}) \\ &= (\vec{g}_0)_{\Sigma} \cdot \text{Grad}_{\Sigma}\,(\vec{N} \cdot \vec{U})\end{aligned}$$

due to the symmetry of $\text{Grad}_{\Sigma}\,\vec{N}$, see (B.16). As previously stated, $\vec{g}_0$ is not continuous along Σ_F but, correct to first order in $|\vec{U}|$, it can be considered continuous in the product $(\vec{g}_0)_{\Sigma} \cdot \text{Grad}_{\Sigma}\,(\vec{N} \cdot \vec{U})$. Moreover, using the continuity of the normal component of displacement along Σ_F yields

$$\Big[(\vec{g}_0)_{\Sigma} \cdot \text{Grad}_{\Sigma}\,(\vec{N} \cdot \vec{U})\Big]_{-}^{+} = 0 \qquad \text{on } \Sigma_F. \tag{9.122}$$

Consequently, the sum of the last two terms on the left-hand side of (9.121) is equal to zero and the jump condition (9.121) at a discontinuity Σ_F reduces to the jump condition (9.118) valid at a discontinuity Σ_S. We conclude that for the both types of discontinuity, the linearised form of the jump condition for gravitation is

$$\Big[\vec{N} \cdot \text{Grad}\,(\Phi^L + \vec{U} \cdot \vec{g}_0) + 4\pi G\varrho_0(\vec{N} \cdot \vec{U})\Big]_{-}^{+} = 0 \qquad \text{on } \Sigma. \tag{9.123}$$

The corresponding result for ϕ^E,

Table 9.3 Summary of linearised gravitational jump conditions

Discontinuity type	Linearised jump conditions
Σ_S: solid–solid	$[\Phi^L]_-^+ = 0 \qquad$ (exact)
	$[\phi^E]_-^+ = 0$
	$\left[\vec{N} \cdot \text{Grad}\,(\Phi^L + \vec{U} \cdot \vec{g}_0) + 4\pi G\varrho_0(\vec{N} \cdot \vec{U})\right]_-^+ = 0$
	$\left[\vec{N} \cdot \text{Grad}\,\phi^E + 4\pi G\varrho_0(\vec{N} \cdot \vec{U})\right]_-^+ = 0$
Σ_F: fluid–solid	$\left[\Phi^L + \vec{U} \cdot \vec{g}_0\right]_-^+ = 0$
	$[\phi^E]_-^+ = 0$
	$\left[\vec{N} \cdot \text{Grad}\,(\Phi^L + \vec{U} \cdot \vec{g}_0) + 4\pi G\varrho_0(\vec{N} \cdot \vec{U})\right]_-^+ = 0$
	$\left[\vec{N} \cdot \text{Grad}\,\phi^E + 4\pi G\varrho_0(\vec{N} \cdot \vec{U})\right]_-^+ = 0$

$$\left[\vec{N} \cdot \text{Grad}\,\phi^E + 4\pi G\varrho_0(\vec{N} \cdot \vec{U})\right]_-^+ = 0 \qquad \text{on } \Sigma, \tag{9.124}$$

is obtained from the relation (9.90) between the increments ϕ^E and Φ^L. The complete set of linearised jump conditions at both types of discontinuities is summarised in Table 9.3.

9.10 Equations of Motion for a Self-gravitating Body

We now deduce the linearised equations of motion describing the infinitesimal deformation of a *self-gravitating body*, which is a body whose deformations are large enough to change the gravitation of the body, and, in turn, changes in gravitation are capable of deforming the body itself. For this purpose, the general theory presented in the preceding sections will be applied to the special case where the gravitational force is the only force acting in the body. Since the self-gravitating body is initially pre-stressed by its own gravitation, we substitute the initial gravitational attraction $\vec{g}_0$ for the initial body force $\vec{f}_0$ into the static linear momentum equation (9.1),

$$\vec{f}_0 = -\text{Grad}\,\phi_0. \tag{9.125}$$

The Lagrangian increment in body force $\vec{F}^L$, defined by $(9.30)_2$, that occurs in the equation of motion is replaced by the Lagrangian increment in gravitation $\vec{G}^L$, which can be expressed in terms of either the Eulerian increment ϕ^E in gravitational potential by (9.94) or the Lagrangian increment Φ^L in gravitational potential by (9.96):

$$\begin{aligned}\vec{F}^L &= -\mathrm{Grad}\,\phi^E - \vec{U}\cdot\mathrm{Grad}\,\mathrm{Grad}\,\phi_0\\ &= -\mathrm{Grad}\,\Phi^L + \mathrm{Grad}\,\vec{U}\cdot\mathrm{Grad}\,\phi_0.\end{aligned} \tag{9.126}$$

Similarly, the Eulerian increment in body force $\vec{f}^E$, defined by $(9.30)_1$, is replaced by the Eulerian increment in gravitation $\vec{g}^E$, which can be expressed in terms of either the Eulerian increment ϕ^E in gravitational potential by (9.92) or the Lagrangian increment Φ^L in gravitational potential by (9.93):

$$\begin{aligned}\vec{f}^E &= -\mathrm{Grad}\,\phi^E\\ &= -\mathrm{Grad}\,\Phi^L + \mathrm{Grad}\,(\vec{U}\cdot\mathrm{Grad}\,\phi_0).\end{aligned} \tag{9.127}$$

All forms of the equation of motion derived in Sect. 9.6 can be expressed in terms of either Φ^L or ϕ^E using relations (9.125)–(9.127). Since this substitution is straightforward, we need not write the resulting relations explicitly.

Appendix A
Vector and Tensor Differential Identities

Let ϕ and ψ be differentiable scalar functions, $\vec{u}$, $\vec{v}$ and $\vec{w}$ be differentiable vector functions, $\boldsymbol{A}$ and $\boldsymbol{B}$ be differentiable second-order tensor functions, and $\boldsymbol{I}$ represent the second-order unit tensor. The following vector differential identities can be verified:

$$\text{grad}\,(\phi\psi) = \phi\,\text{grad}\,\psi + \psi\,\text{grad}\,\phi \tag{A.1}$$

$$\text{grad}\,(\phi\vec{u}) = \phi\,\text{grad}\,\vec{u} + \text{grad}\,\phi \otimes \vec{u} \tag{A.2}$$

$$\text{grad}\,(\vec{u}\cdot\vec{v}) = \text{grad}\,\vec{u}\cdot\vec{v} + \text{grad}\,\vec{v}\cdot\vec{u} \tag{A.3}$$

$$\text{grad}\,(\vec{u}\times\vec{v}) = \text{grad}\,\vec{u}\times\vec{v} - \text{grad}\,\vec{v}\times\vec{u} \tag{A.4}$$

$$\begin{aligned}\text{grad}\,(\vec{u}\otimes\vec{v}) &= \text{grad}\,\vec{u}\otimes\vec{v} + (\vec{u}\otimes\text{grad}\,\vec{v})^{213} \\ &= \text{grad}\,\vec{u}\otimes\vec{v} + (\text{grad}\,\vec{v}\otimes\vec{u})^{132}\end{aligned} \tag{A.5}$$

$$\vec{w}\cdot\text{grad}\,(\vec{u}\otimes\vec{v}) = (\vec{w}\cdot\text{grad}\,\vec{u})\otimes\vec{v} + \vec{u}\otimes(\vec{w}\cdot\text{grad}\,\vec{v}) \tag{A.6}$$

$$\text{grad}\,(\phi\boldsymbol{A}) = \phi\,\text{grad}\,\boldsymbol{A} + \text{grad}\,\phi\otimes\boldsymbol{A} \tag{A.7}$$

$$\text{grad}\,(\boldsymbol{A}\cdot\vec{u}) = \text{grad}\,\boldsymbol{A}\cdot\vec{u} + \text{grad}\,\vec{u}\cdot\boldsymbol{A}^T \tag{A.8}$$

$$\text{grad}\,(\vec{u}\cdot\boldsymbol{A}) = \text{grad}\,\vec{u}\cdot\boldsymbol{A} + \vec{u}\cdot(\text{grad}\,\boldsymbol{A})^{213} \tag{A.9}$$

$$\text{grad}\,\vec{v}\cdot\vec{u} - \vec{u}\cdot\text{grad}\,\vec{v} = \vec{u}\times\text{rot}\,\vec{v} \tag{A.10}$$

$$\text{grad}\,\vec{u} - (\text{grad}\,\vec{u})^T = -\boldsymbol{I}\times\text{rot}\,\vec{u} \tag{A.11}$$

$$\text{div}\,(\phi\vec{u}) = \phi\,\text{div}\,\vec{u} + \text{grad}\,\phi\cdot\vec{u} \tag{A.12}$$

$$\text{div}\,(\vec{u}\times\vec{v}) = \text{rot}\,\vec{u}\cdot\vec{v} - \vec{u}\cdot\text{rot}\,\vec{v} \tag{A.13}$$

$$\text{div}\,(\vec{u}\otimes\vec{v}) = (\text{div}\,\vec{u})\vec{v} + \vec{u}\cdot\text{grad}\,\vec{v} \tag{A.14}$$

$$\text{div}\,(\phi\boldsymbol{A}) = \phi\,\text{div}\,\boldsymbol{A} + \text{grad}\,\phi\cdot\boldsymbol{A} \tag{A.15}$$

Z. Martinec, *Principles of Continuum Mechanics*, Nečas Center Series,
https://doi.org/10.1007/978-3-030-05390-1

$$\operatorname{div}(\phi \boldsymbol{I}) = \operatorname{grad}\phi \tag{A.16}$$

$$\operatorname{div}(\boldsymbol{A}\cdot\vec{u}) = \operatorname{div}\boldsymbol{A}\cdot\vec{u} + \boldsymbol{A}^T : \operatorname{grad}\vec{u} \tag{A.17}$$

$$\operatorname{div}(\vec{u}\cdot\boldsymbol{A}) = \operatorname{grad}\vec{u} : \boldsymbol{A} + \vec{u}\cdot\operatorname{div}\boldsymbol{A}^T \tag{A.18}$$

$$\operatorname{div}(\boldsymbol{A}\otimes\vec{u}) = \operatorname{div}\boldsymbol{A}\otimes\vec{u} + \boldsymbol{A}^T\cdot\operatorname{grad}\vec{u} \tag{A.19}$$

$$\operatorname{div}(\vec{u}\otimes\boldsymbol{A}) = (\operatorname{div}\vec{u})\boldsymbol{A} + \vec{u}\cdot\operatorname{grad}\boldsymbol{A} \tag{A.20}$$

$$\operatorname{div}(\boldsymbol{A}\times\vec{u}) = \operatorname{div}\boldsymbol{A}\times\vec{u} + \boldsymbol{A}^T \dot{\times} \operatorname{grad}\vec{u} \tag{A.21}$$

$$\operatorname{div}(\vec{u}\times\boldsymbol{A}) = \operatorname{rot}\vec{u}\cdot\boldsymbol{A} - \vec{u}\cdot\operatorname{rot}\boldsymbol{A} \tag{A.22}$$

$$\operatorname{div}(\boldsymbol{I}\times\vec{u}) = \operatorname{rot}\vec{u} \tag{A.23}$$

$$\operatorname{div}\operatorname{rot}\vec{u} = 0 \tag{A.24}$$

$$\operatorname{div}(\boldsymbol{A}\cdot\boldsymbol{B}) = \operatorname{div}\boldsymbol{A}\cdot\boldsymbol{B} + \boldsymbol{A}^T : \operatorname{grad}\boldsymbol{B} \tag{A.25}$$

$$\operatorname{div}(\vec{u}\cdot\operatorname{grad}\boldsymbol{A}) = \operatorname{grad}\vec{u} : \operatorname{grad}\boldsymbol{A} + \vec{u}\cdot\operatorname{grad}\operatorname{div}\boldsymbol{A} \tag{A.26}$$

$$\operatorname{div}[(\operatorname{grad}\vec{u})^T] = \operatorname{grad}\operatorname{div}\vec{u} \tag{A.27}$$

$$\operatorname{rot}(\phi\vec{u}) = \phi\operatorname{rot}\vec{u} + \operatorname{grad}\phi\times\vec{u} \tag{A.28}$$

$$\operatorname{rot}(\vec{u}\times\vec{v}) = \vec{v}\cdot\operatorname{grad}\vec{u} - \vec{u}\cdot\operatorname{grad}\vec{v} + \vec{u}\operatorname{div}\vec{v} - \vec{v}\operatorname{div}\vec{u} \tag{A.29}$$

$$\operatorname{rot}(\vec{u}\otimes\vec{v}) = \operatorname{rot}\vec{u}\otimes\vec{v} - \vec{u}\times\operatorname{grad}\vec{v} \tag{A.30}$$

$$\operatorname{rot}(\phi\boldsymbol{A}) = \phi\operatorname{rot}\boldsymbol{A} + \operatorname{grad}\phi\times\boldsymbol{A} \tag{A.31}$$

$$\operatorname{rot}(\phi\boldsymbol{I}) = \operatorname{grad}\phi\times\boldsymbol{I} \tag{A.32}$$

$$\operatorname{rot}(\vec{u}\times\boldsymbol{I}) = \operatorname{rot}(\boldsymbol{I}\times\vec{u}) = (\operatorname{grad}\vec{u})^T - (\operatorname{div}\vec{u})\boldsymbol{I} \tag{A.33}$$

$$\operatorname{rot}\operatorname{rot}\vec{u} = \operatorname{grad}\operatorname{div}\vec{u} - \operatorname{div}\operatorname{grad}\vec{u} \tag{A.34}$$

$$\operatorname{rot}\operatorname{grad}\phi = \vec{0} \tag{A.35}$$

$$\operatorname{rot}(\operatorname{grad}\vec{u})^T = (\operatorname{grad}\operatorname{rot}\vec{u})^T \tag{A.36}$$

$$\boldsymbol{I} : \operatorname{grad}\vec{u} = \operatorname{div}\vec{u} \tag{A.37}$$

$$\boldsymbol{I}\dot{\times}\operatorname{grad}\vec{u} = \operatorname{rot}\vec{u} \tag{A.38}$$

$$\nabla^2(\phi\psi) = \psi\nabla^2\phi + \phi\nabla^2\psi + 2(\operatorname{grad}\phi)\cdot(\operatorname{grad}\psi) \tag{A.39}$$

$$\nabla^2(\vec{u}\cdot\vec{v}) = (\nabla^2\vec{u})\cdot\vec{v} + \vec{u}\cdot\nabla^2\vec{v} + 2(\operatorname{grad}\vec{u})^T : \operatorname{grad}\vec{v} \tag{A.40}$$

$$\nabla^2(\operatorname{grad}\phi) = \operatorname{grad}(\nabla^2\phi) \tag{A.41}$$

$$\nabla^2(\operatorname{rot}\vec{u}) = \operatorname{rot}(\nabla^2\vec{u}) \tag{A.42}$$

$$\nabla^2(\operatorname{rot}\operatorname{rot}\vec{u}) = \operatorname{rot}\operatorname{rot}(\nabla^2\vec{u})\,. \tag{A.43}$$

The symbol **:** denotes the double-dot product of dyads (see footnote 6 on page 11) and $\dot{\times}$ the dot-cross product of dyads (see footnote 3 on page 65). The symbols

$(\ )^T$, $(\ )^{213}$ and $(\ )^{132}$ denote, respectively, the transpose of a dyad, the left transpose of a triad and the right transpose of a triad, e.g. $(\vec{u} \otimes \vec{v})^T = \vec{v} \otimes \vec{u}$, $(\vec{u} \otimes \vec{v} \otimes \vec{w})^{213} = (\vec{v} \otimes \vec{u} \otimes \vec{w})$, $(\vec{u} \otimes \vec{v} \otimes \vec{w})^{132} = (\vec{u} \otimes \vec{w} \otimes \vec{v})$.

Appendix B
Surface Geometry

In Chap. 9, various jump conditions involving vectors that are normal or tangent to a surface are presented. The aim of this appendix is to show how those vectors are computed and used to define quantities such as surface gradients.

B.1 Local Base Vectors

Let the position of an arbitrary point P on the surface Σ in a three-dimensional space be described by two coordinates, ϑ_1 and ϑ_2, and let the position vector of P be denoted by $\vec{p}_\Sigma(\vartheta_1, \vartheta_2)$. In view of Sect. C.2, two vectors

$$\frac{\partial \vec{p}_\Sigma}{\partial \vartheta_\alpha}, \qquad \alpha = 1, 2, \tag{B.1}$$

are tangent to the surface. Specifically, $\partial \vec{p}_\Sigma / \partial \vartheta_1$ is tangent to a ϑ_1 coordinate line on the surface (i.e., the curve on Σ where ϑ_2 is held fixed), and $\partial \vec{p}_\Sigma / \partial \vartheta_2$ is tangent to a ϑ_2 coordinate line. In general, these vectors are not orthogonal to each other and they are not of unit length. However, we will assume that the ϑ_1 and ϑ_2 coordinate lines coincide with the principal directions of Σ at point P so that the vectors $\partial \vec{p}_\Sigma / \partial \vartheta_\alpha$ are then orthogonal to each other. In addition, they can be scaled to the unit length such that

$$\vec{e}_\alpha := \frac{1}{h_\alpha} \frac{\partial \vec{p}_\Sigma}{\partial \vartheta_\alpha}, \tag{B.2}$$

where h_α are the Lamé coefficients (C.9) associated with the coordinates ϑ_α. Hence,

$$\vec{e}_\alpha \cdot \vec{e}_\beta = \delta_{\alpha\beta}. \tag{B.3}$$

Z. Martinec, *Principles of Continuum Mechanics*, Nečas Center Series,
https://doi.org/10.1007/978-3-030-05390-1

The cross product of $\vec{e}_1$ and $\vec{e}_2$ is orthogonal to both vectors and hence normal to Σ. Since $\vec{e}_1$ and $\vec{e}_2$ are unit vectors, a unit normal vector to the surface is

$$\vec{n} = \vec{e}_1 \times \vec{e}_2. \tag{B.4}$$

Differentiating $\vec{n} \cdot \vec{n} = 1$ with respect to ϑ_α yields

$$0 = \frac{\partial}{\partial \vartheta_\alpha}(\vec{n} \cdot \vec{n}) = 2\vec{n} \cdot \frac{\partial \vec{n}}{\partial \vartheta_\alpha},$$

which shows that two vectors

$$\frac{\partial \vec{n}}{\partial \vartheta_\alpha}, \qquad \alpha = 1, 2,$$

are perpendicular to $\vec{n}$, and thus tangent to Σ in the direction of the ϑ_α coordinate lines. Hence, $\partial \vec{n}/\partial \vartheta_\alpha$ is a multiple of $\vec{e}_\alpha$, say

$$\frac{\partial \vec{n}}{\partial \vartheta_\alpha} = k_\alpha \vec{e}_\alpha \qquad \text{(no summation over } \alpha\text{)}, \tag{B.5}$$

where k_α are proportional to the so-called *principal curvatures* of Σ at point P. The explicit form of k_α is not needed in the following. Note that (B.5) holds under our starting assumption that the ϑ_1 and ϑ_2 coordinate lines at point P coincide with the principal directions of Σ at point P, that is, with the eigenvectors of the matrix $\partial \vec{n}/\partial \vartheta_\alpha$ at P.

B.2 Tangent Vectors and Tensors

A vector defined on the surface Σ with the unit normal vector $\vec{n}$ can be decomposed into a normal and a tangential part,

$$\vec{u} = u_n \vec{n} + \vec{u}_\Sigma, \tag{B.6}$$

where $u_n = \vec{n} \cdot \vec{u}$ and $\vec{u}_\Sigma = \vec{u} \cdot (\boldsymbol{I} - \vec{n} \otimes \vec{n})$ with $\vec{n} \cdot \vec{u}_\Sigma = 0$. The quantity u_n is the normal component of $\vec{u}$, while $\vec{u}_\Sigma$ is referred to as a *tangent vector* to Σ. The component form of $\vec{u}_\Sigma$ is

$$\vec{u}_\Sigma = \sum_\alpha u_\alpha \vec{e}_\alpha, \tag{B.7}$$

where the summation index α takes the values 1 and 2.

A tensor $\boldsymbol{A}_\Sigma$ defined on the surface Σ is called a *tangent tensor* to Σ if

$$\vec{n} \cdot \boldsymbol{A}_\Sigma = \boldsymbol{A}_\Sigma \cdot \vec{n} = \vec{0}. \tag{B.8}$$

A dyadic form of $\boldsymbol{A}_\Sigma$ is

$$\boldsymbol{A}_\Sigma = \sum_{\alpha\beta} A_{\alpha\beta}(\vec{e}_\alpha \otimes \vec{e}_\beta). \tag{B.9}$$

B.3 The Surface Gradient

The three-dimensional gradient operator can be decomposed into its normal and tangential parts,

$$\text{grad} = \vec{n}\frac{\partial}{\partial n} + \text{grad}_\Sigma, \tag{B.10}$$

where $\frac{\partial}{\partial n} := \vec{n} \cdot \text{grad}$ and $\vec{n} \cdot \text{grad}_\Sigma = 0$. The tangential part, grad_Σ, is called the *surface gradient operator* and its component form is

$$\text{grad}_\Sigma = \sum_\alpha \frac{\vec{e}_\alpha}{h_\alpha}\frac{\partial}{\partial \vartheta_\alpha}. \tag{B.11}$$

Since grad_Σ involves only differentiation in directions tangent to the surface Σ, it can be applied to any scalar, vector or tensor field defined on Σ, whether that field is defined elsewhere or not. Similarly, the three-dimensional divergence operator can be decomposed into its normal and tangential parts,

$$\text{div} = \vec{n} \cdot \frac{\partial}{\partial n} + \text{div}_\Sigma, \tag{B.12}$$

where the component form of the *surface divergence operator* div_Σ is

$$\text{div}_\Sigma = \sum_\alpha \frac{\vec{e}_\alpha}{h_\alpha} \cdot \frac{\partial}{\partial \vartheta_\alpha}. \tag{B.13}$$

B.4 Differential Identities

Let ϕ and ψ be differentiable scalar functions, and $\vec{u}$ and $\vec{v}$ be differentiable vector functions. The following vector differential identities can be verified:

$$\begin{aligned}
\text{grad}_\Sigma(\phi\psi) &= \phi\,\text{grad}_\Sigma\psi + \psi\,\text{grad}_\Sigma\phi,\\
\text{grad}_\Sigma(\vec{u}\cdot\vec{v}) &= \text{grad}_\Sigma\vec{u}\cdot\vec{v} + \text{grad}_\Sigma\vec{v}\cdot\vec{u},\\
\text{grad}_\Sigma(\phi\vec{u}) &= \phi\,\text{grad}_\Sigma\vec{u} + (\text{grad}_\Sigma\phi)\otimes\vec{u},
\end{aligned} \tag{B.14}$$

$$\mathrm{div}_\Sigma(\phi\vec{u}) = \phi\,\mathrm{div}_\Sigma\vec{u} + \mathrm{grad}_\Sigma\phi\cdot\vec{u},$$
$$\mathrm{div}_\Sigma(\vec{u}\otimes\vec{v}) = (\mathrm{div}_\Sigma\vec{u})\vec{v} + \vec{u}\cdot\mathrm{grad}_\Sigma\vec{v}.$$

B.5 The Curvature Tensor

The surface gradient applied to the unit normal vector $\vec{n}$, the so-called *surface curvature tensor*, is often used to classify the curvature of the surface. In view of (B.5) and (B.11), we have

$$\mathrm{grad}_\Sigma\vec{n} = \sum_\alpha \frac{k_\alpha}{h_\alpha}(\vec{e}_\alpha\otimes\vec{e}_\alpha). \tag{B.15}$$

The tensor $\mathrm{grad}_\Sigma\vec{n}$ is symmetric,

$$(\mathrm{grad}_\Sigma\vec{n})^T = \mathrm{grad}_\Sigma\vec{n}, \tag{B.16}$$

and is tangent to Σ,

$$\vec{n}\cdot\mathrm{grad}_\Sigma\vec{n} = \mathrm{grad}_\Sigma\vec{n}\cdot\vec{n} = \vec{0}. \tag{B.17}$$

The surface divergence of the unit normal vector $\vec{n}$ is obtained from (B.15) by replacing the dyadic product by the scalar product such that

$$\mathrm{div}_\Sigma\vec{n} = \sum_\alpha \frac{k_\alpha}{h_\alpha}(\vec{e}_\alpha\cdot\vec{e}_\alpha) = \frac{k_1}{h_1} + \frac{k_2}{h_2} =: 2\overline{c}, \tag{B.18}$$

where $\overline{c}$ is the *mean curvature* of Σ. Another relation of interest in this context is

$$\mathrm{div}_\Sigma(\vec{n}\otimes\vec{n}) = (\mathrm{div}_\Sigma\vec{n})\vec{n}, \tag{B.19}$$

where $(\mathrm{B.14})_5$ and (B.15) have been used.

B.6 Divergence of a Vector

When applying the operator (B.12) to a decomposed vector of the form (B.6), it must be remembered that the surface gradient also acts on the unit normal vector $\vec{n}$. For instance, the surface divergence of a vector function is

$$\mathrm{div}_\Sigma\vec{u} = \mathrm{div}_\Sigma(\vec{n}u_n + \vec{u}_\Sigma) = (\mathrm{div}_\Sigma\vec{n})u_n + \mathrm{grad}_\Sigma u_n\cdot\vec{n} + \mathrm{div}_\Sigma\vec{u}_\Sigma,$$

where the identity $(B.14)_4$ has been used. Since the vector $\mathrm{grad}_\Sigma u_n$ is tangent to the surface, the second term on the right is equal to zero, leaving

$$\mathrm{div}_\Sigma \vec{u} = (\mathrm{div}_\Sigma \vec{n}) u_n + \mathrm{div}_\Sigma \vec{u}_\Sigma. \tag{B.20}$$

B.7 Gradient of a Vector

We now wish to show that $\mathrm{grad}_\Sigma \vec{u}$ is **not** in general a tangent tensor. According to (B.11),

$$\vec{n} \cdot \mathrm{grad}_\Sigma \vec{u} = \vec{n} \cdot \sum_\alpha \frac{\vec{e}_\alpha}{h_\alpha} \otimes \frac{\partial \vec{u}}{\partial \vartheta_\alpha} = \sum_\alpha \frac{1}{h_\alpha} (\vec{n} \cdot \vec{e}_\alpha) \frac{\partial \vec{u}}{\partial \vartheta_\alpha},$$

which implies that

$$\vec{n} \cdot \mathrm{grad}_\Sigma \vec{u} = \vec{0} \tag{B.21}$$

in view of (B.4).

To find the other scalar product, $\mathrm{grad}_\Sigma \vec{u} \cdot \vec{n}$, we differentiate $\vec{n} \cdot \vec{u}_\Sigma = 0$ with respect to ϑ_α, obtaining

$$0 = \frac{\partial}{\partial \vartheta_\alpha} (\vec{n} \cdot \vec{u}_\Sigma) = \frac{\partial \vec{n}}{\partial \vartheta_\alpha} \cdot \vec{u}_\Sigma + \vec{n} \cdot \frac{\partial \vec{u}_\Sigma}{\partial \vartheta_\alpha},$$

which yields

$$\vec{n} \cdot \frac{\partial \vec{u}_\Sigma}{\partial \vartheta_\alpha} = -\frac{\partial \vec{n}}{\partial \vartheta_\alpha} \cdot \vec{u}_\Sigma. \tag{B.22}$$

Using (B.5), (B.6) and (B.11), the scalar product $\mathrm{grad}_\Sigma \vec{u} \cdot \vec{n}$ can be successively manipulated to give

$$\begin{aligned} \mathrm{grad}_\Sigma \vec{u} \cdot \vec{n} &= \sum_\alpha \frac{1}{h_\alpha} \Big(\vec{e}_\alpha \otimes \frac{\partial \vec{u}}{\partial \vartheta_\alpha} \Big) \cdot \vec{n} = \sum_\alpha \frac{\vec{e}_\alpha}{h_\alpha} \Big(\frac{\partial \vec{u}}{\partial \vartheta_\alpha} \cdot \vec{n} \Big) \\ &= \sum_\alpha \frac{\vec{e}_\alpha}{h_\alpha} \Big(\frac{\partial u_n}{\partial \vartheta_\alpha} \vec{n} + u_n \frac{\partial \vec{n}}{\partial \vartheta_\alpha} + \frac{\partial \vec{u}_\Sigma}{\partial \vartheta_\alpha} \Big) \cdot \vec{n} = \sum_\alpha \frac{\vec{e}_\alpha}{h_\alpha} \Big(\frac{\partial u_n}{\partial \vartheta_\alpha} + \frac{\partial \vec{u}_\Sigma}{\partial \vartheta_\alpha} \cdot \vec{n} \Big), \end{aligned}$$

which yields

$$\mathrm{grad}_\Sigma \vec{u} \cdot \vec{n} = \sum_\alpha \frac{1}{h_\alpha} \Big(\frac{\partial u_n}{\partial \vartheta_\alpha} - \frac{\partial \vec{n}}{\partial \vartheta_\alpha} \cdot \vec{u}_\Sigma \Big) \vec{e}_\alpha \tag{B.23}$$

in view of (B.22). Hence, the vector $\mathrm{grad}_\Sigma \vec{u} \cdot \vec{n}$ is tangent to the surface,

$$\mathrm{grad}_\Sigma \vec{u} \cdot \vec{n} = (\mathrm{grad}_\Sigma \vec{u} \cdot \vec{n}) \cdot (\boldsymbol{I} - \vec{n} \otimes \vec{n}), \tag{B.24}$$

and is, in general, non-zero,

$$\mathrm{grad}_\Sigma \vec{u} \cdot \vec{n} \neq \vec{0}, \tag{B.25}$$

from which we conclude that, despite (B.21), $\mathrm{grad}_\Sigma \vec{u}$ is not a tangent tensor to the surface. However,

$$\vec{n} \cdot \mathrm{grad}_\Sigma \vec{u} \cdot \vec{n} = 0. \tag{B.26}$$

B.8 Spatial and Surface Invariants

Applying the operators (B.10) and (B.12) to a vector $\vec{u}$, the three-dimensional gradient of $\vec{u}$ and the three-dimensional divergence of $\vec{u}$ are

$$\mathrm{grad}\, \vec{u} = \mathrm{grad}_\Sigma \vec{u} + \vec{n} \otimes (\vec{n} \cdot \mathrm{grad}\, \vec{u}), \tag{B.27}$$

$$\mathrm{div}\, \vec{u} = \mathrm{div}_\Sigma \vec{u} + (\vec{n} \cdot \mathrm{grad}\, \vec{u} \cdot \vec{n}). \tag{B.28}$$

Multiplying (B.27) scalarly by $\vec{n}$ from the right leads to

$$(\mathrm{grad}\, \vec{u}) \cdot \vec{n} = (\mathrm{grad}_\Sigma \vec{u}) \cdot \vec{n} + \vec{n}(\vec{n} \cdot \mathrm{grad}\, \vec{u} \cdot \vec{n}). \tag{B.29}$$

Subtracting (B.28) multiplied by $\vec{n}$ from (B.29) gives

$$(\mathrm{grad}\, \vec{u}) \cdot \vec{n} - (\mathrm{div}\, \vec{u})\vec{n} = (\mathrm{grad}_\Sigma \vec{u}) \cdot \vec{n} - (\mathrm{div}_\Sigma \vec{u})\vec{n}, \tag{B.30}$$

which shows that the quantity

$$(\mathrm{grad}\, \vec{u}) \cdot \vec{n} - (\mathrm{div}\, \vec{u})\vec{n}$$

is an invariant in the sense that it is not altered when the three-dimensional gradient operator is replaced by the surface gradient operator. Similarly, it can be shown that

$$\vec{n} \cdot \mathrm{rot}\, \vec{u} = \vec{n} \cdot \mathrm{rot}_\Sigma \vec{u}. \tag{B.31}$$

The normal component of the invariant equality (B.30) is

$$\vec{n} \cdot \mathrm{grad}\, \vec{u} \cdot \vec{n} = \mathrm{div}\, \vec{u} - \mathrm{div}_\Sigma \vec{u}, \tag{B.32}$$

where (B.26) has been used.

Appendix C
Orthogonal Curvilinear Coordinates

We recall that the basic field equations and constitutive equations developed in Chaps. 1–9 exclusively the Cartesian coordinate system. This is the simplest coordinate system in which the three Cartesian unit base vectors have a fixed orientation in space. It is suitable for solving problems with rectangular symmetry, however, we often need to treat finite configurations with spherical, spheroidal, ellipsoidal or other types of symmetry. We must therefore work with vectors and tensors in a wider class of coordinate systems. In this appendix, we present the methods used to express vectors, tensors and their derivatives in orthogonal curvilinear coordinates. The formulae derived in arbitrary orthogonal curvilinear coordinates are then specifically expressed in spherical coordinates.

C.1 Coordinate Transformations

In a three-dimensional space, we define a system of *curvilinear coordinates* x_k by specifying three *coordinate transformation* functions $\hat{x}_k$ of the Cartesian[1] coordinates y_l,

$$x_k = \hat{x}_k(y_1, y_2, y_3), \qquad k = 1, 2, 3. \tag{C.1}$$

We assume that the three functions $\hat{x}_k$ have continuous first-order partial derivatives and the Jacobian determinant does not vanish,

$$j := \det\left(\frac{\partial \hat{x}_k}{\partial y_l}\right) \neq 0; \tag{C.2}$$

[1]More generally, the coordinate transformation functions can be introduced between any two curvilinear coordinates.

Z. Martinec, *Principles of Continuum Mechanics*, Nečas Center Series,
https://doi.org/10.1007/978-3-030-05390-1

exceptions may occur at singular points or curves, but never throughout a volume. Then, the correspondence between x_k and y_l is one-to-one and there exists a unique inverse of (C.1),

$$y_k = \hat{y}_k(x_1, x_2, x_3), \qquad k = 1, 2, 3. \tag{C.3}$$

If x_1 is held fixed, the three equations (C.3) define parametrically a surface, giving its Cartesian coordinates as a function of the two parameters x_2 and x_3. The first equation in (C.1), $x_1 = \hat{x}_1(y_1, y_2, y_3)$, defines the same surface implicitly. Similarly, we obtain the other two surfaces for fixed values of x_2 and x_3. The three surfaces so obtained are the *curvilinear coordinate surfaces*. Each pair of coordinate surfaces intersects at a *curvilinear coordinate line*, along which only one of the three parameters x_k varies. As opposed to Cartesian coordinate lines, curvilinear coordinate lines are space curves. The three coordinate surfaces (and also all three coordinate lines) intersect each other at a single point P marked with specific values of x_1, x_2 and x_3. We may take the values of x_1, x_2 and x_3 as the *curvilinear coordinates* of point P (Fig. C.1). If the coordinate lines through each point P are mutually orthogonal, the curvilinear coordinates x_k are called *orthogonal*.

Example 1 The *spherical coordinates* r, ϑ, $\lambda (r \equiv x_1, \vartheta \equiv x_2, \lambda \equiv x_3)$ are defined by their relations to the Cartesian coordinates y_k as

$$y_1 = r \sin\vartheta \cos\lambda, \qquad y_2 = r \sin\vartheta \sin\lambda, \qquad y_3 = r \cos\vartheta, \tag{C.4}$$

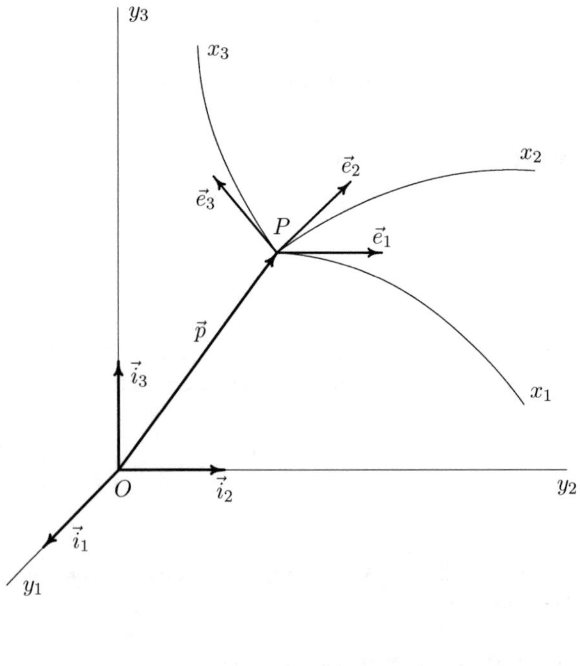

Fig. C.1 Curvilinear coordinates

or, conversely,

$$r = \sqrt{y_1^2 + y_2^2 + y_3^2}, \qquad \vartheta = \arctan\left(\frac{\sqrt{y_1^2 + y_2^2}}{y_3}\right), \qquad \lambda = \arctan\left(\frac{y_2}{y_1}\right). \tag{C.5}$$

The ranges of values are $0 \le r < \infty$, $0 \le \vartheta \le \pi$, and $0 \le \lambda \le 2\pi$. The inverse of the Jacobian j is

$$j^{-1} = \begin{vmatrix} \sin\vartheta\cos\lambda & \sin\vartheta\sin\lambda & \cos\vartheta \\ r\cos\vartheta\cos\lambda & r\cos\vartheta\sin\lambda & -r\sin\vartheta \\ -r\sin\vartheta\sin\lambda & r\sin\vartheta\cos\lambda & 0 \end{vmatrix} = r^2\sin\vartheta. \tag{C.6}$$

Hence, the unique inverse of (C.4) exists everywhere, except for $r = 0$, $\vartheta = 0$ and $\vartheta = \pi$. The coordinate surfaces are the concentric spheres $r =$ const. centred at the origin, the circular cones $\vartheta =$ const. centred on the y_3-axis and the half planes $\lambda =$ const. passing through the y_3-axis. The r-coordinate lines are half straights, the ϑ-coordinate lines are circles or *meridians* and the λ-coordinate lines are circles or *parallels*.

C.2 Base Vectors

We now introduce three base vectors $\vec{e}_k$ in curvilinear coordinates. We observe that if we move along a curvilinear coordinate line, only one of the three curvilinear coordinates varies.
For instance, along the x_1-coordinate line only x_1 varies while x_2 and x_3 are kept fixed. By analogy with the definition of the base vectors in Cartesian coordinates, the unit base vector $\vec{e}_1$ in curvilinear coordinates will be defined as the tangent vector to the x_1-coordinate line. Similarly, $\vec{e}_2$ and $\vec{e}_3$ are the unit tangent vectors to the curvilinear coordinate lines of varying x_2 and x_3, respectively. The direction cosines of the base vector $\vec{e}_1$ are

$$\cos\alpha = \frac{1}{h_1}\frac{\partial \hat{y}_1}{\partial x_1}, \quad \cos\beta = \frac{1}{h_1}\frac{\partial \hat{y}_2}{\partial x_1}, \quad \cos\gamma = \frac{1}{h_1}\frac{\partial \hat{y}_3}{\partial x_1},$$

where, for instance, α is the angle between the positive x_1 direction and the positive y_1 direction (Fig. C.2). A scale factor h_1 is introduced to satisfy the orthonormality condition for the direction cosines,

$$\cos^2\alpha + \cos^2\beta + \cos^2\gamma = 1,$$

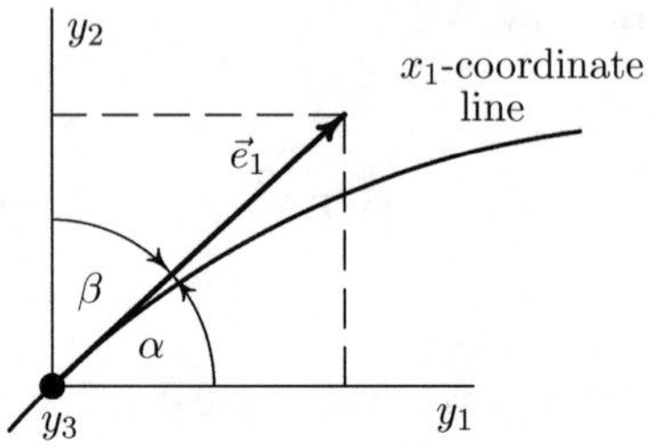

Fig. C.2 The unit base vector $\vec{e}_1$

which gives

$$h_1 = \sqrt{\sum_{l=1}^{3} \left(\frac{\partial \hat{y}_l}{\partial x_1} \right)^2}.$$

Throughout this appendix, we will suspend the Einstein summation convention and explicitly indicate, when necessary, all summations, which range from 1 to 3. The unit base vector $\vec{e}_1$ can now be written as

$$\vec{e}_1 = \frac{1}{h_1} \left(\frac{\partial \hat{y}_1}{\partial x_1} \vec{i}_1 + \frac{\partial \hat{y}_2}{\partial x_1} \vec{i}_2 + \frac{\partial \hat{y}_3}{\partial x_1} \vec{i}_3 \right),$$

where $\vec{i}_1$, $\vec{i}_2$ and $\vec{i}_3$ are the Cartesian unit base vectors. Expressing the position vector $\vec{p}$ of point P in the form

$$\vec{p} = \sum_k y_k \vec{i}_k, \tag{C.7}$$

where the summation index k ranges from 1 to 3, gives the curvilinear coordinate unit base vector $\vec{e}_1$,

$$\vec{e}_1 = \frac{1}{h_1} \frac{\partial \vec{p}}{\partial x_k}.$$

Similar considerations can be made for the unit base vectors $\vec{e}_2$ and $\vec{e}_3$, hence,

$$\vec{e}_k = \frac{1}{h_k} \frac{\partial \vec{p}}{\partial x_k} \qquad \text{(no summation over } k) \tag{C.8}$$

for $k = 1, 2, 3$. The non-negative functions

$$h_k := \sqrt{\frac{\partial \vec{p}}{\partial x_k} \cdot \frac{\partial \vec{p}}{\partial x_k}} \qquad \text{(no summation over } k) \tag{C.9}$$

are called *scale factors* or *Lamé coefficients*.

In what follows, we will consider only orthogonal curvilinear coordinates. We order the three curvilinear coordinates such that each base vector $\vec{e}_k$ points in the direction of increasing x_k and the three base vectors form a right-handed orthonormal system,

$$\vec{e}_k \cdot \vec{e}_l = \delta_{kl}, \qquad \vec{e}_k \times \vec{e}_l = \varepsilon_{klm}\vec{e}_m, \tag{C.10}$$

where δ_{kl} and ε_{klm} are the Kronecker delta symbols and the Levi–Cività alternating symbols, respectively.

An infinitesimal vector at point P can be expressed as

$$d\vec{p} = \sum_k \frac{\partial \vec{p}}{\partial x_k} dx_k = \sum_k h_k \vec{e}_k dx_k, \tag{C.11}$$

where we have substituted $\partial \vec{p}/\partial x_k$ from (C.8). The scalar product $d\vec{p} \cdot d\vec{p}$ gives the square of the distance between two neighbouring points, that is, the infinitesimal element of arc length ds on an arbitrary curve passing through point P,

$$(ds)^2 = (h_1 dx_1)^2 + (h_2 dx_2)^2 + (h_3 dx_3)^2. \tag{C.12}$$

Along the x_k coordinate line, only the increment dx_k differs from zero. The elementary distance along the x_k coordinate line is then

$$ds_k = h_k dx_k \qquad \text{(no summation over } k\text{)}. \tag{C.13}$$

Note that the three curvilinear coordinates x_k need not be lengths and the scale factors h_k may have dimensions. The products $h_k dx_k$ must, however, have dimensions of length.

From (C.13) we immediately determine the area element of the coordinate surface

$$da_{kl} = ds_k ds_l = h_k h_l dx_k dx_l, \tag{C.14}$$

and the volume element

$$dV = ds_1 ds_2 ds_3 = h_1 h_2 h_3 dx_1 dx_2 dx_3. \tag{C.15}$$

Example 2 The scale factors in spherical coordinates r, ϑ, λ are

$$h_r = 1, \qquad h_\vartheta = r, \qquad h_\lambda = r \sin\vartheta, \tag{C.16}$$

while the square of the arc length is

$$(ds)^2 = (dr)^2 + r^2 (d\vartheta)^2 + r^2 \sin^2\vartheta (d\lambda)^2. \tag{C.17}$$

The unit base vectors $\vec{e}_r$, $\vec{e}_\vartheta$ and $\vec{e}_\lambda$ pointing in the direction of increasing r, ϑ, λ, respectively, form a local right-handed orthonormal basis related to the Cartesian unit base vectors $\vec{i}_1$, $\vec{i}_2$ and $\vec{i}_3$ by

$$\begin{aligned}
\vec{e}_r &= \sin\vartheta\cos\lambda\,\vec{i}_1 + \sin\vartheta\sin\lambda\,\vec{i}_2 + \cos\vartheta\,\vec{i}_3,\\
\vec{e}_\vartheta &= \cos\vartheta\cos\lambda\,\vec{i}_1 + \cos\vartheta\sin\lambda\,\vec{i}_2 - \sin\vartheta\,\vec{i}_3,\\
\vec{e}_\lambda &= -\sin\lambda\,\vec{i}_1 + \cos\lambda\,\vec{i}_2,
\end{aligned} \tag{C.18}$$

or, conversely,

$$\begin{aligned}
\vec{i}_1 &= \sin\vartheta\cos\lambda\,\vec{e}_r + \cos\vartheta\cos\lambda\,\vec{e}_\vartheta - \sin\lambda\,\vec{e}_\lambda,\\
\vec{i}_2 &= \sin\vartheta\sin\lambda\,\vec{e}_r + \cos\vartheta\sin\lambda\,\vec{e}_\vartheta + \cos\lambda\,\vec{e}_\lambda,\\
\vec{i}_3 &= \cos\vartheta\,\vec{e}_r - \sin\vartheta\,\vec{e}_\vartheta.
\end{aligned} \tag{C.19}$$

C.3 Derivatives of Unit Base Vectors

Since the unit base vectors $\vec{e}_k$ are functions of position, they vary in direction as the curvilinear coordinates vary. Hence, they cannot be treated as constants in differentiation and we must now evaluate the partial derivatives $\partial\vec{e}_k/\partial x_l$. As any other vector quantity, they can be represented as a linear combination of the three base vectors $\vec{e}_m$,

$$\frac{\partial\vec{e}_k}{\partial x_l} = \sum_{m=1}^{3} \binom{m}{k\,l}\vec{e}_m, \tag{C.20}$$

where the expansion coefficients $\binom{m}{k\,l}$ are known as *Christoffel symbols*.[2] Their meaning is the mth component of the derivative of the kth unit base vector along the lth coordinate. Another common notation for the Christoffel symbols is

$$\Gamma^m_{kl} \equiv \binom{m}{k\,l},$$

[2]In general curvilinear coordinates, we usually introduce the Christoffel symbols of the first kind, $[kl, m]$, and the second kind, $\{kl, m\}$. In orthogonal curvilinear coordinates, we can use only the Christoffel symbols $\binom{m}{k\,l}$, which may correspond to either $[kl, m]$ or $\{kl, m\}$. If the base vectors are also normalised, the symbols $\binom{m}{k\,l}$ correspond to $[kl, m]$.

which we do not use because the symbols Γ^m_{kl} resemble mixed components of a third-order tensor, which they are not.

Each Christoffel symbol is a function of the scale factors only. To find its explicit functional form, we take the scalar product of (C.20) with the unit base vector $\vec{e}_n$, use the orthonormality relation $(\mathrm{C}.10)_1$ and subsequently replace the index n by m. We then obtain the explicit expression for the Christoffel symbol

$$\binom{m}{k\,l} = \frac{\partial \vec{e}_k}{\partial x_l} \cdot \vec{e}_m. \tag{C.21}$$

Substituting for the unit base vectors from (C.8) yields

$$\begin{aligned}\binom{m}{k\,l} &= \frac{\partial}{\partial x_l}\left(\frac{1}{h_k}\frac{\partial \vec{p}}{\partial x_k}\right) \cdot \frac{1}{h_m}\frac{\partial \vec{p}}{\partial x_m} \\ &= -\frac{1}{h_k^2 h_m}\frac{\partial h_k}{\partial x_l}\left(\frac{\partial \vec{p}}{\partial x_k} \cdot \frac{\partial \vec{p}}{\partial x_m}\right) + \frac{1}{h_k h_m}\left(\frac{\partial^2 \vec{p}}{\partial x_l \partial x_k} \cdot \frac{\partial \vec{p}}{\partial x_m}\right),\end{aligned}$$

or

$$\binom{m}{k\,l} = -\frac{1}{h_k}\frac{\partial h_k}{\partial x_l}\delta_{km} + \frac{1}{h_k h_m}\left(\frac{\partial^2 \vec{p}}{\partial x_l \partial x_k} \cdot \frac{\partial \vec{p}}{\partial x_m}\right), \tag{C.22}$$

where $(\mathrm{C}.10)_1$ has been used. To find an explicit form of the second term on the right, we differentiate the orthonormality relation

$$\frac{\partial \vec{p}}{\partial x_k} \cdot \frac{\partial \vec{p}}{\partial x_m} = h_k h_m \delta_{km}$$

with respect to x_l to obtain

$$\frac{\partial^2 \vec{p}}{\partial x_l \partial x_k} \cdot \frac{\partial \vec{p}}{\partial x_m} + \frac{\partial \vec{p}}{\partial x_k} \cdot \frac{\partial^2 \vec{p}}{\partial x_l \partial x_m} = \frac{\partial (h_k h_m)}{\partial x_l}\delta_{km}. \tag{C.23}$$

Cyclical permutations of k, l, m give two other forms of (C.23),

$$\frac{\partial^2 \vec{p}}{\partial x_m \partial x_l} \cdot \frac{\partial \vec{p}}{\partial x_k} + \frac{\partial \vec{p}}{\partial x_l} \cdot \frac{\partial^2 \vec{p}}{\partial x_m \partial x_k} = \frac{\partial (h_l h_k)}{\partial x_m}\delta_{lk}, \tag{C.24}$$

$$\frac{\partial^2 \vec{p}}{\partial x_k \partial x_m} \cdot \frac{\partial \vec{p}}{\partial x_l} + \frac{\partial \vec{p}}{\partial x_m} \cdot \frac{\partial^2 \vec{p}}{\partial x_k \partial x_l} = \frac{\partial (h_m h_l)}{\partial x_k}\delta_{ml}. \tag{C.25}$$

Adding (C.23) and (C.25), and subtracting (C.24) yields

$$2\,\frac{\partial^2 \vec{p}}{\partial x_k \partial x_l} \cdot \frac{\partial \vec{p}}{\partial x_m} = \frac{\partial (h_k h_m)}{\partial x_l}\delta_{km} + \frac{\partial (h_m h_l)}{\partial x_k}\delta_{ml} - \frac{\partial (h_l h_k)}{\partial x_m}\delta_{lk}. \tag{C.26}$$

Substituting this expression into (C.22), we obtain

$$\left(\begin{matrix} m \\ k\,l \end{matrix}\right) = -\frac{1}{h_k}\frac{\partial h_k}{\partial x_l}\delta_{km} + \frac{1}{2h_k h_m}\left[\frac{\partial (h_k h_m)}{\partial x_l}\delta_{km} + \frac{\partial (h_m h_l)}{\partial x_k}\delta_{ml} - \frac{\partial (h_l h_k)}{\partial x_m}\delta_{lk}\right], \tag{C.27}$$

which, after simple manipulation, finally results in

$$\left(\begin{matrix} m \\ k\,l \end{matrix}\right) = \frac{1}{h_k}\frac{\partial h_l}{\partial x_k}\delta_{lm} - \frac{1}{h_m}\frac{\partial h_k}{\partial x_m}\delta_{kl}. \tag{C.28}$$

If $k \neq l \neq m$, then

$$\left(\begin{matrix} m \\ k\,l \end{matrix}\right) = \left(\begin{matrix} k \\ k\,k \end{matrix}\right) = \left(\begin{matrix} k \\ k\,l \end{matrix}\right) = 0, \tag{C.29}$$

where the last equality is a consequence of the fact that $\partial \vec{e}_k / \partial x_l$ is orthogonal to the x_k coordinate line and, thus, has no component in the direction of $\vec{e}_k$ (but may have components in both directions orthogonal to $\vec{e}_l$). In view of (C.29), at most 12 of the 27 Christoffel symbols are non-zero,

$$\left(\begin{matrix} l \\ k\,l \end{matrix}\right) = \frac{1}{h_k}\frac{\partial h_l}{\partial x_k}, \qquad \left(\begin{matrix} l \\ k\,k \end{matrix}\right) = -\frac{1}{h_l}\frac{\partial h_k}{\partial x_l} \qquad \text{if } k \neq l, \tag{C.30}$$

of which only six can be independent because

$$\left(\begin{matrix} l \\ k\,l \end{matrix}\right) = -\left(\begin{matrix} k \\ l\,l \end{matrix}\right) \tag{C.31}$$

holds. We have shown that the Christoffel symbols are fully defined in terms of the scale factors. Equation (C.30) shows that all Christoffel symbols vanish in a Cartesian coordinate system (rectangular or skew) because the scale factors are constant.

Example 3 Only six Christoffel symbols are non-zero in the spherical coordinates r, ϑ, λ:

$$\begin{gathered} \left(\begin{matrix} \vartheta \\ r\,\vartheta \end{matrix}\right) = 1, \quad \left(\begin{matrix} \lambda \\ r\,\lambda \end{matrix}\right) = \sin\vartheta, \quad \left(\begin{matrix} \lambda \\ \vartheta\,\lambda \end{matrix}\right) = \cos\vartheta, \\ \left(\begin{matrix} r \\ \vartheta\,\vartheta \end{matrix}\right) = -1, \left(\begin{matrix} r \\ \lambda\,\lambda \end{matrix}\right) = -\sin\vartheta, \left(\begin{matrix} \vartheta \\ \lambda\,\lambda \end{matrix}\right) = -\cos\vartheta. \end{gathered} \tag{C.32}$$

The partial derivatives of the unit base vectors $\vec{e}_r$, $\vec{e}_\vartheta$ and $\vec{e}_\lambda$ are

$$
\begin{aligned}
&\frac{\partial \vec{e}_r}{\partial r} = 0, &&\frac{\partial \vec{e}_\vartheta}{\partial r} = 0, &&\frac{\partial \vec{e}_\lambda}{\partial r} = 0,\\
&\frac{\partial \vec{e}_r}{\partial \vartheta} = \vec{e}_\vartheta, &&\frac{\partial \vec{e}_\vartheta}{\partial \vartheta} = -\vec{e}_r, &&\frac{\partial \vec{e}_\lambda}{\partial \vartheta} = 0,\\
&\frac{\partial \vec{e}_r}{\partial \lambda} = \vec{e}_\lambda \sin\vartheta, &&\frac{\partial \vec{e}_\vartheta}{\partial \lambda} = \vec{e}_\lambda \cos\vartheta, &&\frac{\partial \vec{e}_\lambda}{\partial \lambda} = -\vec{e}_r \sin\vartheta - \vec{e}_\vartheta \cos\vartheta.
\end{aligned}
\tag{C.33}
$$

C.4 Derivatives of Vectors and Tensors

The preceding formulae for differentiating the unit base vectors are now used to derive expressions for the partial derivatives of a vector and tensor.

Let $\vec{v}$ be a differentiable vector function represented in component form as $\vec{v} = \sum_k v_k \vec{e}_k$. Then, the partial derivative of $\vec{v}$ is

$$
\frac{\partial \vec{v}}{\partial x_l} = \frac{\partial}{\partial x_l}\left(\sum_k v_k \vec{e}_k\right) = \sum_k \left(\frac{\partial v_k}{\partial x_l}\vec{e}_k + v_k \frac{\partial \vec{e}_k}{\partial x_l}\right) = \sum_k \left[\frac{\partial v_k}{\partial x_l}\vec{e}_k + v_k \sum_m \left(\begin{matrix} m \\ k\,l \end{matrix}\right)\vec{e}_m\right].
$$

Interchanging the summation indices k and m in the last term, the vector $\partial\vec{v}/\partial x_l$ can be expressed in a compact form as

$$
\frac{\partial \vec{v}}{\partial x_l} = \sum_k v_{k;l}\vec{e}_k,
\tag{C.34}
$$

where

$$
v_{k;l} := \frac{\partial v_k}{\partial x_l} + \sum_m \left(\begin{matrix} k \\ m\,l \end{matrix}\right) v_m
\tag{C.35}
$$

is the *balanced* or *neutral derivative* of v_k with respect to x_l.

The partial derivatives of tensors are evaluated in a similar manner. For example, representing a differentiable tensor function $\boldsymbol{A}$ as a dyadic, that is, as a linear combination of the nine dyads formed by three curvilinear coordinate unit base vectors,[3]

$$
\boldsymbol{A} = \sum_{kl} A_{kl}(\vec{e}_k \otimes \vec{e}_l),
\tag{C.36}
$$

[3]This is Gibbs' dyadic notation applied to orthogonal curvilinear coordinates.

the partial derivatives of $\boldsymbol{A}$ are

$$\frac{\partial \boldsymbol{A}}{\partial x_m} = \sum_{kl} \left[\frac{\partial A_{kl}}{\partial x_m} (\vec{e}_k \otimes \vec{e}_l) + A_{kl} \frac{\partial \vec{e}_k}{\partial x_m} \otimes \vec{e}_l + A_{kl} \vec{e}_k \otimes \frac{\partial \vec{e}_l}{\partial x_m} \right]$$

$$= \sum_{kl} \left[\frac{\partial A_{kl}}{\partial x_m} (\vec{e}_k \otimes \vec{e}_l) + A_{kl} \sum_n \binom{n}{k\,m} (\vec{e}_n \otimes \vec{e}_l) + A_{kl} \sum_n \binom{n}{l\,m} (\vec{e}_k \otimes \vec{e}_n) \right],$$

where we have substituted for the derivatives of the unit base vectors from (C.20). We interchange the summation indices n and k in the second term, and n and l in the last term, to obtain

$$\frac{\partial \boldsymbol{A}}{\partial x_m} = \sum_{kl} A_{kl;m} (\vec{e}_k \otimes \vec{e}_l), \tag{C.37}$$

where

$$A_{kl;m} = \frac{\partial A_{kl}}{\partial x_m} + \sum_n \binom{k}{n\,m} A_{nl} + \sum_n \binom{l}{n\,m} A_{kn} \tag{C.38}$$

is the *balanced* or *neutral derivative* of A_{kl} with respect to x_m. Higher-order tensors can be treated in the same way by representing them as polyadics.

C.5 Invariant Differential Operators

The results of the preceding section on the derivatives of unit base vectors, vectors and tensors are now used to derive expressions for the gradient, divergence and curl in the orthogonal curvilinear coordinates.

C.5.1 Gradient of a Scalar

We first find the components of the nabla operator in orthogonal curvilinear coordinates. This operator in Cartesian coordinates is defined by

$$\vec{\nabla} \bullet := \sum_k \vec{i}_k \frac{\partial \bullet}{\partial y_k}. \tag{C.39}$$

In view of (C.1), (C.3) and the chain rule of differentiation,

$$\vec{\nabla}\bullet = \sum_{kl} \vec{i}_l \frac{\partial \hat{x}_k}{\partial y_l} \frac{\partial \bullet}{\partial x_k}.$$

To find the explicit expression for $\partial \hat{x}_k/\partial y_l$, we use (C.7) and (C.8) in the orthogonality relation $(C.10)_1$ to obtain

$$\sum_m \frac{\partial \hat{y}_m}{\partial x_k} \frac{\partial \hat{y}_m}{\partial x_l} = h_k h_l \delta_{kl},$$

which, after multiplication by $\partial \hat{x}_l/\partial y_n$ and summation over l, gives

$$\sum_m \frac{\partial \hat{y}_m}{\partial x_k} \sum_l \frac{\partial \hat{y}_m}{\partial x_l} \frac{\partial \hat{x}_l}{\partial y_n} = \sum_l h_k h_l \delta_{kl} \frac{\partial \hat{x}_l}{\partial y_n}. \tag{C.40}$$

Since (C.1) is invertible,

$$\sum_l \frac{\partial \hat{y}_m}{\partial x_l} \frac{\partial \hat{x}_l}{\partial y_n} = \delta_{mn},$$

which reduces (C.40) to

$$\frac{\partial \hat{y}_n}{\partial x_k} = h_k^2 \frac{\partial \hat{x}_k}{\partial y_n}. \tag{C.41}$$

The nabla operator can now be expressed as

$$\vec{\nabla}\bullet = \sum_{kl} \frac{\vec{i}_l}{h_k^2} \frac{\partial \hat{y}_l}{\partial x_k} \frac{\partial \bullet}{\partial x_k} = \sum_k \frac{1}{h_k^2} \frac{\partial \vec{p}}{\partial x_k} \frac{\partial \bullet}{\partial x_k}.$$

Substituting $\vec{e}_k$ from (C.8) results in an expression for the nabla operator in orthogonal curvilinear coordinates:

$$\vec{\nabla}\bullet = \sum_k \frac{\vec{e}_k}{h_k} \frac{\partial \bullet}{\partial x_k}. \tag{C.42}$$

Now, let $\phi(x_1, x_2, x_3)$ be a differentiable scalar function. The gradient of ϕ, denoted by $\text{grad}\,\phi$, is defined by a product of the nabla operator with ϕ,

$$\text{grad}\,\phi := \vec{\nabla}\phi. \tag{C.43}$$

Note that this definition makes no reference to a coordinate system. Substituting the nabla operator from (C.42) yields grad ϕ in orthogonal curvilinear coordinates:

$$\operatorname{grad}\phi = \sum_k \frac{\vec{e}_k}{h_k}\frac{\partial\phi}{\partial x_k}. \tag{C.44}$$

Example 4 The gradient of a scalar function ϕ in the spherical coordinates r, ϑ, λ is

$$\operatorname{grad}\phi = \frac{\partial\phi}{\partial r}\vec{e}_r + \frac{1}{r}\frac{\partial\phi}{\partial\vartheta}\vec{e}_\vartheta + \frac{1}{r\sin\vartheta}\frac{\partial\phi}{\partial\lambda}\vec{e}_\lambda. \tag{C.45}$$

C.5.2 Divergence of a Vector

The divergence of a vector function $\vec{v}$ is a scalar function, defined in any orthogonal curvilinear coordinates as the scalar product of the nabla operator and $\vec{v}$,

$$\operatorname{div}\vec{v} := \vec{\nabla}\cdot\vec{v}. \tag{C.46}$$

This definition is manipulated to obtain

$$\begin{aligned}
\operatorname{div}\vec{v} &= \sum_k \frac{\vec{e}_k}{h_k}\cdot\frac{\partial\vec{v}}{\partial x_k} = \sum_k \frac{\vec{e}_k}{h_k}\cdot\sum_l v_{l;k}\vec{e}_l = \sum_{kl}\frac{v_{l;k}}{h_k}\delta_{kl} = \sum_k \frac{v_{k;k}}{h_k} \\
&= \sum_k \frac{1}{h_k}\left[\frac{\partial v_k}{\partial x_k} + \sum_m \binom{k}{m\,k} v_m\right] \\
&= \sum_k \frac{1}{h_k}\left(\frac{\partial v_k}{\partial x_k} + \sum_{\substack{m \\ m\neq k}} \frac{1}{h_m}\frac{\partial h_k}{\partial x_m}v_m\right) = \frac{1}{h_1}\frac{\partial v_1}{\partial x_1} + \frac{1}{h_1 h_2}\frac{\partial h_1}{\partial x_2}v_2 + \frac{1}{h_1 h_3}\frac{\partial h_1}{\partial x_3}v_3 \\
&\quad + \frac{1}{h_2}\frac{\partial v_2}{\partial x_2} + \frac{1}{h_2 h_3}\frac{\partial h_2}{\partial x_3}v_3 + \frac{1}{h_2 h_1}\frac{\partial h_2}{\partial x_1}v_1 \\
&\quad + \frac{1}{h_3}\frac{\partial v_3}{\partial x_3} + \frac{1}{h_3 h_1}\frac{\partial h_3}{\partial x_1}v_1 + \frac{1}{h_3 h_2}\frac{\partial h_3}{\partial x_2}v_2,
\end{aligned}$$

or, in a compact form,

$$\operatorname{div}\vec{v} = \frac{1}{h_1 h_2 h_3}\left[\frac{\partial}{\partial x_1}(h_2 h_3 v_1) + \frac{\partial}{\partial x_2}(h_3 h_1 v_2) + \frac{\partial}{\partial x_3}(h_1 h_2 v_3)\right]. \tag{C.47}$$

Example 5 The divergence of a vector function $\vec{v}$ in the spherical coordinates r, ϑ, λ is

$$\operatorname{div}\vec{v} = \frac{1}{r^2}\frac{\partial}{\partial r}(r^2 v_r) + \frac{1}{r\sin\vartheta}\frac{\partial}{\partial\vartheta}(\sin\vartheta\; v_\vartheta) + \frac{1}{r\sin\vartheta}\frac{\partial v_\lambda}{\partial\lambda}. \tag{C.48}$$

The divergence of the spherical unit base vectors is

$$\operatorname{div}\vec{e}_r = \frac{2}{r}, \qquad \operatorname{div}\vec{e}_\vartheta = \frac{1}{r}\cot\vartheta, \qquad \operatorname{div}\vec{e}_\lambda = 0. \tag{C.49}$$

C.5.3 Curl of a Vector

The curl of a vector function $\vec{v}$ is a vector function defined as the cross-product of the nabla operator and $\vec{v}$,

$$\operatorname{rot}\vec{v} := \vec{\nabla}\times\vec{v}. \tag{C.50}$$

In orthogonal curvilinear coordinates, this definition is manipulated to obtain

$$\begin{aligned}
\operatorname{rot}\vec{v} &= \sum_k \frac{\vec{e}_k}{h_k}\times\frac{\partial\vec{v}}{\partial x_k} = \sum_k \frac{\vec{e}_k}{h_k}\times\sum_l v_{l;k}\vec{e}_l = \sum_{klm}\frac{v_{l;k}}{h_k}\varepsilon_{klm}\vec{e}_m \\
&= \sum_{klm}\frac{\varepsilon_{klm}}{h_k}\left[\frac{\partial v_l}{\partial x_k} + \sum_n \left(\begin{smallmatrix} l \\ n\,k\end{smallmatrix}\right)v_n\right]\vec{e}_m \\
&= \sum_{klm}\frac{\varepsilon_{klm}}{h_k}\left[\frac{\partial v_l}{\partial x_k} + \left(\begin{smallmatrix} l \\ k\,k\end{smallmatrix}\right)v_k\right]\vec{e}_m + \sum_{klm}\frac{\varepsilon_{klm}}{h_k}\sum_{\substack{n \\ n\neq k}}\left(\begin{smallmatrix} l \\ n\,k\end{smallmatrix}\right)v_n\vec{e}_m \\
&= \sum_{klm}\frac{\varepsilon_{klm}}{h_k}\left(\frac{\partial v_l}{\partial x_k} - \frac{1}{h_l}\frac{\partial h_k}{\partial x_l}v_k\right)\vec{e}_m \\
&= \sum_{klm}\left(\frac{\varepsilon_{klm}}{h_k}\frac{\partial v_l}{\partial x_k} + \frac{\varepsilon_{lkm}}{h_k h_l}\frac{\partial h_k}{\partial x_l}v_k\right)\vec{e}_m = \sum_{klm}\left(\frac{\varepsilon_{klm}}{h_k}\frac{\partial v_l}{\partial x_k} + \frac{\varepsilon_{klm}}{h_k h_l}\frac{\partial h_l}{\partial x_k}v_l\right)\vec{e}_m,
\end{aligned}$$

or, in a compact form,

$$\operatorname{rot}\vec{v} = \sum_m\left[\sum_{kl}\frac{\varepsilon_{klm}}{h_k h_l}\frac{\partial(h_l v_l)}{\partial x_k}\right]\vec{e}_m. \tag{C.51}$$

For example, the coefficient of $\vec{e}_1$ is

$$(\operatorname{rot}\vec{v})_1 = \frac{1}{h_2 h_3}\left[\frac{\partial(h_3 v_3)}{\partial x_2} - \frac{\partial(h_2 v_2)}{\partial x_3}\right].$$

The other components can be expressed by the cyclic permutation of indices. It is often convenient to write the curl of a vector in the determinant form

$$\operatorname{rot}\vec{v} = \frac{1}{h_1 h_2 h_3}\begin{vmatrix} h_1\vec{e}_1 & h_2\vec{e}_2 & h_3\vec{e}_3 \\ \dfrac{\partial}{\partial x_1} & \dfrac{\partial}{\partial x_2} & \dfrac{\partial}{\partial x_3} \\ h_1 v_1 & h_2 v_2 & h_3 v_3 \end{vmatrix}. \tag{C.52}$$

Example 6 The curl of a vector function $\vec{v}$ in the spherical coordinates r, ϑ, λ is

$$\operatorname{rot}\vec{v} = \left[\frac{1}{r\sin\vartheta}\frac{\partial(\sin\vartheta\, v_\lambda)}{\partial\vartheta} - \frac{1}{r\sin\vartheta}\frac{\partial v_\vartheta}{\partial\lambda}\right]\vec{e}_r + \left[\frac{1}{r\sin\vartheta}\frac{\partial v_r}{\partial\lambda} - \frac{1}{r}\frac{\partial(r\, v_\lambda)}{\partial r}\right]\vec{e}_\vartheta$$
$$+ \left[\frac{1}{r}\frac{\partial(r\, v_\vartheta)}{\partial r} - \frac{1}{r}\frac{\partial v_r}{\partial\vartheta}\right]\vec{e}_\lambda, \tag{C.53}$$

or, in the determinant form,

$$\operatorname{rot}\vec{v} = \frac{1}{r^2\sin\vartheta}\begin{vmatrix} \vec{e}_r & r\,\vec{e}_\vartheta & r\sin\vartheta\,\vec{e}_\lambda \\ \dfrac{\partial}{\partial r} & \dfrac{\partial}{\partial\vartheta} & \dfrac{\partial}{\partial\lambda} \\ v_r & r\, v_\vartheta & r\sin\vartheta\, v_\lambda \end{vmatrix}. \tag{C.54}$$

The curl of the spherical unit base vectors is then

$$\operatorname{curl}\vec{e}_r = 0, \qquad \operatorname{curl}\vec{e}_\vartheta = \frac{1}{r}\vec{e}_\lambda, \qquad \operatorname{curl}\vec{e}_\lambda = \frac{1}{r}\cot\vartheta\,\vec{e}_r - \frac{1}{r}\vec{e}_\vartheta. \tag{C.55}$$

C.5.4 Gradient of a Vector

The gradient of a vector function $\vec{v}$ is a non-symmetric, second-order tensor function defined as the left dyadic product of the nabla operator with $\vec{v}$,

$$\operatorname{grad}\vec{v} := \vec{\nabla}\otimes\vec{v}. \tag{C.56}$$

In orthogonal curvilinear coordinates, this definition is manipulated to obtain

$$\begin{aligned}
\operatorname{grad}\vec{v} &= \sum_k \frac{\vec{e}_k}{h_k} \otimes \frac{\partial \vec{v}}{\partial x_k} = \sum_k \frac{\vec{e}_k}{h_k} \otimes \sum_l v_{l;k}\vec{e}_l \\
&= \sum_{kl} \frac{1}{h_k}\left[\frac{\partial v_l}{\partial x_k} + \sum_m \left(\begin{matrix} l \\ m\,k \end{matrix}\right) v_m\right](\vec{e}_k \otimes \vec{e}_l) \\
&= \sum_k \frac{1}{h_k}\left[\frac{\partial v_k}{\partial x_k} + \sum_m \left(\begin{matrix} k \\ m\,k \end{matrix}\right) v_m\right](\vec{e}_k \otimes \vec{e}_k) \\
&\quad + \sum_k \sum_{\substack{l \\ l \neq k}} \frac{1}{h_k}\left[\frac{\partial v_l}{\partial x_k} + \sum_m \left(\begin{matrix} l \\ m\,k \end{matrix}\right) v_m\right](\vec{e}_k \otimes \vec{e}_l) \\
&= \sum_k \frac{1}{h_k}\left(\frac{\partial v_k}{\partial x_k} + \sum_{\substack{m \\ m \neq k}} \frac{1}{h_m}\frac{\partial h_k}{\partial x_m} v_m\right)(\vec{e}_k \otimes \vec{e}_k) \\
&\quad + \sum_k \sum_{\substack{l \\ l \neq k}} \frac{1}{h_k}\left(\frac{\partial v_l}{\partial x_k} - \frac{1}{h_l}\frac{\partial h_k}{\partial x_l} v_k\right)(\vec{e}_k \otimes \vec{e}_l).
\end{aligned}$$

Hence, the orthogonal curvilinear components of $\operatorname{grad}\vec{v}$ are

$$(\operatorname{grad}\vec{v})_{kl} = \begin{cases} \dfrac{1}{h_k}\left(\dfrac{\partial v_k}{\partial x_k} + \displaystyle\sum_{\substack{m \\ m \neq k}} \dfrac{1}{h_m}\dfrac{\partial h_k}{\partial x_m} v_m\right) & \text{if } l = k, \\ \dfrac{1}{h_k}\left(\dfrac{\partial v_l}{\partial x_k} - \dfrac{1}{h_l}\dfrac{\partial h_k}{\partial x_l} v_k\right) & \text{if } l \neq k. \end{cases} \tag{C.57}$$

In particular, the symmetric part of $\operatorname{grad}\vec{v}$, i.e., $\frac{1}{2}[\operatorname{grad}\vec{v} + (\operatorname{grad}\vec{v})^T]$, has components

$$\begin{aligned}
&\frac{1}{2}[\operatorname{grad}\vec{v} + (\operatorname{grad}\vec{v})^T]_{kl} \\
&= \begin{cases} \dfrac{1}{h_k}\left(\dfrac{\partial v_k}{\partial x_k} + \displaystyle\sum_{\substack{m \\ m \neq k}} \dfrac{1}{h_m}\dfrac{\partial h_k}{\partial x_m} v_m\right) & \text{if } l = k, \\ \dfrac{1}{2}\left(\dfrac{1}{h_k}\dfrac{\partial v_l}{\partial x_k} + \dfrac{1}{h_l}\dfrac{\partial v_k}{\partial x_l} - \dfrac{1}{h_k h_l}\dfrac{\partial h_k}{\partial x_l} v_k - \dfrac{1}{h_k h_l}\dfrac{\partial h_l}{\partial x_k} v_l\right) & \text{if } l \neq k. \end{cases}
\end{aligned} \tag{C.58}$$

Example 7 The symmetric part of the gradient of $\vec{v}$ in the spherical coordinates r, ϑ, λ is

$$
\begin{aligned}
\frac{1}{2}[\operatorname{grad}\vec{v} + (\operatorname{grad}\vec{v})^T] &= \frac{\partial v_r}{\partial r}(\vec{e}_r \otimes \vec{e}_r) + \frac{1}{r}\left(\frac{\partial v_\vartheta}{\partial \vartheta} + v_r\right)(\vec{e}_\vartheta \otimes \vec{e}_\vartheta) \\
&+ \frac{1}{r}\left(\frac{1}{\sin\vartheta}\frac{\partial v_\lambda}{\partial \lambda} + v_r + \cot\vartheta\, v_\vartheta\right)(\vec{e}_\lambda \otimes \vec{e}_\lambda) \\
&+ \frac{1}{2}\left(\frac{\partial v_\vartheta}{\partial r} + \frac{1}{r}\frac{\partial v_r}{\partial \vartheta} - \frac{v_\vartheta}{r}\right)(\vec{e}_r \otimes \vec{e}_\vartheta + \vec{e}_\vartheta \otimes \vec{e}_r) \qquad \text{(C.59)} \\
&+ \frac{1}{2}\left(\frac{\partial v_\lambda}{\partial r} + \frac{1}{r\sin\vartheta}\frac{\partial v_r}{\partial \lambda} - \frac{v_\lambda}{r}\right)(\vec{e}_r \otimes \vec{e}_\lambda + \vec{e}_\lambda \otimes \vec{e}_r) \\
&+ \frac{1}{2r}\left(\frac{\partial v_\lambda}{\partial \vartheta} + \frac{1}{\sin\vartheta}\frac{\partial v_\vartheta}{\partial \lambda} - \cot\vartheta\, v_\lambda\right)(\vec{e}_\vartheta \otimes \vec{e}_\lambda + \vec{e}_\lambda \otimes \vec{e}_\vartheta).
\end{aligned}
$$

C.5.5 Divergence of a Tensor

The divergence of a tensor function $\boldsymbol{A}$ is a vector function defined as the left scalar product of the nabla operator with $\boldsymbol{A}$,

$$\operatorname{div}\boldsymbol{A} := \vec{\nabla}\cdot\boldsymbol{A}. \qquad \text{(C.60)}$$

In orthogonal curvilinear coordinates, this definition is manipulated to obtain

$$
\begin{aligned}
\operatorname{div}\boldsymbol{A} &= \sum_m \frac{\vec{e}_m}{h_m}\cdot\frac{\partial \boldsymbol{A}}{\partial x_m} = \sum_m \frac{\vec{e}_m}{h_m}\cdot\sum_{kl} A_{kl;m}(\vec{e}_k \otimes \vec{e}_l) \\
&= \sum_{klm} \frac{A_{kl;m}}{h_m}(\vec{e}_m\cdot\vec{e}_k)\vec{e}_l = \sum_{kl}\frac{A_{kl;k}}{h_k}\vec{e}_l \\
&= \sum_{kl}\frac{1}{h_k}\left[\frac{\partial A_{kl}}{\partial x_k} + \sum_m \binom{k}{m\,k} A_{ml} + \sum_m \binom{l}{m\,k} A_{km}\right]\vec{e}_l \\
&= \sum_{kl}\frac{1}{h_k}\left[\frac{\partial A_{kl}}{\partial x_k} + \sum_{\substack{m \\ m\neq k}} \binom{k}{m\,k} A_{ml} + \binom{l}{k\,k} A_{kk} + \sum_{\substack{m \\ m\neq k}} \binom{l}{m\,k} A_{km}\right]\vec{e}_l \\
&= \sum_{kl}\frac{1}{h_k}\left[\frac{\partial A_{kl}}{\partial x_k} + \sum_{\substack{m \\ m\neq k}} \binom{k}{m\,k} A_{ml}\right]\vec{e}_l + \sum_l \sum_{\substack{k \\ k\neq l}} \frac{1}{h_k}\binom{l}{k\,k} A_{kk}\vec{e}_l
\end{aligned}
$$

$$+\sum_{k}\sum_{\substack{m\\ m\neq k}}\frac{1}{h_k}\binom{k}{m\,k}A_{km}\vec{e}_k$$

$$=\sum_{kl}\frac{1}{h_k}\left[\frac{\partial A_{kl}}{\partial x_k}+\sum_{\substack{m\\ m\neq k}}\binom{k}{m\,k}A_{ml}\right]\vec{e}_l$$

$$+\sum_{l}\sum_{\substack{k\\ k\neq l}}\left[\frac{1}{h_k}\binom{l}{k\,k}A_{kk}+\frac{1}{h_l}\binom{l}{k\,l}A_{lk}\right]\vec{e}_l$$

$$=\sum_{kl}\frac{1}{h_k}\left(\frac{\partial A_{kl}}{\partial x_k}+\sum_{\substack{m\\ m\neq k}}\frac{1}{h_m}\frac{\partial h_k}{\partial x_m}\right)\vec{e}_l+\sum_{l}\sum_{\substack{k\\ k\neq l}}\frac{1}{h_k h_l}\left(\frac{\partial h_l}{\partial x_k}A_{lk}-\frac{\partial h_k}{\partial x_l}A_{kk}\right)\vec{e}_l.$$

The sum of the first two terms can be arranged in the same way as the divergence of a vector function (see Sect. C.5.2). The lth component of div $\boldsymbol{A}$ is

$$(\operatorname{div}\boldsymbol{A})_l=\frac{1}{h_1h_2h_3}\left[\frac{\partial}{\partial x_1}(h_2h_3A_{1l})+\frac{\partial}{\partial x_2}(h_3h_1A_{2l})+\frac{\partial}{\partial x_3}(h_1h_2A_{3l})\right]$$
$$+\sum_{k}\frac{1}{h_k h_l}\left(\frac{\partial h_l}{\partial x_k}A_{lk}-\frac{\partial h_k}{\partial x_l}A_{kk}\right). \tag{C.61}$$

Note that the tensor $\boldsymbol{A}$ is not assumed to be symmetric.

Example 8 The divergence of a tensor function $\boldsymbol{A}$ in the spherical coordinates r, ϑ, λ is

$$\operatorname{div}\boldsymbol{A}=\left[\frac{1}{r^2}\frac{\partial}{\partial r}(r^2A_{rr})+\frac{1}{r\sin\vartheta}\frac{\partial}{\partial\vartheta}(\sin\vartheta\,A_{\vartheta r})+\frac{1}{r\sin\vartheta}\frac{\partial A_{\lambda r}}{\partial\lambda}\right.$$
$$\left.-\frac{1}{r}(A_{\vartheta\vartheta}+A_{\lambda\lambda})\right]\vec{e}_r$$
$$+\left[\frac{1}{r^2}\frac{\partial}{\partial r}(r^2A_{r\vartheta})+\frac{1}{r\sin\vartheta}\frac{\partial}{\partial\vartheta}(\sin\vartheta\,A_{\vartheta\vartheta})+\frac{1}{r\sin\vartheta}\frac{\partial A_{\lambda\vartheta}}{\partial\lambda}\right.$$
$$\left.+\frac{1}{r}(A_{\vartheta r}-\cot\vartheta\,A_{\lambda\lambda})\right]\vec{e}_\vartheta$$

$$+\left[\frac{1}{r^2}\frac{\partial}{\partial r}(r^2 A_{r\lambda}) + \frac{1}{r\sin\vartheta}\frac{\partial}{\partial\vartheta}(\sin\vartheta\, A_{\vartheta\lambda}) + \frac{1}{r\sin\vartheta}\frac{\partial A_{\lambda\lambda}}{\partial\lambda}\right.$$
$$\left. + \frac{1}{r}(A_{\lambda r} + \cot\vartheta\, A_{\lambda\vartheta})\right]\vec{e}_\lambda. \tag{C.62}$$

C.5.6 Laplacian of a Scalar and a Vector

The Laplacian of a scalar function ϕ is obtained by substituting $\vec{v} = \text{grad}\,\phi$ into (C.44) and using (C.47),

$$\nabla^2\phi \equiv \text{div grad}\,\phi$$
$$= \frac{1}{h_1h_2h_3}\left[\frac{\partial}{\partial x_1}\left(\frac{h_2h_3}{h_1}\frac{\partial\phi}{\partial x_1}\right) + \frac{\partial}{\partial x_2}\left(\frac{h_3h_1}{h_2}\frac{\partial\phi}{\partial x_2}\right) + \frac{\partial}{\partial x_3}\left(\frac{h_1h_2}{h_3}\frac{\partial\phi}{\partial x_3}\right)\right]. \tag{C.63}$$

Example 9 The Laplacian of a scalar function ϕ in the spherical coordinates r, ϑ, λ is

$$\nabla^2\phi = \frac{1}{r^2}\frac{\partial}{\partial r}\left(r^2\frac{\partial\phi}{\partial r}\right) + \frac{1}{r^2\sin\vartheta}\frac{\partial}{\partial\vartheta}\left(\sin\vartheta\frac{\partial\phi}{\partial\vartheta}\right) + \frac{1}{r^2\sin^2\vartheta}\frac{\partial^2\phi}{\partial\lambda^2}. \tag{C.64}$$

In curvilinear coordinates, the Laplacian of a vector function is more difficult to derive than the Laplacian of a scalar function due to the spatial dependence of the unit base vectors. One way to obtain it is to express the Laplacian of a vector by the vector differential identity (A.34). Without providing a detailed derivation,

$$\begin{aligned}
(\nabla^2\vec{v})_r &= \nabla^2 v_r - \frac{2}{r^2}v_r - \frac{2}{r^2}\frac{\partial v_\vartheta}{\partial\vartheta} - \frac{2}{r^2}\frac{\cos\vartheta}{\sin\vartheta}v_\vartheta - \frac{2}{r^2\sin\vartheta}\frac{\partial v_\lambda}{\partial\lambda},\\
(\nabla^2\vec{v})_\vartheta &= \nabla^2 v_\vartheta - \frac{1}{r^2\sin^2\vartheta}v_\vartheta + \frac{2}{r^2}\frac{\partial v_r}{\partial\vartheta} - \frac{2}{r^2}\frac{\cos\vartheta}{\sin^2\vartheta}\frac{\partial v_\lambda}{\partial\lambda},\\
(\nabla^2\vec{v})_\lambda &= \nabla^2 v_\lambda - \frac{1}{r^2\sin^2\vartheta}v_\lambda + \frac{2}{r^2\sin\vartheta}\frac{\partial v_r}{\partial\lambda} + \frac{2}{r^2}\frac{\cos\vartheta}{\sin^2\vartheta}\frac{\partial v_\vartheta}{\partial\lambda}
\end{aligned} \tag{C.65}$$

in spherical coordinates.

Appendix D
Selected References for General Reading

The literature on Continuum Mechanics is very extensive, ranging from books oriented towards the more practical aspects of this discipline to those providing an exact mathematical treatment. The literature listed in References was used during the preparation of this book and preceding lecture series. Moreover, the table below should aid the reader to find supplementary literature related to a particular chapter of this book. However, I must emphasise that the provided list is by no means exhaustive (Table D.1).

Table D.1 A selection of supplementary literature related to particular chapters of this book

Reference	Chapter 1	Chapter 2	Chapter 3	Chapter 4
Brdička (1959)	4.1–4.2	12.1–12.4	3.1–3.5	13.1–13.5
Chadwick (1999)	2.1–2.3	2.4–2.6	3.1–3.3	3.4–3.6
Dahlen and Tromp (1998)	2.1–2.3	2.4	2.5	2.6–2.7
Eringen (1962)	2–14	17–23	25, 29–31	26, 34, 37–41
Eringen (1967)	1.2–1.12	2.2–2.7	3.2–3.4	2.8–2.9, 3.5–3.9
Greve (2003)	1.1–1.3	1.1–1.3	2.1, 2.8	2.3–2.7
Gurtin (1981)	6–8	9–11	14–15	12–13
Hutter and Jöhnk (2004)	1.1–1.3, 1.5	1.4, 2.2.1	2.1.2	2.1–2.3
Liu (2002)	1.1–1.5	1.6	2.3	2.1–2.5
Malvern (1969)	4.2, 4.5	4.3–4.4, 4.6	3.1–3.3	5.1–5.6
Maršík (1999)	4.1–4.5	4.7–4.10	4.1–4.5	6.1–6.6
Marsden and Hughes (1983)	1.1–1.4	1.6	2.2	2.1–2.5
Mase and Mase (1991)	4.1–4.4, 4.6–4.9	4.5, 4.10, 4.11, 5.2	3.1–3.7	5.3–5.7
Smith (1993)	2.1, 3.1, 3.2	2.2	4.1–4.6	2.3–2.5
Truesdell (1991)	1.2, 1.7, 1.10, 2.3–2.5	2.6, 2.9, 2.13	1.5, 3.1–3.4	1.4, 3.5–3.6

(continued)

Z. Martinec, *Principles of Continuum Mechanics*, Nečas Center Series,
https://doi.org/10.1007/978-3-030-05390-1

Table D.1 (continued)

Reference	Chapter 5	Chapter 6	Chapter 7	Chapter 8	Chapter 9
Biot (1965)					5.1–5.7
Brdička (1959)		5.2–5.3, 11.1		5.1–5.7	
Chadwick (1999)	4.3	4.1–4.2, 4.4–4.8			
Dahlen and Tromp (1998)				2.10	3.1–3.6
Eringen (1962)	27	44–49		51–61	
Eringen (1967)	2.10	5.2–5.10	4.2–4.8	6.1–6.6	
Eringen and Suhubi (1974)					4.2
Greve (2003)	1.4	5.1–5.6	6.1–6.2	3.1–3.3	
Gurtin (1981)	20–21	16–19, 22		30–35	
Hutter and Jöhnk (2004)	4.1–4.3	5.1–5.4, 5.6	5.7–5.8		
Liu (2002)	1.7	3.1–3.7, 4.1–4.3	7.1–7.4	6.1, 6.5	
Malvern (1969)		6.7–6.8		8.1–8.4	
Maršík (1999)		7.1–7.6	8.1–8.5		
Marsden and Hughes (1983)		3.1–3.5		4.1–4.4	
Mase and Mase (1991)				6.1–6.9	
Nečas and Hlaváček (1981)				1.1–3.6	
Smith (1993)	1.1–1.3	5.1–5.7, 6.1–6.3, 8.1–8.4		10.1–10.3	
Truesdell (1991)	1.9, 1.11	6.1–6.8, 6.14–6.17			

The numbers refer to specific sections of the literature

References

Biot, M. A. (1965). *Mechanics of incremental deformations*. New York: Wiley.

Brdička, M. (1959). *Continuum mechanics*. Prague: NČSAV (in Czech).

Chadwick, P. (1999). *Continuum mechanics: Concise theory and problems*. New York: Dower.

Coleman, B. D., & Noll, W. (1960). An approximation theorem for functionals with applications in continuum mechanics. *Archive for Rational Mechanics and Analysis, 6*, 355–374.

Dahlen, F. A., & Tromp, J. (1998). *Theoretical global seismology*. Princeton, NJ: Princeton University Press.

Eringen, A. C. (1962). *Nonlinear theory of continuous media*. New York: McGraw-Hill.

Eringen, A. C. (1967). *Mechanics of continua*. New York: Wiley.

Eringen, A. C., & Suhubi, E. S. (1974). *Elastodynamics. Vol. 1: Finite motions*. New York: Academic.

Greve, R. (2003). *Kontinuumsmechanik. Ein Grundkurs für Ingenieure und Physiker*. Berlin: Springer (in German).

Gurtin, M. E. (1981). *An introduction to continuum mechanics*. New York: Academic.

Hutter, K., & Jöhnk, K. (2004). *Continuum methods of physical modeling. Continuum mechanics, dimensional analysis, turbulence*. Berlin: Springer.

Liu, I.-S. (1972). Method of Lagrange multipliers for exploitation of the entropy principle. *Archive for Rational Mechanics and Analysis, 46*, 131–148.

Liu, I.-S. (2002). *Continuum mechanics*. Berlin: Springer.

Liu, I.-S., & Sampaio, R. (2014). Remarks on material frame-indifference controversy. *Acta Mechanica, 225*, 331–348.

Malvern, L. E. (1969). *Introduction to the mechanics of a continuous medium*. Englewood Cliffs, NJ: Prentice-Hall.

Marsden, J. E., & Hughes, T. J. R. (1983). *Mathematical foundations of elasticity*. New York: Dower.

Maršík, F. (1999). *Thermodynamics of continua*. Prague: Academia (in Czech).

Mase, G. E., & Mase, G. T. (1991). *Continuum mechanics for engineers*. New York: CRC Press.

Müller, I. (1967). On the entropy inequality. *Archive for Rational Mechanics and Analysis, 26*, 118–141.

Nečas, J., & Hlaváček, I. (1981). *Mathematical theory of elastic and elasto-plastic bodies: An introduction*. Amsterdam: Elsevier.

Noll, W. (1958). A mathematical theory of the mechanical behavior of continuous media. *Archive for Rational Mechanics and Analysis, 2*, 197–226.

Rajagopal, K. R., & Srinivasa, A. R. (2004). On the thermomechanics of materials that have multiple natural configurations, part I: Viscoelasticity and classical plasticity. *Zeitschrift für angewandte Mathematik und Physik ZAMP, 55*, 861–893.

Smith, D. R. (1993). *An introduction to continuum mechanics – After Truesdell and Noll*. Dordrecht: Kluwer.

Truesdell, C. (1991). *A first course in rational continuum mechanics, vol 1: General concepts* (2nd ed.). New York: Academic.

Truesdell, C., & Noll, W. (1965). The nonlinear field theories of mechanics. In S. Flügge (Ed.), *Handbook der Physik* (Vol. III/3, pp. 359–366). Berlin: Springer.

Index

Z. Martinec, *Principles of Continuum Mechanics*, Nečas Center Series,
https://doi.org/10.1007/978-3-030-05390-1